Plant Nutrient Dynamics in Stressful Environments

Plant Nutrient Dynamics in Stressful Environments

Special Issue Editors

Urs Feller
Stanislav Kopriva
Valya Vassileva

MDPI • Basel • Beijing • Wuhan • Barcelona • Belgrade

MDPI

Special Issue Editors
Urs Feller
University of Bern
Switzerland

Stanislav Kopriva
University of Cologne
Germany

Valya Vassileva
Bulgarian Academy of Sciences
Bulgaria

Editorial Office
MDPI
St. Alban-Anlage 66
Basel, Switzerland

This is a reprint of articles from the Special Issue published online in the open access journal *Agriculture* (ISSN 2077-0472) from 2017 to 2018 (available at: http://www.mdpi.com/journal/agriculture/special_issues/plant_nutrients_dynamics)

For citation purposes, cite each article independently as indicated on the article page online and as indicated below:

LastName, A.A.; LastName, B.B.; LastName, C.C. Article Title. *Journal Name* **Year**, *Article Number*, Page Range.

ISBN 978-3-03897-063-7 (Pbk)
ISBN 978-3-03897-064-4 (PDF)

Contents

About the Special Issue Editors

Urs Feller is a Professor emeritus at the University of Bern. As a Professor for Plant Physiology he was involved in teaching, research, and management at the Institute of Plant Sciences (IPS) and at the Oeschger Centre for Climate Change Research (OCCR). As an author or coauthor of more than 100 publications he addressed aspects of nutrient and pollutant dynamics in crop plants including nitrogen metabolism during leaf senescence, solute transport via xylem and phloem, and responses to extreme environmental conditions. The mobility of heavy metals in the plant was considered as an important basis for evaluating mechanisms influencing final contents in harvested plant products. During the last two decades he became more and more interested in impacts of extreme climatic events such es heat waves or extended drought periods on the performance of trees, grasslands, and annual crops with special reference to the heat sensitivity of Rubisco activase and the regulation of stomatal opening by these abiotic stresses.

Stanislav Kopriva is Professor of Plant Biochemistry at the Botanical Institute of University of Cologne and deputy speaker of the Cluster of Excellence on Plant Sciences (CEPLAS). The topic of his research is plant mineral nutrition. SK's research concentrates on understanding the molecular basis of control of nutrient uptake and assimilation, integration of sulfate assimilation with primary and secondary metabolism and stress response, and evolution and biodiversity of nutrient uptake and assimilation. This is achieved by a combination of molecular, physiological, and biochemical methods with genetic approaches including association mapping. In the last few years his research started to address the contribution of microorganisms to plant nutrition and the mechanisms by which plant modulate their microbiome.

Valya Vassileva is a Professor at the Institute of Plant Physiology and Genetics (IPPG), Bulgarian Academy of Sciences. She is a Group leader of the team "Regulation of gene expression" that focuses on elucidation of the genetic and epigenetic aspects of the gene expression regulation in higher plants. Particular attention is given to identifying the molecular mechanisms of plant adaptation to abiotic stress in the search for novel strategies for improvement of plant resistance to unfavorable environmental factors. She published more than 70 research papers focused on the nitrogen-fixing symbiosis between rhizobia and legume plants, development of novel molecular markers for plant susceptibility to abiotic stress, and the cellular and molecular mechanisms involved in the regulation of lateral root initiation and development. She also examines the root patterning mechanisms impacting the efficiency of water uptake and acquisition of nutrients.

agriculture

MDPI

Editorial

Plant Nutrient Dynamics in Stressful Environments: Needs Interfere with Burdens

Urs Feller [1,*], Stanislav Kopriva [2] and Valya Vassileva [3]

1   Institute of Plant Sciences and Oeschger Centre for Climate Change Research, University of Bern, Altenbergrain 21, CH-3013 Bern, Switzerland; urs.feller@ips.unibe.ch
2   Botanical Institute, Cluster of Excellence on Plant Sciences (CEPLAS), University of Cologne, 50674 Cologne, Germany; skopriva@uni-koeln.de
3   Institute of Plant Physiology and Genetics, Bulgarian Academy of Sciences, Sofia, Bulgaria; valyavassileva@bio21.bas.bg
*   Correspondence: urs.feller@ips.unibe.ch; Tel.: +41-31-631-4911

Received: 20 June 2018; Accepted: 22 June 2018; Published: 1 July 2018

1. Introduction

Several biotic and abiotic stresses influence plant growth, yield of agricultural crops, and the quality of plant products harvested for human or animal nutrition [1–6]. Abiotic stresses include nutrient starvation, unbalanced nutrient supply, and pollution, as well as climatic factors such as drought, flooding, heat waves, or low temperatures [7,8]. Climatic factors (especially extreme events) become more relevant in the course of global change, and may affect plant nutrition. Besides responses to individual stresses, combinations of stresses (e.g., drought and heat, drought and pests, drought, and unbalanced nutrient supply) must be borne in mind [5,9]. The severity of stresses and the timing of stress periods (phase of plant development, duration) are relevant for the impacts on various plant species or various varieties of a given species. Furthermore, recovery phases following stress periods must be also considered when evaluating stress impacts in a comprehensive manner [10]. This special issue addresses impacts of various stresses on plant nutrient acquisition, translocation, and accumulation in the harvested plant parts; however, only a limited number of stress impacts can be presented in detail.

2. Nutrient Availability and Acquisition

Liebig's law (initially focused on the availability of mineral macronutrients, and later also on micronutrients) was extended and integrated into a new concept presented in the review by Haneklaus et al. [5]. A balanced supply of macro- and micronutrients is essential to decrease the susceptibility of plants to biotic and abiotic stresses. Silicon—an element not belonging to the essential nutrients—plays a key role in the response of plants to biotic stresses (e.g., fungal attack) and a series of abiotic stresses such as drought or heavy metal stress [5,11–13]. For example, silicon can decrease the toxic effects of heavy metals or play a protective role against fungal diseases by positively influencing the structure and function of plant cell walls [5].

A research article contributed to this special issue by Bouranis et al. [14] addresses interactions between various nutrients and forms of nutrients with special reference to elemental sulfur and to iron in calcareous soils. Iron nutrition of crops can be improved by the addition of elemental sulfur to a standard fertilizer mixture. This was demonstrated for durum wheat on a calcareous soil. The addition of elemental sulfur lead to heavier vegetative plant parts and ears. Iron and organic sulfur contents were increased in all plant parts by this treatment, and yield quantity was also positively influenced. This paper nicely demonstrates the necessity to consider various aspects of plant nutrition in a broad context, including plant/soil interactions, in order to optimize yield quantity and quality, as well as to decrease the susceptibility of plants to various stresses.

Interactions between magnesium supply and light levels with respect to photosynthetic efficiency were investigated by Dias et al. [15]. Experiments with coffee plants grown under controlled conditions provide evidence that optimized magnesium concentrations in the nutrient medium are important for maximal CO_2 assimilation rate, as well as for highest water use efficiency. Stomatal conductance depends on light and magnesium supply, illustrating the interaction between these two factors. Furthermore, increased magnesium concentrations in the nutrient solution led to increased magnesium and decreased potassium in leaf dry matter. Physiological processes (e.g., growth, photosynthesis, transpiration, seed filling) and the composition of collected plant material (e.g., hay, cereal grains) must be borne in mind when addressing aspects of plant nutrition in stressful environments.

The three papers mentioned above clearly illustrate the importance of addressing nutrient disorders in a broader context. Various stresses may affect nutrient acquisition by plants and their distribution within plants. In contrast, the nutritional status of plants is relevant for the responses to stresses. Appropriate fertilization is a key aspect, but fertilizers must be applied before or at the beginning of the main growth period when environmental conditions throughout this period and during the subsequent maturation phase cannot yet be known. As a consequence, plant growth, yield, and nutrient consumption may be negatively influenced, and deviate from average seasons. In such situations corrections in agronomic practices may become necessary during the following season(s). It remains a challenge to further integrate nutritional aspects and stress responses. In this context, it must be borne in mind that agricultural practices and the genotype spectrum available for crop production are permanently evolving, and may bring additional complexity to a comprehensive network of regulatory interactions.

3. Nutrient Redistribution within Plants and Accumulation in Harvested Plant Parts

Besides nutrient availability in soils, nutrient distribution and redistribution within the plant are important for the final contents of the various plant parts. Such transport processes and their regulation allow an accumulation of nutrients in harvested vegetative [16] or reproductive plant parts [17,18]. The mobility of an element or of certain forms of an element in the phloem is crucial for redistribution processes within the plant [13,19,20]. Such redistribution processes are crucial for heavy metal homeostasis [21], hyperaccumulation [22], and toxicity [11].

Possibilities to increase phosphorus use efficiency in wheat and Arabidopsis were reviewed by Kisko et al. [19]. The optimized use of phosphorus is highly relevant from an ecological (e.g., risk of eutrophication), as well as from an economical (e.g., fertilizer costs, quality of yield) point of view. Inorganic phosphate transport is emphasized in this review with respect to optimal use of phosphorus fertilizers and phosphorus contents in edible plant parts. A list of genes involved in sensing, uptake, transport, and signaling of inorganic phosphate in *Arabidopsis thaliana* and wheat documents the actual state of knowledge.

Biofortification is an important keyword in this context. Low zinc contents in plant products are a serious issue for human nutrition, and may cause zinc deficiency for a large percentage of the worldwide population [16,23]. Biofortification was, in the past, mainly investigated in staple crops (e.g., wheat, potato). The research article by White et al. [16] is focused on the zinc biofortification in leafy brassicas such as broccoli or cabbage. These plants, grown worldwide, are directly consumed by humans, and improved zinc contents in their leaves may help to provide sufficient zinc for the population in regions with inadequate zinc supplies. However, an excess of zinc can be phytotoxic. Therefore, there are limits for increasing zinc levels in collected plant parts without negatively influencing the plant itself. Based on these facts, zinc biofortification of leafy brassicas must be optimized, taking into account the zinc demands for human nutrition, as well as the possible toxic effects of elevated zinc on plant growth and/or metabolism [11,16,22].

4. Impact of Global Change on Plant Nutrient Dynamics

Global climate change with a string of extreme events (e.g., heat waves, droughts) cannot be ignored when addressing plant nutrient dynamics these days. The availability of nutrients in the soil, their acquisition, assimilation, distribution/redistribution within the plants, and the nutrient balance sheets for fields can be severely disturbed by climatic stress factors [13,24–27]. Such effects may not be restricted to the actual growing season, and may be relevant for the subsequent years(s).

Etienne et al. [13] contributed a review focused on senescence of vegetative plant parts and nutrient redistribution in annual crops. Drought effects on nutrient dynamics within the shoot, including remobilization from senescing leaves and transport to reproductive structures via xylem and phloem, are discussed in this paper. It must be assumed that the points addressed will become more relevant in the course of climate change, with predicted more frequent and/or more severe drought periods in many regions. Water fluxes, assimilatory activities, and the redistribution of inorganic nutrients and of assimilates are affected by this stress. The relative mobilities within plants vary between different macro- and micronutrients. Therefore, the drought effects on redistribution must be considered in an element-specific manner, avoiding unjustified generalizations. Nutrient deficiencies, or other stresses such as drought or heat, may influence the life span of leaves, as well as the composition of the harvest (e.g., grains) or of the stover.

The impact of drought on nitrogen, phosphorus, and potassium nutrition was investigated in young maize plants and documented in a research article by Studer et al. [25]. These three elements are frequently limiting, and represent classical fertilizer components. Changes in root and shoot growth, as documented with altered root/shoot ratios for dry matter, were identified as an important mechanism to improve stress tolerance in crops. In a broader context, structural responses to stresses may be as important as physiological changes to improve overall susceptibility [26–28]. Drought impact on symbiotic interactions, such as nitrogen fixation in legumes [29], or on mycorrhizal symbiosis [30], is a further important mechanism to influence the overall performance of crop plants.

5. Genetics and Breeding for Tolerant Crops

The elucidation of relevant genes and of genetic potentials are an important basis for breeding crop varieties with improved performance in stressful environments [31–35]. Besides traditional breeding programs, new tools to improve stress responses became available over recent decades, and may allow more direct genetic improvements to be made. Such programs could further improve well performing varieties when exposed to stresses.

Mastrodomenico et al. [31] started a research program in the corn belt based on old maize varieties (*expired plant variety protection germplasm*) and aimed at identifying possibilities of improving the performance under nitrogen stress (poor nitrogen availability). The paper is based on data from field experiments from four years at eight locations. Large numbers of inbreds (53 non-stiff, stalk synthetic and 36 stiff stalk synthetic) were included in this program to improve nitrogen-use efficiency. The authors present a perspective to identify the genetic potential, and to breed genotypes with a good overall performance under low nitrogen supply.

The research article by Chietera et al. [32] is also focused on nitrogen stress, and proposes new breeding targets including morphological and physiological properties. This research article is based on experiments with the hydroponically grown model plant *Arabidopsis thaliana*. A broad range of nitrate concentrations in the nutrient medium allowed us to undertake comprehensive analyses of four lines differing in their nitrate utilization properties. The findings were integrated into a statistical model predicting biomass production from nitrate supply. The article may serve as a basis for refining breeding programs for crops.

The possibilities of using all technologies available today—including genetic engineering tools—are presented in a comprehensive review by Roberts and Mattoo [33]. Sustainable intensification in agriculture to provide adequate food for an ever-growing worldwide population is a key aspect in this paper. The authors emphasize the challenge for the scientific community to provide the basis and suitable techniques for breeding programs, taking into account changing environmental conditions (e.g., global change), potentials of changes in cropping systems, as well as biotic stress impacts in the future. Biotechnological approaches must be envisaged to provide cultivars with improved stress tolerance (abiotic and biotic stresses) [33].

6. Conclusions

The papers included in this special issue cover a broad range of aspects ranging from genetics and breeding to crop production in the field. Climate change, intensified agriculture, modifications of land use, or pollution are often accompanied by larger fluctuations including extreme events. The growing world's population and nutrient deficiencies in agricultural products for human or animal nutrition, or pollutants in harvested products in some regions (quality of yield), are important points to be integrated in a comprehensive analysis aimed at supporting agriculture on the way into a challenging future. It is therefore necessary to develop suitable models to identify potentials and risks. Instabilities (e.g., caused by climatic factors or pests) should be detected as early as possible to initiate corrections in the nutrient supply or in other growth conditions. Sensitive detection systems for nutrient disorders in the field can facilitate this task, and are therefore, highly desirable [36].

Author Contributions: The three coeditors equally contributed to organizing the special issue, to editorial work, and to writing this editorial.

Funding: No external funding was obtained for editing this special issue and for writing the editorial.

Acknowledgments: We thank the authors for submitting manuscripts of high quality and their willingness to further improve them after peer review, the reviewers for their careful evaluations aimed at eliminating weaknesses and their suggestions to optimize the manuscripts and the editorial staff of MDPI for the professional support and the rapid actions taken when necessary throughout the editorial process.

Conflicts of Interest: The authors declare no conflict of interest.

References

1. Rose, T.J.; Raymond, C.A.; Bloomfield, C.; King, G.J. Perturbation of nutrient source-sink relationships by post-anthesis stresses in differential accumulation of nutrients in wheat grain. *J. Plant Nutr.* **2015**, *178*, 89–98. [CrossRef]
2. Mittler, R. Abiotic stress, the field environment and stress combination. *Trends Plant Sci.* **2006**, *11*, 15–19. [CrossRef] [PubMed]
3. DaMatta, F.M.; Grandis, A.; Arenque, B.C.; Buckeridge, M.S. Impacts of climate changes on crop physiology and food quality. *Food Res. Int.* **2010**, *43*, 1814–1823. [CrossRef]
4. Zhou, J.; Ju, R.; Li, B.; Wu, J. Responses of soil biota and nitrogen availability to an invasive plant under aboveground herbivory. *Plant Soil* **2017**, *415*, 479–491. [CrossRef]
5. Haneklaus, S.H.; Bloem, E.; Schnug, E. Hungry Plants—A short treatise on how to feed crops under stress. *Agriculture* **2018**, *8*, 43. [CrossRef]
6. Helfensstein, J.; Pawlowski, M.L.; Hill, C.B.; Stewart, J.; Lagos-Kutz, D.; Bowen, C.R.; Frossard, E.; Hartmann, G.L. Zinc deficiency alters soybean susceptibility to pathogens and pests. *J. Plant Nutr. Soil Sci.* **2015**, *178*, 896–903. [CrossRef]
7. IPCC. *Climate Change. 2014: Impacts, Adaptation, and Vulnerability. Part B: Regional Aspects. Contribution of Working Group II to the Fifth Assessment Report of the Intergovernmental Panel on Climate Change*; Cambridge University Press: Cambridge, UK; New York, NY, USA, 2014.
8. Grant, K.; Kreyling, J.; Heilmeier, H.; Beierkuhnlein, C.; Jentsch, A. Extreme weather events and plant-plant interactions: Shifts between competition and facilitation among grassland species in the face of drought and heavy rainfall. *Ecol. Res.* **2014**, *29*, 991–1001. [CrossRef]
9. Simova-Stoilova, L.; Vassileva, V.; Feller, U. Selection and breeding of suitable crop genotypes for drought and heat periods in a changing climate: Which morphological and physiological properties should be considered? *Agriculture* **2016**, *6*, 26. [CrossRef]

10. Kaufmann, I.; Schulze-Till, T.; Schneider, H.U.; Zimmermann, U.; Jakob, P.; Wegner, L.H. Functional repair of embolized vessels in maize roots after temporal drought stress, as demonstrated by magnetic resonance imaging. *New Phytol.* **2009**, *184*, 245–256. [CrossRef] [PubMed]

11. Clemens, S. Toxic metal accumulation, responses to exposure and mechanisms of tolerance in plants. *Biochimie* **2006**, *88*, 1707–1719. [CrossRef] [PubMed]

12. Chaves, M.M.; Maroco, J.P.; Pereira, J.S. Understanding plant responses to drought—From genes to the whole plant. *Funct. Plant Biol.* **2003**, *30*, 239–264. [CrossRef]

13. Etienne, P.; Diquelou, S.; Prudent, M.; Salon, C.; Maillard, A.; Ourry, A. Macro and micronutrient storage in plants and their remobilization when facing scarcity: The case of drought. *Agriculture* **2018**, *8*, 14. [CrossRef]

14. Bouranis, D.L.; Chorianopoulou, S.N.; Margetis, M.; Saridis, G.I.; Sigalas, P.P. Effect of elemental sulfur as fertilizer ingredient on the mobilization of iron from the iron pools of a calcareous soil cultivated with durum wheat and the crop's iron and sulfur nutrition. *Agriculture* **2018**, *8*, 20. [CrossRef]

15. Dias, K.G.L.; Guimarães, P.T.G.; Neto, A.E.F.; Silveira, H.R.O.; Lacerda, J.J.J. Effect of magnesium on gas exchange and photosynthetic efficiency of coffee plants grown under different light levels. *Agriculture* **2017**, *7*, 85. [CrossRef]

16. White, P.J.; Pongrac, P.; Sneddon, C.C.; Thompson, J.A.; Wright, G. Limits to the biofortification of leafy brassicas with zinc. *Agriculture* **2018**, *8*, 32. [CrossRef]

17. Welch, R.M.; Graham, R.D. Breeding crops for enhanced micronutrient content. *Plant Soil* **2002**, *245*, 205–214. [CrossRef]

18. Zheng, L.Q.; Yamaji, N.; Yokosho, K.; Ma, J.F. YSL16 is a phloem-localized transporter of the copper-nicotianamine complex that is responsible for copper distribution in rice. *Plant Cell* **2012**, *24*, 3767–3782. [CrossRef] [PubMed]

19. Kisko, M.; Shukla, V.; Kaur, M.; Bouain, N.; Chaiwong, N.; Lacombe, B.; Pandey, A.K.; Rouached, H. Phosphorus transport in Arabidopsis and wheat: Emerging strategies to improve P pool in seeds. *Agriculture* **2018**, *8*, 27. [CrossRef]

20. Hazama, K.; Nagata, S.; Fujimori, T.; Yanagisawa, S.; Yoeneyama, T. Concentrations of metals and potential metal-binding compounds and speciation of Cd, Zn and Cu in phloem and xylem saps from castor bean plants (*Ricinus communis*) treated with four levels of cadmium. *Physiol. Plant.* **2015**, *154*, 243–255. [CrossRef] [PubMed]

21. Grotz, N.; Guerinot, M.L. Molecular aspects of Cu, Fe and Zn homeostasis in plants. *Biochim. Biophys. Acta Mol. Cell Res.* **2006**, *1763*, 595–608. [CrossRef] [PubMed]

22. Kramer, U. Metal hyperaccumulation in plants. *Annu. Rev. Plant Biol.* **2010**, *61*, 517–534. [CrossRef] [PubMed]

23. White, P.J.; Broadley, M.R. Biofortification of crops with seven mineral elements often lacking in human diets—Iron, zinc, copper, calcium, magnesium, selenium and iodine. *New Phytol.* **2009**, *182*, 49–84. [CrossRef] [PubMed]

24. Si, L.; Xie, Y.; Ma, Q.; Wu, L. The short-term effects of rice straw biochar, nitrogen and phosphorus fertilizer on rice yield and soil properties in a cold waterlogged paddy field. *Sustainability* **2018**, *10*, 537. [CrossRef]

25. Studer, C.; Hu, Y.; Schmidhalter, U. Interactive effects of N-, P- and K-nutrition and drought stress on the development of maize seedlings. *Agriculture* **2017**, *7*, 90. [CrossRef]

26. Hoffmann, C.M. Adaptive responses of *Beta vulgaris* L. and *Cichorium intybus* L. root and leaf forms to drought stress. *J. Agric. Crop Sci.* **2014**, *200*, 108–118. [CrossRef]

27. Grieder, C.; Trachsel, S.; Hund, A. Early vertical distribution of roots and its association with drought tolerance in tropical maize. *Plant Soil* **2014**, *377*, 295–308. [CrossRef]

28. Caser, M.; D'Angiolillo, F.; Chitarra, W.; Lovisolo, C.; Ruffoni, B.; Pistelli, L.; Pistelli, L.; Scariot, V. Ecophysiological and phytochemical responses of *Salvia sinaloensis* Fern. to drought stress. *Plant Growth Regul.* **2018**, *84*, 383–394. [CrossRef]

29. González, E.M.; Larrainzar, E.; Marino, D.; Wienkoop, S.; Gil-Quintana, E.; Arrese-Igor, C. Physiological responses of N_2-fixing legumes to water limitation. In *Legume Nitrogen Fixation in a Changing Environment*; Springer International Publishing: Berlin/Heidelberg, Germany, 2015; pp. 5–33.

30. Auge, R.M.; Toler, H.D.; Saxton, A.M. Arbuscular mycorrhizal symbiosis alters stomatal conductance of host plants more under drought than under amply watered conditions: A meta-analysis. *Mycorrhiza* **2015**, *25*, 13–24. [CrossRef] [PubMed]

31. Mastrodomenico, A.T.; Hendrix, C.C.; Below, F.E. Nitrogen Use Efficiency and the genetic variation of maize expired plant variety protection germplasm. *Agriculture* **2018**, *8*, 3. [CrossRef]
32. Chietera, G.; Chaillou, S.; Bedu, M.; Marmagne, A.; Masclaux-Daubresse, C.; Chardon, F. Impact of the genetic–environment interaction on the dynamic of nitrogen pools in arabidopsis. *Agriculture* **2018**, *8*, 28. [CrossRef]
33. Roberts, D.P.; Mattoo, A.K. Sustainable agriculture—Enhancing environmental benefits, food nutritional quality and building crop resilience to abiotic and biotic stresses. *Agriculture* **2018**, *8*, 8. [CrossRef]
34. Krannich, C.T.; Maletzki, L.; Kurowsky, C.; Horn, R. Network candidate genes in breeding for drought tolerant crops. *Int. J. Mol. Sci.* **2015**, *16*, 16378–16400. [CrossRef] [PubMed]
35. Rasheed, A.; Mujeeb-Kazi, A.; Ogbonnaya, F.C.; He, Z.H.; Rajaram, S. Wheat genetic resources in the post-genomics era: Promise and challenges. *Ann. Bot.* **2018**, *121*, 603–616. [CrossRef] [PubMed]
36. Masseroni, D.; Ortuani, B.; Corti, M.; Gallina, P.M.; Cocetta, G.; Ferrante, A.; Facchi, A. Assessing the reliability of thermal and optical imaging techniques for detecting crop water status under different nitrogen levels. *Sustainability* **2017**, *9*, 1548. [CrossRef]

agriculture

MDPI

Review

Hungry Plants—A Short Treatise on How to Feed Crops under Stress

Silvia H. Haneklaus *, Elke Bloem and Ewald Schnug

Institute for Crop and Soil Science, Julius Kühn-Institut, Bundesallee 69, D-38116 Braunschweig, Germany;
elke.bloem@julius-kuehn.de (E.B.); ewald.schnug@julius-kuehn.de (E.S.)
* Correspondence: silvia.haneklaus@julius-kuehn.de; Tel.: +49-531-596-2121

Received: 31 January 2018; Accepted: 15 March 2018; Published: 17 March 2018

Abstract: Fertilisation is as old as is the cultivation of crops. In the 19th century, plant nutrition became an area of research in the field of agricultural chemistry. *Liebig*'s "Law of the Minimum" (1855) is still the basis for plant nutrition. It states that the exploitation of the genetically fixed yield potential of crops is limited by that variable, which is insufficiently supplied to the greatest extent. With a view to abiotic and biotic stress factors, this postulation should be extended by the phrase "and/or impaired by the strongest stress factor". Interactions between mineral elements and plant diseases are well known for essential macro- and micronutrients, and silicon. In comparison, the potential of fertilisation to alleviate abiotic stress has not been compiled in a user-orientated manner. It is the aim of this chapter to summarise the influence of nutrient deficiency in general, and the significance of sodium, potassium, and silicon, in particular, on resistance of crop plants to abiotic stress factors such as drought, salinity, and heavy metal stress. In addition, the significance of seed priming with various nutrients and water to provide tolerance against abiotic stress is discussed. Underlying physiological mechanisms will be elaborated, and information on fertiliser application rates from practical experiences provided.

Keywords: drought; heavy metal pollution; no-effect value; potassium; salinity; seed priming; silicon; sodium

1. Introduction

Instinctive management practices feeding plants can be traced back to the Neolithic agricultural revolution [1]. But it was the development of mineral fertilisers, in particular nitrogen (N), after World War I which revolutionised agricultural production. *Liebig*'s "Law of the Minimum" (1855) is still the basic concept of plant nutrition. It states that the "exploitation of the genetically fixed yield potential of crops is limited by that nutrient, which is insufficiently supplied to the greatest extent" (Figure 1). This principle assumes that all other growth factors, such as water supply and temperature, are optimum. The significance of regularly occurring abiotic and biotic stress factors is neglected, but deserves attention where they significantly affect crop production. Thus, *Liebig*'s law of the minimum should be extended by the phrase "and/or is impaired by the strongest stress factor".

Though the significance of individual nutrients for maintaining or promoting plant health saw some interest in the 1960s and 1970s [2], research in the field of nutrient induced resistance mechanisms has been scarce because of its complexity and limited practical significance, due to the availability of effective pesticides. Recent omics approaches have enabled identification of underlying physiological mechanisms of nutrient induced resistance against diseases [3] and tolerance against abiotic stress [4]. Abiotic stress factors comprise nutrient and water deficiency, soil pH, temperature, oxygen supply, mechanical pressure, injury, chemical compounds, and heavy metals [5].

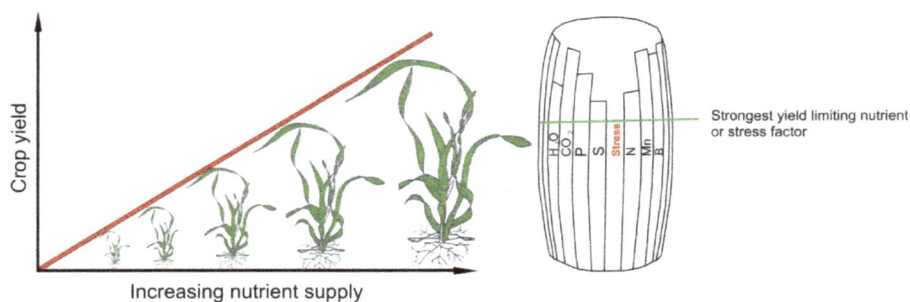

Figure 1. *Liebig*'s law of the minimum (diagrammatic illustration).

Sulphate and phosphate fertiliser applications have been tested to fix Sr^{90} in soil, but even excessively high rates yielded no significant effect on Sr^{90} uptake of plants in field experimentation [6]. Likewise, no data exist on a quantitatively relevant detoxification of heavy metals in plants by sulphur (S) fertilisation through the formation of phytochelatins and metallothioneins. Both metabolites are S-containing secondary compounds that bind, for instance, As, Cd, Cr, Cu, and Ni. The question remains open as to whether graded S fertiliser rates yield a parallel increase in these secondary components as has been shown, for example, for glucosinolates [7]. Rather, Ernst [8] stated that phytochelatin synthesis is an unspecific metabolic reaction to increased heavy metal concentrations, which bind only small amounts of metal, and that resistance of the biomembrane is a much more efficient mechanism against heavy metal stress. The amelioration and reclamation of acid, saline, and polluted soils by fertiliser practices are not subjects of this review, as these topics have been discussed comprehensively before [9,10].

The phenomenon that mild stress induces yield gains and quality improvements is well known from the cultivation and processing of herbal plants [11–13]. In both cases, the effect is explained by a stimulation of metabolic processes in order to defy oxidative stress. By contrast, worldwide severe stress conditions are a serious threat to crop productivity, and measures to alleviate temporal and permanent abiotic stress are an important contribution to food security. Though various essential macro- and micronutrients are involved in tolerance mechanisms against abiotic stress, only a limited number of beneficial elements proved to alleviate stress conditions under field conditions. The aim of this chapter is to compile fertiliser practices which have proved to significantly mitigate abiotic and biotic stress under field conditions. In addition, underlying physiological mechanisms have been assessed, and guidelines for fertilisation will be provided.

2. Balanced Nutrient Supply—Essential to Secure Productivity and Pivotal Barrier against Abiotic and Biotic Stress

A balanced nutrient supply is a basic requirement to protect plants against all forms of stress. The depletion of nutrients, soil organic matter, and erosion are the principal forms of soil degradation. Nutrient deficiency is an abiotic stress factor, whereby a limited nutrient stock needs to be distinguished from restricted nutrient availability. In the first case, the deficiency can be balanced by adequate soil-applied fertiliser rates, in the second case, the mobility can be enhanced, for instance, by increasing or decreasing the soil pH value by applying lime and acidifying fertilisers, respectively. Another alternative is the use of foliar fertiliser applications. The ideal causal chain to avoid stress induced by nutrient deficiency is assessment of the nutrient supply status–establishment of fertiliser response curves–targeted fertiliser application.

The nutrient status of plants follows the typical *Mitscherlich* growth functions from severe over moderate deficiency, to optimum supply, and finally, toxicity (Figure 2a). The so-called upper boundary lines represent the mathematical, usually, 4th order polynomial functions of the impact of increasing nutrient concentrations in the plant tissue on crop yield at defined growth stages, if no other growth factor is yield limiting [14]. The mathematical procedures to determine upper boundary line functions have been developed by Schnug et al. [15]. Boundary lines describe the "pure effect of a nutrient" on crop yield under *ceteris paribus* conditions [16–18], as the line describes the highest yields observed over the range of nutrient values measured. Data points below this line relate to samples where some other factor limited the crop's response to the nutrient. It is important to note that the slopes of the growth functions of essential macro- and micronutrients increase steeply with decreasing critical elemental concentrations in the plant tissue (Figure 2a).

So far, it is not known what impact low input systems on their own, or combined with stress conditions, have on curve progression (Figure 2). If another variable, for instance, a stress factor (drought, salinity) and stress relieving minerals, such as sodium (Na) and silicon (Si), are having a significant effect on the response to the nutrient, their presence will be indicated by two or more distinct concentrations of points, each with its own boundary line response to the nutrient (Figure 2b). Then, the data can be classified on the basis of stress factor and concentration of essential plant nutrient and stress relieving minerals; then boundary lines can be determined separately for each class (Figure 2, [14]).

The physiological nutrient demand can be higher under stress conditions than needed for high yield. This has been shown for S in order to trigger *SIR* (sulphur induced resistance) against fungal pathogens [19], and also will be shown for the application of silicon (Si) against fungal diseases in this chapter.

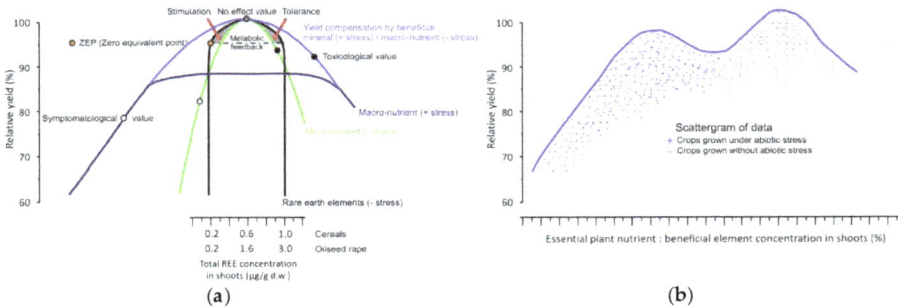

Figure 2. Common, abstract upper boundary line progression for essential macro- and micronutrients and beneficial minerals in relation to abiotic stress, and predicted progression of upper boundary lines for rare earth elements (REEs) (**a**); scattergram showing the relationship between essential plant nutrients and beneficial elements on crop yield indicating two categories (+/−) of the discriminating variable abiotic stress (**b**) (adapted from [20]).

Minerals are rated as beneficial for plants if they promote growth and yield, replace the function of essential nutrients in plant metabolism, or strengthen the natural resistance of plants against abiotic and biotic stress. Prominent examples are, for example, Na and Si. If abiotic stress reduces crop yield, for instance, by 15%, a mineral should have the potential to compensate this gap. As a result, the boundary line progression of the essential plant nutrient will not change under stress (Figure 2a). Such stress-compensating effect has been postulated for rare earth elements (REEs), too, but data suggest that REEs yield, most likely, a hormetic effect ([20], Figure 2a). It is difficult to come to a final decision as to whether REEs are beneficial for plant growth and act as mediators against abiotic stress, as contradictory findings exist.

Critical values are indispensable for evaluating the nutritional status of a crop. Important threshold markers are the symptomatological value, which reflects the nutrient concentration below which deficiency symptoms become visible; the no-effect value, which stands for the nutrient concentration above which the plant is sufficiently supplied for achieving the maximum potential yield, the critical nutrient value for realising a yield level which is 5, 10, and 20% lower than the maximum yield [21]; and the toxicological value, which indicates the nutrient concentration above which toxicity symptoms can be observed (Figure 2a). The sufficiency range indicates the span of concentrations within which any change of the nutrient content will not influence crop yield [22].

It is important to note that there is no one exclusive critical nutrient value for each crop, as this depends on the growth conditions, the developmental stage of the plant at sampling, the specific plant part, the determined nutrient species, the targeted yield and mathematical approach for calculating it. This also implies that it is more or less impossible to compare results from different experiments, particularly as critical values are often based on not more than a single experiment [23]. Consequently, a sheer innumerable amount of critical nutrient values can be found in literature for different crops. For these reasons, Smith and Loneragan [22] stressed that it is only possible to define ranges, and not specific values, for different nutritional levels. As an example, Haneklaus et al. [7] compiled and categorised available individual data with varying experimental conditions from the literature, for the variables total S and sulphate concentrations, and N/S ratios in relation to different crop species, in order to facilitate an easy and appropriate evaluation of the S status. Plant groups were assembled by morphogenetic and physiological features. Because of the wide heterogeneity of results for similar classes of S supply and for a better comparability of results, concentrations were combined in three major categories: deficient, adequate, and high, irrespective of the sampled plant part during vegetative growth. The results of the approach chosen by [7,14] were in close agreement with individual no-effect values and ranges that the authors determined independently by employing BOLIDES ((Boundary Line Development System), Table 1). The data sets interpreted by [14] comprised several thousand entries for cereals and oilseed rape, and more than 500 for sugar beet from field surveys, field trials, or pot experiments, which cover a wide range of growth factor combinations. In Table 1, the no-effect values for macro- and micronutrients, and Na contents in cereals, sugar beet, and oilseed rape are summarised. Nutrient concentrations which equal or are higher than the no-effect values are required to achieve maximum crop yields.

Table 1. No-effect values for essential macro- and micronutrients, and Na concentrations in cereals, oilseed rape, and sugar beet (extracted from [14]).

Nutrient	Cereals	Oilseed Rape	Sugar Beet
(mg/g d.w.)			
N	35	40	46
P	4	4.2	4.5
S	4	6.5	3.5
K	35	35	42
Ca	4	22.5	4.2
Mg	1.1	1.5	1.8
Na	-	-	2.0
(μg/g d.w.)			
Fe	60	100	100
Mn	28	30	30
Zn	25	33	50
Cu	4	4.5	15
Cl	100	100	-
B	3	25	30
Mo	0.2	0.3	-

For sugar beet, the question arises as to whether the no-effect values vary in relation to the reference value of productivity, namely root and sugar yield. The sugar beet root contains, on average, 18% sugar with a range of variation from 15 to 23%, depending on soil type, cultivar, and climate. The upper boundary line analysis revealed that relevant differences in the sufficiency nutrient ranges existed for Mg and Na with respect to root and sugar yield [14]. In the case of these elements, the concentration range for a sufficient supply with respect to the sugar yield was 2.2–4.0 mg/g for Mg and 4.0–6.5 mg/g for Na. This means that for maximising the sugar yield, Mg and Na concentrations that equal the no-effect values are not sufficient for maximising the sugar yield. This fact needs to be taken into account when the nutrient status is assessed by plant analysis, and considered in fertiliser management, respectively.

Where resilient critical nutrient values, sufficiency ranges or no-effect values are not available for a specific crop species, it is recommended that these are calculated site-specifically for each location by employing on-farm experimentation techniques, such as yield monitoring and directed sampling [24]. The overall advantage is that threshold values are tailor-made. The last step is the development of algorithms for a variable rate application of nutrients and stress relieving minerals, in relation to stress. The technology for continuous, variable rate application of fertilisers is available, and the best option for realising the site-specific yield potential under stress conditions. In low input systems, in particular, including organic farming, the systematic variable rate application of farmyard manure offers a solution to the key problem of strictly limited N and phosphorus (P) sources. A local resource management will harmonise the N and P supply. Where the technology is not available, "local knowledge" may be an adequate alternative [24]. In comparison, in high input systems, the main objective of variable rate mineral and organic fertilisation is to adjust rates in such way that crop productivity is maintained, whilst negative environmental side effects, for instance, by leaching and run-off of N and P, are avoided. Independently of the land use system, variable rate application of lime seems an appropriate measure to reduce the availability of undesirable heavy metals in soils and their uptake by plants.

3. Silicon—Multifunctional Mediator against Biotic, Drought, Salinity, and Heavy Metal Stress

Amongst all minerals, it is Si which manifests significant effects against biotic and abiotic stress under field conditions. The Si content in plants varies between 0.1 and 10% d.w. [25]. Plants show an active Si uptake, a passive Si uptake by diffusion, and a rejective Si uptake. Based on this differentiation one can distinguish between Si accumulators (>1.0% Si), intermediates, and excluders (<0.5% Si). Accumulating plants are sugarcane and rice, intermediates dryland *Gramineae*, while *Brassica* crops and potatoes are Si rejective [26]. Intrinsically, positive effects of Si applications can be expected for Si accumulating crop plants. Plants take up Si in the form of the uncharged molecule H_4SiO_4 (orthosilicic acid) by roots, while no hard evidence exists that plants can absorb Si through their leaves. This implies, in the case of biotic stress, that foliar sprays with Si have a direct impact on the pathogen, for instance, an osmotic effect on spores. Alternatively, the alkalisation of condensate water in branches of the crop may yield a fungitoxic effect [27]. Comprehensive data about Si in plants, the role of Si against biotic and abiotic stress, and fertiliser recommendations have been published in textbooks and reviews, and will provide detailed information about different aspects [3,28–30].

There is a green line through the mode of action of Si in response to abiotic and biotic stress, which is a fortification of the structural integrity of cell walls, the stimulation of the synthesis of defence components, and the contribution to the osmotic adjustment together with ion balance (homeostasis). The key functions in relation to the various stress factors are summarised next.

Si proved to be highly efficient against biotrophic and hemi-biotrophic fungi, such as mildew, *Septoria tritici*, and *Fusarium*, after soil application [31]. Several modes of action of Si against fungal diseases seem to be involved in resistance to infection. Hereby, it can be seen that Si acts as a natural activator of plant disease resistance. Si increases the structural integrity of cells by incorporating Si (amorphous $(SiO_2)m\cdot n(H_2O)$) in cell walls and intercellular spaces [32]. In the shoot, orthosilicic acid is polymerised to amorphous silica $((SiO_2)m\cdot n(H_2O))$ which is deposited in specific cells. This deposited Si in plants functions as a physical barrier. In addition, soluble silicic acid may act prophylactic against pathogens. Besides this function, Si induces resistance pathways, for instance, phytoalexin synthesis [32] and synthesis of phenolic compounds [33]. Si fertilisation at rates of 5–15 t/ha have been most efficient in reducing the disease incidence of plants infected by biotrophic, hemi-biotrophic, and necrotrophic fungi by 40–70% in Si hyper- and semi-hyperaccumulating crops [3]. The disease intensities of some diseases for Si-supplied plants were lowered to the same level achieved with fungicides [28]. A similar efficacy has been observed for elemental S applications, and the disease incidence and severity of cereals with *Fusarium* head blight [34].

Si mitigates phytotoxicity of heavy metals, such as Mn, Al, Cu, Zn, Cd, As, Cr, and Pb [29]. Most studies concentrated on the influence of Si applications on alleviating Mn and Al toxicity. In principle, two mechanisms can be distinguished. Firstly, *ex planta* mechanisms in form of a pH effect, which results in reduced availability of the heavy metal and formation of more stable heavy metal fractions. The latter involves the formation of hydroxy-aluminosilicate (HAS) and binding of heavy metals to organic matter and crystalline Fe-oxides instead of labile heavy metal pools. Secondly, are *in planta* mechanisms that inhibit Mn uptake [35], restrict root to shoot transport, and induce a more even co-distribution of Si and Mn. Other studies showed an enhanced binding of Mn in cell walls and the detoxification of apoplastic Mn by soluble Si [36], and the co-precipitation and/or co-deposition of Si and Al, which reduces the Al transport into the symplast [37].

The percentage of arable land worldwide, which is adversely affected by salinity, amounts to some 20%, so any agrotechnical measure to tackle the problem deserves attention. Si increases the tolerance against salt stress significantly [29]. Research suggests that phytoliths (SiO_2)m·n(H_2O)) in the apoplast enhances water retention by reducing transpirational water loss. Next, soluble Si in the symplast is involved in the biosynthesis of hormones, antioxidant defence enzymes, H^+-pumps, and osmolytes to rebalance ion stoichiometry, reduce membrane permeability and losses of electrolytes, and improve membrane structure and stability [28,30]. Other studies showed that Si reduced Na and increased K uptake/content of shoots and roots, in vivo, through the influence of proton pumps on plasma membranes [29]. Thus, Si increased the photosynthetic activity, leaf area, chlorophyll content chloroplast structure, and biomass production.

About 30% of land worldwide is arid or semi-arid, and climate change will enforce the problem of drought in the near future. Si increases the tolerance to drought, low temperature, and UV-B radiation stress [28,29,38]. Si decreases cuticle transpiration mechanically, and is involved in osmotic adjustment in plant metabolism. Physiologically and biochemically, Si maintains membrane stability and functions, decreases oxidative damage, and increases antioxidant defence. Thus, Si improves water retention, root water uptake root growth, and increases photosynthesis and crop growth.

The positive effects of Si on plant metabolism, particularly against abiotic and biotic stress, are numerous and impressive. Next, an attempt was made to attribute the various physiological effects of Si to that of the corresponding essential plant nutrient and detrimental minerals, respectively, in order to obtain an improved overview of the reaction behaviour of Si in plants (Table 2).

Sterner and Elser [39] have provided a comprehensive overview of homeostasis in the context of stoichiometry in the animate and inanimate world. Homeostasis assumes that complex physiological processes maintain a steady state in the plant organism, unless any severe imbalance of ions taken up by the plant will impair, for instance, the functionality of chloroplasts and reduce biomass production [40]. Homeostasis is maintained most efficiently by balanced nutrient ratios in plant tissues, which then foster crop productivity, quality, and plant health, while excessive loads of minerals will reduce yield. Si seems to be able to mediate nutrient imbalances by biomineralisation and directed de-swelling of Si(OH)$_4$ after uptake by roots. A unique characteristic of Si is that it is taken up as a weak acid, Si(OH)$_4$, and by biomineralisation through a successive loss of water; heavy metals are bound in phytoliths, and SiO_2 is finally deposited in cell walls.

Table 2. Silicon, essential plant nutrients, and pollutants involved in the regulation of some stress related physiological processes.

Physiological Action	Essential Nutrient/Pollutant	Interaction	Stimulating (+) Repressive (−) Si Effect
Antagonistic uptake	As, Sb, Mn, Na, Al, Cr	⇌	−
Transpirational bypass flow change	Na, Cl	⇌	−
Biological silicification—heavy metals bound in in cell walls and phytoliths	Cd, Zn	⇌	+
Stimulation of phytoalexin synthesis	$CuSO_4$		+
Osmotic adjustment	K, Na, Cl		+
Membrane stability (physical barrier)	Ca		+
Ion homeostasis	K/Na balance (salt stress)		+
Signalling oxidative stress	K, SO_4-S, Ca		+
For references see [19,30,39,41–46]			

Debona et al. [28] concluded in their review that a minimum Si concentration in the roots and/or shoots of higher plants, especially monocots, is needed to effectively combat stress conditions that would otherwise reduce yield. The beneficial effects of Si suggest a yield increase of potatoes by 22%, rice by 30%, and sugar cane by 45% [47,48]. Si showed its highest potency to increase yield of various grasses under drought stress [49]. Similar positive effects were observed in upland rice by [50]. Favourable effects of Si on yield and oil content of canola were most probably indirectly induced through a strengthened resistance against disease, as this crop species actively excludes Si from uptake [51].

In high input systems, Si fertilisation against abiotic and biotic stress will gain relevance if suitable agrochemicals are not available and there is an augmentation of stress incidences that reduce crop productivity. Currently, the net return of prophylactic Si applications is not predictable, even if Si-containing fertiliser materials are available.

Though Si is the second most abundant element in the earth's crust [25], it is hardly available to crop plants. Basically, the following soluble Si sources are commercially available for fertilisation: Ca-metasilicate; wollastonite ($CaSiO_3$), which is an electric furnace calcium silicate slag from elemental P processing; sodium silicate (water glass) for soil and foliar application; and by-product slags (blast furnace slags, silico-manganese slags, stainless steel slags, converter slags). The major disadvantages of the three sources are the putative radon (Rn) contamination of wollastonite, the high price of water glass, and the limited solubility of by-products slags, which requires higher application rates.

4. Potassium and Sodium—Nutritional Associates against Drought

Positive yield effects of Na fertilisation are known since long [52,53]. Natrophilic crops, such as sugar beet, spinach, celery, and cabbage, contain 1–3% Na in their leaves, and the prospect of beneficial effects of Na fertilisation on crop growth are highest if the K supply is limited [54,55]. Marschner [55] and Broadley et al. [56] summarised the physiological effects of Na as follows: Na regulates the water supply by a faster closure of stomata; Na increases leaf area and number of stomata while it reduces the chlorophyll content per unit leaf area; Na stimulates the assimilate transport to roots.

NaCl fertilisation decreased the K uptake in favour of Na and Cl, which induced a higher water content in sugar beet leaves at growth stage BBCH33 and 39 ([57,58], Figure 3). Broadley et al. [56] determined thicker, more succulent leaves, which stored more water per unit leaf area after Na fertilisation.

Shabala and Cuin [59] see the maintenance of K-specific enzyme functions as the key role of an adequate K supply. With respect to drought and salinity stress, a sufficiently high K supply is indispensable for osmoregulation and stomatal functioning [43,60]. A balanced K and Na supply proved to be essential for achieving maximum yields; compensation of K by Na was obvious in the range of moderate K deficiency, but for maximum yield, a minimum K concentration in leaves of 36 mg/g proved to be necessary [58]. For maximum root yield, a concentration of 2.0 mg/g Na was required, and for maximum sugar yield, 4.0 mg/g Na was required (see Section 1). However, a higher Na uptake lowers the quality of the beet root. Haneklaus et al. [58] determined that the Na content in the beet root increased by 0.29 meq/kg f.w. per 10 kg/ha Na applied to the soil.

In 1995, a separate evaluation of the K and Na supply in northern Germany and Denmark revealed that 38% of all Danish samples and only 1% of the German samples were in the range of an insufficient K supply, with concentrations <35 mg/g K in the leaf tissue at row closing [58]. In the German samples, 4% revealed Na contents <2 mg/g, indicating a severe undersupply of this mineral. The distinctively better Na supply of Danish samples can be explained by the fact that Na fertilisation at a rate of 60 k/ha Na is a standard production technique in the country [61]. The results showed with respect to K/Na ratios that, in total, 62% of the German, but only 10% of the Danish samples showed an unfavourable nutrient relation. Highest yields could only be obtained if plants contained at least 35 mg/g K and 6 mg/g Na. Yield losses due to an imbalanced K/Na nutrition were as high as 60% or 37 t/ha [58,62]. In northern Germany, NaCl fertilisation increased beet root yield by 8.7 t/ha on a clayey soil, and 4.1 and 7.3 t/ha on loamy sand soils. The sugar yield remained unaffected on all sites [62].

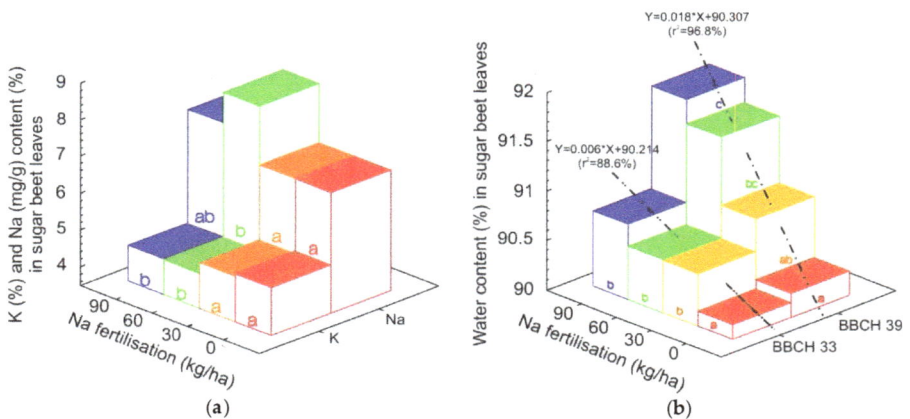

Figure 3. Influence of graded rates of Na (kg/ha) applied as NaCl on the K (%) and Na (mg/g) concentration (**a**), and water content (%) (**b**) of younger leaves of sugar beet at row closing (BBCH33 and BBCH39) at three experimental sites in northern Germany (extracted from [58,62]).

5. Seed Priming—Promoter for Improved Development at Early Stages Combats Salinity, Drought, and Nutrient Deficiency Stress

Seed priming has been applied since the 1970s, and positive effects on germination rate, plant vigour and development in early growth stages, and resistance against abiotic stress have been described under field conditions. In general, priming techniques involve imbibitions of water, hormones, chemicals, biota, and salts during the first reversible stages of germination [63]. These are summarised under the terms osmo-, hormo-, chemical, bio-, and halopriming [63]. Seed priming yields more homogenous development of seedlings and an advanced physiological status over a certain period of time [63]. Other suggested improvements are an increased efficiency of nutrient use and improved regulation of the plant water status. Seed priming not only accelerates germination, but also mechanisms which improve seed vigour. Seed priming implies an osmotic (e.g., 0.1% copper sulphate, 0.1% zinc sulphate, and 0.1% sodium sulphate) resulting in a water potential of −0.5 to −2.0 MPa, the temperature varying between 15 and 20 °C, and a duration that ranges between hours and weeks in relation to crop type. Afterwards, seeds are rinsed and dried [64]. It seems to be important during priming that the process is aerated, which improves seed performance [65]. Other factors are temperature and solution concentration. It is not possible to provide general recommendations for seed priming, as the best method varies in relation to variety, species, and seed lots [63].

Most field studies on seed priming were carried out in low input system in India and Pakistan. Varier et al. [66] suggested that priming provides extra time for the repair of damaged DNA, and an error free template for replication and transcription. Practically, the treatments improved seedling vigour, ion homeostasis (higher uptake of beneficial than toxic minerals), and starch metabolism (starch hydrolysis yields a higher content of soluble sugars) under normal and saline conditions. Primed seeds show a higher capacity for osmotic adjustment because roots take up more Na and Cl, while in leaves, the sugar and organic acid content is higher than in non-primed seeds [41]. Calcium (Ca) salts were suggested to enhance oxygen uptake and promote α-amylase activity, resulting in a higher content of total and reducing sugars [67]. Chilling stress delayed the germination of rice by more than 3 days; priming with selenium (Se) enhanced the tolerance of rice against chilling through an enhanced starch metabolism, respiration rate, anti-oxidative defence system (glutathione), and lower lipid peroxidation [68].

Main stress factors are limited soil moisture, high temperatures, and soil-crust formation in semi-arid areas, which impair germination and seedling emergence [69]. Under severe stress conditions, less than 10% of the sown seeds establish successfully [69]. Seed priming with water has been described by Wilkinson [70], and its effect on germination, crop establishment, growth, and yield has been intensely studied and reported [69,71,72]. Crop establishment in marginal rainfed areas was significantly enhanced by seed priming with water and/or nutrient solutions [69]. Seed priming with water and micronutrients was beneficial, in particular where stress affected juvenile plants [69]. As could be expected, seed priming with micronutrients enhanced their concentration in the seeds, while the content in the progeny seeds remained unaffected [73]. Calcium sulphate and calcium chloride proved to be more effective at enhancing germination of wheat under salt stress than sodium chloride [74].

Farooq et al. [75] postulate the equality of fertilisers applied to soil, foliage, and seeds, but this seems highly questionable, since under conditions of severe nutrient deficiency, particularly that of macronutrients, the rates required can only be applied to the soil. In addition, the risk of burning the plant tissue needs to be taken into account if the nutrient concentrations in foliar fertilisers are too high [76]. The same applies to in seed priming solutions, since these may result in negative yield effects [75].

The physiologically positive effects of seed priming transfer directly into yield increase. In on-farm studies, wheat yield varied between 1.2–1.4 to 4.2 t/ha in India and Pakistan, and 2.3 t/ha in Nepal [69]. Priming increased yield by 152–505 kg/ha with a mean value of 270 kg/ha. In another study, wheat grain yield increased by 200 kg/ha, and straw yield by 400 kg/ha, after seed priming with water [77]. With seed priming, 50 kg/ha N resulted in the same yield level as a crop that received 75 kg/ha N without priming [69]. These data show that seed priming enhanced N uptake of the crop plants. In addition, primed seeds showed an improved resistance to pest and diseases [69].

In the literature, results are expressed as increases in percentages, because the total yield increase is rather low, as is the yield level in general (Table 3). Maximum yield increases of 34.9 to 53.7% have been reported for seed priming of wheat with Zn, and common bean with Mo, respectively (Table 3).

To recap, it can be stated that seed priming is suitable for all crops, and proved to be effective against moderate and severe abiotic stress. Seed priming with nutrients or water seems a practical solution to alleviate nutrient deficiencies, drought, and salinity stress on marginal soils where crop production levels are low.

Table 3. Yield increase by seed priming [1].

Element	Crop	Yield Increase (%)
Zn	rice	6.8–29.6
	wheat	14.0–34.9
	chickpea	17.7–36.0
B	oats	8.4
	cowpea	37.3
Mo	chickpea	20.0
	French bean	34.8
Mn	wheat	12.8
Mo	common bean	11.6–53.7
Co.	common bean	5.0–52.5 [2]

[1] references in [75]; [2] economic yield.

6. Significance of Fertiliser Practices against Abiotic Stress in Practice—a Critical Assessment

A proven positive effect of an essential plant nutrient or beneficial element on abiotic stress factors under field conditions is a required premise for the recommendation and implementation of fertiliser practices on production fields. Another important aspect is the positive long-term performance of fertiliser applications in practice under stress conditions. For example, Na will increase yield of sugar beet under conditions of transient drought, but putatively negative side effects have to be taken into account, such as silting, and an increase in soluble salts in soils if Na is applied regularly. In addition, a regular, prophylactic application of Na will only be profitable if drought is a regular phenomenon in the production zone. In the case of Si fertiliser costs, efficacy of the fertiliser and practicability of the handling and application of the fertiliser materials are the main obstacles for being part of a routine fertiliser scheme. On-site experimentation seems the best solution for targeted fertiliser applications against biotic and abiotic stress, though the possibility of alleviating abiotic stress by fertilisation is strictly limited in high input systems. Rather, a balanced nutrient management seems important for counteracting soil degradation, maintaining the soil organic matter content on a site-specific level, and avoiding physical and chemical deterioration by regular liming. The implementation of biological know-how into fertiliser strategies, for example, a crop-specific S and Si fertilisation in combination with threshold applications of fungicides, would significantly limit the input of agrochemicals distinctly. In low input systems seriously affected by abiotic stress factors, seed priming with water and nutrients is a sensible measure to counteract drought and nutrient deficiencies.

Acknowledgments: The authors wish to express their most sincere gratitude to Mr. Herbert Daybell (Agrimedia, Bottesford, UK) for the linguistic revision of this chapter.

Author Contributions: All three authors collected the relevant literature and jointly performed the interpretation of the data including drawing of graphics and generation of tables; the first author wrote the paper.

Conflicts of Interest: The authors declare no conflict of interest.

References

1. Matoyer, M.; Roudart, L. *A History of World Agriculture. From the Neolithic Age to the Current Crisis*; Earthscan: London, UK; Sterling, VA, USA, 2006; p. 525.
2. Bergmann, H. *Ernährungsstörungen bei Kulturpflanzen*; Gustav Fischer Verlag: Stuttgart, Germany, 1983.
3. Datnoff, L.; Rodrigues, F.A.; Seebold, K.W. Silicon and Plant Disease. In *Mineral Nutrition and Plant Diseases*; Datnoff, L., Elmer, W., Huber, D., Eds.; APS Press: St. Paul, MN, USA, 2007; pp. 233–246.
4. Shanker, A.K.; Venkateswarlu, B. *Abiotic Stress in Plants—Mechanisms and Adaptations*; InTech: Rijeka, Croatia, 2011; p. 428. ISBN 978-953-307-394-1.
5. Weiler, E.W.; Nover, L. *Allgemeine und Molekulare Botanik*; Georg Thieme Verlag KG: Stuttgart, Germany, 2008.
6. Haneklaus, S.; Schnug, E. Impact of Agro-Technical Measures on the Strontium Uptake of Agricultural Crops. *Landbauforschung Völkenrode* **2001**, *51*, 77–86.

7. Haneklaus, S.; Bloem, E.; Schnug, E.; De Kok, L.; Stulen, I. Sulphur. In *Handbook of Plant Nutrition*; Barker, A.V., Pilbeam, D.J., Eds.; CRC Press: Boca Raton, FL, USA, 2006; pp. 183–238.

8. Ernst, W.H.O. Ecological aspects of sulphur metabolism. In *Sulfur Nutrition and Sulphur Assimilation in Higher Plants—Fundamental, Environmental and Agricultural Aspects*; Rennenberg, H., Brunold, C., De Kok, L.J., Stulen, I., Eds.; SPB Academic Publishing: Leiden, The Netherlands, 1989; pp. 131–144.

9. Prapagar, K.; Premanandharajah, P.; Indraratne, S. *Reclamation of Saline-Sodic Soils: Gypsum Amended Organic Materials*; LAP LAMBERT Academic Publishing: Riga, Latvia, 2013; p. 112.

10. Ondrasek, G.; Rengel, Z.; Veres, S. Soil Salinisation and Salt Stress in Crop Production. In *Abiotic Stress in Plants—Mechanisms and Adaptations*; Shanker, A.K., Venkateswarlu, B., Eds.; InTech: Rijeka, Croatia, 2011; pp. 171–190.

11. Bloem, E.; Haneklaus, S.; Keinwächter, M.; Paulsen, J.; Schnug, E.; Selmar, D. Stress-induced changes of bioactive compounds in *Tropaeolum majus* L. *Ind. Crops Prod.* **2014**, *60*, 349–359. [CrossRef]

12. Bloem, E.; Haneklaus, S.; Schnug, E.; Paulsen, J.; Kleinwächter, M.; Selmar, D. Steigerung der Produktqualität von Arznei- und Gewürzpflanzen durch induzierten leichten Stress—Möglichkeiten und Grenzen. *Z. Arznei Gewürzpflanzen* **2017**, *22*, 115–120.

13. Selmar, D.; Kleinwächter, M. Influencing the product quality by deliberately applying drought stress during the cultivation of medicinal plants. *Ind. Crops Prod.* **2013**, *42*, 558–566. [CrossRef]

14. Schnug, E.; Haneklaus, S. Evaluation of the Relative Significance of Sulfur and Other Essential Mineral Elements in Oilseed Rape, Cereals, and Sugar Beet Production. In *Sulfur: A Missing Link Between Soils, Crops, and Nutrition*; Crops, B., Sulfur, J.J., Eds.; CSSA-ASA-SSSA: Madison, WI, USA, 2008; pp. 219–233.

15. Schnug, E.; Heym, J.; Achwan, F. Establishing critical values for soil and plant analysis by means of the boundary line development system (BOLIDES). *Commun. Soil Sci. Plant Anal.* **1996**, *27*, 2739–2748. [CrossRef]

16. Evanylo, G.K.; Sumner, M.E. Utilization of the boundary line approach in the development of soil nutrient norms for soybean production. *Commun. Soil Sci. Plant Anal.* **1987**, *18*, 1355–1377. [CrossRef]

17. Moeller-Nielsen, P.; Frijs-Nielsen, H. Evaluation and control of the nutritional status of cereals. II. Pure effects of a nutrient. *Plant Soil* **1976**, *45*, 339–351. [CrossRef]

18. Walworth, J.L.; Letzsch, W.S.; Sumner, M.E. Use of boundary lines in establishing diagnostic norms. *Soil Sci. Soc. Am. J.* **1986**, *50*, 123–128. [CrossRef]

19. Haneklaus, S.; Bloem, E.; Schnug, E. Sulfur and plant disease. In *Mineral Nutrition and Plant Diseases*; Datnoff, L., Elmer, W., Huber, D., Eds.; APS Press: St. Paul, MN, USA, 2007; pp. 101–118.

20. Haneklaus, S.; Schnug, E.; Lottermoser, B.; Hu, Z. Lanthanides. In *Handbook of Plant Nutrition*; Barker, A.V., Pilbeam, D.J., Eds.; CRC Press: Boca Raton, FL, USA, 2015; pp. 625–649.

21. Reuter, D.J.; Robinson, J.B. *Plant Analysis—An Interpretation Manual*; CSIRO Publishing: Collingwood, Australia, 1997.

22. Smith, F.W.; Loneragan, J.F. Interpretation of plant analysis: Concepts and principles. In *Plant Analysis—An Interpretation Manual*; Reuter, D.J., Robinson, J.B., Eds.; CSIRO Publishing: Collingwood, Australia, 1997; pp. 3–26.

23. Vielemeyer, H.-P.; Neubert, P.; Hundt, I.; Vanselow, G.; Weissert, P. Ein neues Verfahren zur Ableitung von Pflanzenanalyse-Grenzwerten fuer die Einschaetzung des Ernaehrungszustandes landwirtschaftlicher Kulturpflanzen. *Arch. Acker Pflanzenbau Bodenkde* **1983**, *27*, 445–453.

24. Haneklaus, S.; Schnug, E. Site-specific nutrient management—Objectives, current status and future research needs. In *Precision Farming—A Global Perspective*; Srinivasan, A., Ed.; Marcel Dekker: New York, NY, USA, 2006; pp. 91–151.

25. McGinnity, P. Silicon and Its Role in Crop Production—A Review. 2015. Available online: http://planttuff.com/wp-content/uploads/2015/12/silicon-agriculture-iiterature-rvw-1.pdf (accessed on 13 January 2018).

26. Tubana, B.S.; Heckman, J.R. Silicon in soils and plants. In *Silicon and Plant Diseases*; Rodrigues, F.A., Datnoff, L.E., Eds.; Springer: Cham, Switzerland, 2015; pp. 7–51.

27. Schnug, E.; von Franck, E. Bedeutung nützlicher Silizium-Effekte für intensiv angebaute landwirtschaftliche Kulturpflanzen. *Mitt. Dt. Bodenkd. Ges.* **1984**, *39*, 47–52.

28. Debona, D.; Rodrigues, F.A.; Datnoff, L.E. Silicon's role in abiotic and biotic plant stresses. *Annu. Rev. Phytopathol.* **2017**, *55*, 85–107. [CrossRef] [PubMed]

29. Liang, Y.; Nikolic, M.; Bélanger, R.; Gong, H.; Song, A. *Silicon in Agriculture—From Theory to Practice*; Springer: Heidelberg, Germany, 2015.

30. Meharg, C.; Meharg, A.A. Silicon, the silver bullet for mitigating biotic and abiotic stress, and improving grain quality in rice? *Environ. Exp. Bot.* **2015**, *120*, 8–17. [CrossRef]

31. Rodgers-Gray, B.S.; Shaw, M.W. Substantial reductions in winter wheat diseases caused by addition of straw but not manure to soil. *Plant Pathol.* **2000**, *49*, 590–599. [CrossRef]

32. Rodrigues, F.A.; McNally, D.J.; Datnoff, L.E.; Jones, J.B.; Labbé, C.; Benhamou, N.; Bélanger, R.R. Silicon enhances the accumulation of diterpenoid phytoalexins in rice: A potential mechanism for blast resistance. *Phytopathology* **2004**, *94*, 177–183. [CrossRef] [PubMed]

33. Carver, T.L.W.; Robbins, M.P.; Thomas, B.J.; Raistrick, N.; Zeyen, R.J. Silicon deprivation enhances localized autofluorescent responses and phenylalanine ammonia-lyase activity in oat attacked by *Blumeria graminis*. *Physiol. Mol. Plant Pathol.* **1998**, *52*, 245–257. [CrossRef]

34. Haneklaus, S.; Bloem, E.; Funder, U.; Schnug, E. Effect of foliar-applied elemental sulphur on Fusarium infections in barley. *Landbauforschung Völkenrode* **2007**, *57*, 213–217.

35. Tavvakkoli, E.; Lyons, G.; English, P.; Guppy, C.N. Silicon nutrition of rice is affected by soil pH, weathering and silicon fertilization. *J. Plant Nutr. Soil Sci.* **2011**, *174*, 437–446. [CrossRef]

36. Iwasaki, K.; Maier, P.; Fecht, M.; Horst, W.J. Leaf apoplastic silicon enhances manganese tolerance of cowpea (*Vigna unguiculata*). *J. Plant Physiol.* **2002**, *159*, 167–173. [CrossRef]

37. Cocker, K.M.; Evans, D.E.; Hodson, M.J. The amelioration of aluminium toxicity by silicon in wheat (*Triticum aestivum* L.): Malate exudation as evidence for an in planta mechanism. *Planta* **1998**, *204*, 318–323. [CrossRef]

38. Shen, X.; Li, X.; Li, Z.; Li, J.; Duan, L.; Eneji, A.E. Growth, Physiological Attributes and Antioxidant Enzyme Activities in Soybean Seedlings Treated With or Without Silicon Under UV-B Radiation. *Stress J. Agron. Crop Sci.* **2010**, *196*, 431–439. [CrossRef]

39. Sterner, R.W.; Elser, J.J. *Ecological Stoichiometry: The Biology of Elements from Molecules to the Biosphere*; Princeton University Press: Princeton, NJ, USA, 2002; pp. 1–584.

40. Solymosi, K.; Bertrand, M. Soil metals, chloroplast, and secure crop production: A review. *Agron. Sustain. Dev.* **2010**, *32*, 245–272. [CrossRef]

41. Cayuela, E.; Perez-Alfocea, F.; Caro, M.; Bolarin, M.C. Priming of seeds with NaCl induces physiological changes in tomato plants grown under salt stress. *Physiol. Plant.* **1996**, *96*, 231–236. [CrossRef]

42. Kuroyanagi, M.; Arakaya, Y.; Mikami, K.; Yoshida, N.; Kawahar, T.; Hayashi, H.; Ishimaru, H. Phytoalexins from hairy root culture of *Hyoscyamus albus* treated with methyl jasmonate. *J. Nat. Prod.* **1998**, *61*, 1516–1519. [CrossRef] [PubMed]

43. Osterhuis, D.M.; Loka, D.A.; Raper, T.B. Potassium and stress alleviation: Physiological functions and management of cotton. *J. Plant Nutr. Soil Sci.* **2013**, *176*, 331–343. [CrossRef]

44. Sewelam, N.; Kazan, K.; Schenk, P.M. Global plant stress signalling: Reactive oxygen species at the cross-road. *Front. Plant Sci.* **2016**, *7*, 187. [CrossRef] [PubMed]

45. Zhu, J.-K. Regulation of ion homeostasis under salt stress. *Curr. Opin. Plant Biol.* **2003**, *5*, 441–445. [CrossRef]

46. Whitehead, I.M.; Threlfall, D.R. Production of phytoalexins by plant tissue cultures. *J. Biotechnol.* **1992**, *26*, 63–81. [CrossRef]

47. Luz, J.M.Q.; Rodrigues, C.R.; Goncalves, M.V.; Coelho, L. The effect of silicate on potatoes in Minas Gerais, Brazil. In Proceedings of the 4th International Conference on Silicon in Agriculture Port Edward, South Africa, 31 October 2008; Universidade Federal de Uberlandia: Amazonas 4C 127 Uberlandia, Brazil, 2008; p. 67.

48. Kingston, G. Silicon Fertilisers—Requirements and field experiences. In Proceedings of the 4th International Conference on Silicon in Agriculture, Port Edward, South Africa, 31 October 2008; p. 52.

49. Eneji, A.E.; Inanaga, S.; Muranaka, S.; Li, J.; Hattori, T.; An, P.; Tsuji, W. Growth and nutrient use in four grasses under drought stress as mediated by silicon fertilizers. *J. Plant Nut.* **2008**, *31*, 355–365. [CrossRef]

50. Nolla, A.; de Faria, R.J.; Korndörfer, G.H.; da Silva, T.R.B. Effect of silicon on drought tolerance of upland rice. *J. Food Agric. Environ.* **2012**, *10*, 269–272.

51. Lynch, M. Silicates in contemporary Australian farming: A 20 year review. In Proceedings of the 4th International Conference on Silicon in Agriculture, Port Edward, South Africa, 31 October 2008; North Coast Testing Services: Bellingen, NSW, Australia; p. 49.

52. Harmer, P.M.; Benne, E.J. Sodium as a crop nutrient. *Soil Sci.* **1945**, *60*, 137–148. [CrossRef]

53. Truog, E.; Berger, K.C.; Attoe, O.J. Response of nine economic plants to fertilization with sodium. *Soil Sci.* **1953**, *76*, 41–50. [CrossRef]

54. Finck, A. *Pflanzenernährung in Stichworten*; Hirt Verlag: Kiel, Germany, 1978; p. 200.
55. Marschner, H. *Mineral Nutrition of Higher Plants*; Academic Press: London, UK, 1986; p. 341.
56. Broadley, M.; Brown, P.; Cakmak, I.; Ma, J.F.; Rengel, Z.; Zhao, F. Beneficial elements. In *Mineral Nutrition of Higher Plants*; Marschner, H., Ed.; Elsevier Academic Press: Amsterdam, The Netherlands, 2012; pp. 249–269.
57. Stauß, R.; Bleiholder, H.; Van der Boom, T.; Buhr, L.; Hack, H.; Hess, M.; Klose, R.; Meier, U.; Weber, E. *Einheitliche Codierung der phänologischen Entwicklungsstadien mono- und Dikotyler Pflanzen*; Ciba Geigy AG: Basel, Switzerland, 1994.
58. Haneklaus, S.; Knudsen, L.; Schnug, E. Minimum factors in the mineral nutrition of field grown sugar beets in northern Germany and eastern Denmark. *Asp. Appl. Biol.* **1998**, *52*, 57–64.
59. Shabala, S.; Cuin, T.A. Potassium transport and plant salt tolerance. *Physiol. Plant.* **2007**, *133*, 651–669. [CrossRef]
60. Zörb, C.; Senbayaram, M.; Peiter, E. Potassium in agriculture—Status and perspectives. *J. Plant Physiol.* **2014**, *171*, 656–669. [CrossRef] [PubMed]
61. Knudsen, L. SEGES, Skejby, Denmark. Personal communication, 2018.
62. Haneklaus, S.; Knudsen, L.; Schnug, E. Relationship between potassium and sodium in sugar beet. *Commun. Soil Sci. Plant Anal.* **1998**, *29*, 1793–1798. [CrossRef]
63. Lutts, S.; Benincasa, P.; Wojtyla, L.; Kubala, S.S.; Pace, R.; Lechowska, K.; Quinet, M.; Garnczarska, M. Seed Priming: New Comprehensive Approaches for an Old Empirical Technique. In *New Challenges in Seed Biology—Basic and Translational Research Driving Seed Technology*; InTech: Rijeka, Croatia, 2016; pp. 1–47. [CrossRef]
64. Taylor, A.G.; Harman, G.E. Concepts and technologies of selected seed treatments. *Annu. Rev. Phytopathol.* **1990**, *28*, 321–339. [CrossRef]
65. Farooq, M.; Aziz, T.; Rehman, H.; Rehman, A.; Cheema, S.A.; Aziz, T. Evaluating surface drying and re-drying for wheat seed priming with polyamines: Effects of emergence, early seedling growth and starch metabolism. *Acta Physiol. Plant.* **2011**, *33*, 1707–1713. [CrossRef]
66. Varier, A.; Vari, A.K.; Dadlani, M. The subcellular basis of seed priming. *Curr. Sci.* **2010**, *99*, 450–456.
67. Lee, S.S.; Song, K. Total sugars, alpha-amylase activity, and germination after priming of normal and aged rice seeds. *Korean J. Crop Sci.* **2000**, *45*, 108–111.
68. Hussain, S.; Khan, F.; Hussain, H.A.; Nie, L. Physiological and biochemical mechanisms of seed priming-induced chilling tolerance in rice cultivars. *Front. Plant Sci.* **2016**, *7*, 116. [CrossRef] [PubMed]
69. Harris, D. Development and testing of "on-farm" seed priming *Adv. Agron.* **2006**, *90*, 129–178.
70. Wilkinson, A.E. Soaking seeds before planting. *Mark. Grow. J.* **1918**, *6*, 22–26.
71. Hegarty, T.W. The physiology of seed hydration and dehydration, and the relation between water stress and the control of germination: A review. *Plant Cell Environ.* **1978**, *1*, 101–119. [CrossRef]
72. Ashraf, M.; Foolad, M.R. Pre-sowing seed treatment: A shotgun approach to improve germination, plant growth and crop yield under saline and non-saline conditions. *Adv. Agron.* **2006**, *88*, 223–271.
73. Johnson, S.E.; Lauren, J.G.; Welch, R.M.; Duxbury, J.M. A comparison of the effects of micronutrient seed priming and soil fertilization on the mineral nutrition of chickpea (*Cicer arietinum*), lentil (*Lens culinaris*), rice (*Oryza sativa*) and wheat (*Triticum aestivum*) in Nepal. *Exp. Agric.* **2005**, *41*, 427–448. [CrossRef]
74. Afzal, I.; Rauf, S.; Basra, S.M.A.; Murtaza, G. Halopriming improves vigor, metabolism of reserves and ionic contents in wheat seedlings under salt stress. *Plant Soil Environ.* **2008**, *54*, 382–388. [CrossRef]
75. Farooq, M.; Wahid, A.; Kadambot Siddique, H.M. Micronutrient application through seed treatments—A review. *J. Soil Sci. Plant Nutr.* **2012**, *12*, 125–142. [CrossRef]
76. Fageria, N.K.; Barbosa, M.P.; Moreira, A.; Guimaraes, C.M. Foliar Fertilization of Crop Plants. *J. Plant Nutr.* **2009**, *32*, 1044–1064. [CrossRef]
77. Dhingra, K.K.; Gill, G.S.; Kaul, J.N. Agronomic studies on the late-sown wheat. *J. Res.* **1978**, *11*, 262–268.

agriculture

MDPI

Article

Effect of Elemental Sulfur as Fertilizer Ingredient on the Mobilization of Iron from the Iron Pools of a Calcareous Soil Cultivated with Durum Wheat and the Crop's Iron and Sulfur Nutrition

Dimitris L. Bouranis [1,*], Styliani N. Chorianopoulou [1], Miltiadis Margetis [1], Georgios I. Saridis [2] and Petros P. Sigalas [1]

[1] Plant Physiology Laboratory, Crop Science Department, Agricultural University of Athens, 75 Iera Odos, 11855 Athens, Greece; s.chorianopoulou@aua.gr (S.N.C.); miltosmargetis@hotmail.com (M.M.); sigalas-peter@hotmail.com (P.P.S.)
[2] Botanical Institute, Cologne Biocenter, University of Cologne, D–50674 Cologne, Germany; georg.saridis@gmail.com
* Correspondence: bouranis@aua.gr

Received: 28 October 2017; Accepted: 29 January 2018; Published: 1 February 2018

Abstract: The granules of conventional fertilizers have been enriched recently with 2% elemental sulfur (S^0) via a binding material of organic nature and such fertilizers are suitable for large scale agriculture. In a previous work, we demonstrated that a durum wheat crop that received the enriched fertilization scheme (FBS^0-crop) accumulated a higher amount of Fe compared to the durum wheat crop fertilized by the corresponding conventional fertilization scheme (F-crop). In this study, we investigated the effect of S^0 on the contingent mobilization of iron from the iron pools of the calcareous field that affiliated the durum wheat crop and the corresponding effect on the crop's iron nutrition and sulfur nutrition. A sequential extraction of Fe from root zone soil (rhizosoil) was applied and the fluctuations of these fractions during crop development were monitored. The fertilization with FBS^0 at sowing affected the iron fractions of the rhizosoil towards iron mobilization, thus providing more iron to the crop, which apart from the iron nutrition fortified the crop's sulfur nutrition, too. No iron was found as iron attached to carbonates of the rhizosoil. Fluctuations of the iron pool, bound or adsorbed to the organic matter, were exactly the opposite to those of the iron pool associated with the clay particles in both treatments, suggesting iron exchange between the two pools. Replenishment of the F-crop's Fe content and a deficit in the FBS^0-crop's Fe content in the rhizosoil were found at the end of the cultivation period. Furthermore, the initiation of the fast stem elongation stage (day 125) constituted a turning point. Before day 125, the use of FBS^0 increased the iron concentration in the main stems and this was an early fortification effect, followed by an increase in the organic S concentration. Following day 125, the FBS^0-crop consisted of plants with higher main stems and less tillers. A late fortification effect was observed in the iron concentration of the main stems and their heads after the stage of complete flowering. Prior to harvesting in the FBS^0-crop, all plant parts were heavier, with more iron and organic sulfur accumulated in these plant parts, and the obtained commercial yield of the FBS^0-crop was higher by 27.3%.

Keywords: iron dynamics; sulfur; sulfate; durum wheat crop; calcareous soil; elemental sulfur; enriched fertilizer; rhizosoil iron fractions

---

1. Introduction

Sulfur (S) fertilizers with very different physical and chemical forms, have been developed to alleviate S deficiency, and they are distinguished by two categories: (1) S fertilizers containing sulfate forms, i.e., the directly available S chemical form for plant uptake and (2) S fertilizers containing non-sulfate forms [1]. The S content of the latter is not directly available for plant uptake and requires oxidative conversion to sulfate. Being effective in the rapid alleviation of crop S deficiency, sulfate S fertilizers are widely used [1]. The factors affecting the oxidation of elemental S (S^0) in soils have been reviewed [2,3]. From the non-sulfate fertilizers, the ones based on S^0 are still a challenge. S^0 is an emerging fertilizer product and market, which apart from the S^0 itself includes micronized granular S^0, nitrogen and/or phosphate fertilizers enhanced/enriched with S^0 and S^0-coated fertilizers [4].

Recently, S^0 has been attached successfully onto the surface of the granules of commercial fertilizers (F) via a binder (B) by Sulphur Hellas S.A., under the commercial name "Sulfogrow" (FBS^0) [5] for use in large scale agriculture. The F granules act as a core, effectively covered by an amount of 2% (w/w) of S^0 yellow dust. Does such a small amount of S^0 contribute to the efficacy of the granules or is it negligible? In a previous work [6] on the nutritional dynamics of a durum wheat (*Triticum durum*, Poaceae) commercial crop during development, we reported that the FBS^0-treated crop presented denser plantation with more robust plants in comparison to the conventional F-treated crop, whilst the accumulated amounts of iron per plant were found to be significantly increased at day 61 after sowing in the above ground crop part, comprising an early effect. After day 100, the accumulated dry mass per plant was twice that of the control and iron accumulation curves per plant were statistically higher than the control ones. It was suggested that the FBS^0 product was more effective in comparison to its core fertilizer, as regards the examined parameters, i.e., the dry mass, sulfate, total sulfur, organic nitrogen, and iron concentrations in the above ground plant part, highlighting the fact that higher early mobilization of Fe coincided with a decrease in rhizosoil content of humic substances [6]. Given that the crop was established and developed on a calcareous soil, in which Fe availability is highly restricted due to Fe^{3+} precipitation and immobilization, the fact that the FBS^0-crop's plants accumulated greater amounts of Fe implies that soil reserves of Fe were of adequate amounts to support the enhanced growth of FBS^0-crop's plants and that they had been mobilized.

In calcareous soils, calcium carbonate buffers soil solution pH in the range 7.5–8.5 [7] and elevated bicarbonate concentration is present in the solution [8]. The solubility of Fe in well aerated soils is controlled by both the Fe oxides [9] and the pH of calcareous soils. The most soluble Fe oxide limits total soluble Fe concentration at around 10^{-10} M, much lower than that required (10^{-8} M) for optimal plant growth [10]. In addition, bicarbonate may hinder Fe uptake and its translocation in plants [11]. In calcareous soils Fe not only is little soluble [12–14], but usually the concentrations of soluble and exchangeable Fe are much lower than those necessary for adequate plant growth and plants have also developed mechanisms to make other forms of Fe available [13]. Most Fe in soil is in the ferric form. Being a Poaceae species, wheat follows the Strategy II to take up the ferric iron, by the action of phytosiderophores [15].

The aim of this work was to elaborate on the effect of S^0 as a fertilizer ingredient on the mobilization of iron from the several iron pools of the calcareous field that affiliated the durum wheat crop. To this end, a fractionation scheme was adopted to distinguish the various fractions of rhizosoil iron, in order to depict the iron dynamics in the rhizosoil of FBS^0-treated crop. The allocation of iron among the various pools within the rhizosoil was correlated with the allocation of iron in the main stem, the tillers, and the corresponding heads. Furthermore, given the strong relationship between iron and sulfur metabolisms, iron allocation in the above ground plant parts was studied in relation to the corresponding allocations of sulfate and organic (reduced) sulfur.

2. Materials and Methods

2.1. Experimental Field Trial

A durum wheat (*Triticum durum*, cv. SIMETO) commercial crop was established in Lefktra (latitude 38.25° N, longitude 23.18° E, 352 m a.s.l.) at Viotia county, central Greece, in a production field of 2.2 ha with calcareous soil. The area was divided into two parts of 1.1 ha each; one of them was subject to conventional F-treatment according to the local agricultural practices (F-crop), whilst the other one received the corresponding FBS^0-treatment (FBS^0-crop, Figure 1). In order to ensure comparable soil conditions, prior to crop establishment each area was arbitrarily divided into plots 15 m × 7 m (105 m^2) each (105 plots in total). All 44 perimetric plots were excluded, whilst the internal 60 plots were grouped into five successive groups of 12 plots each. Within each group of plots, soil quality was tested in a random fashion by analyzing one composite sample per plot collected at the depth of ca 20 cm, until a set of (5 + 5) plots with comparable quality was secured, one plot within each group (Table 1). Then, for each of these plots, two more composite soil samples per plot were collected and analyzed. At each sampling day, sampling took place from the same plots.

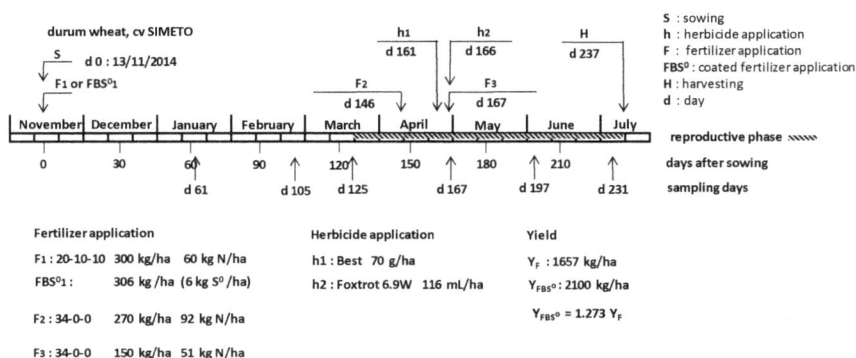

Figure 1. Overview of experimental work. On day 61 initiation of tillering took place. Stem elongation proceeded between days 105 to 146, whilst on day 197 the crop was in the process of grain filling.

Table 1. Soil quality of the experimental fields.

Field	Clay %	Loam %	Sand %	Class	CaCO$_3$ %	NO$_3^-$ ppm	P-Olsen ppm	K_{exch} ppm	Mn-DTPA ppm	Cu-DTPA ppm	Zn-DTPA ppm
F-crop	42	28	30	C	13.1	10.51	18.2	170	7.51	1.74	1.1
FBS^0-crop	41	31	28	C	16.7	11.68	19.7	240	7.84	1.5	0.92

F-crop: the crop that was subject to conventional fertilization (F-) treatment according to the local agricultural practices. FBS^0-crop: the crop which received the corresponding FBS^0-treatment. K_{exch}: soil exchangable potassium. DTPA: diethylenetriaminepentaacetic acid

Sowing day and fertilizer application took place in 13 November 2014 (day 0). The control crop was fertilized with a commercial 20-10-10 fertilizer (nitrogen was provided as ammonium sulfate, phosphorus as triple superphosphate, whilst potassium was provided as potassium sulfate) at a rate of 300 kg ha^{-1}. The FBS^0-treated crop received the equivalent fertilization with the corresponding "Sulfogrow" 20-10-10 commercial fertilizer at the same rate, carrying 2% S^0 (306 kg ha^{-1}). At days 146 and 167 after sowing, additional fertilizations with commercial ammonium nitrate fertilizer took place at the rates of 270 and 150 kg ha^{-1}, respectively. At days 161 and 166 after sowing, herbicide applications took place at the rates of 70 g ha^{-1} (Best) and 1.1 L ha^{-1} (Foxtrot 6.9W), respectively. Both crops received no irrigation. For the determination of morphometric characteristics, 20 plants per plot were used. For the chemical analyses, at least five plants per plot were collected with their root system and the surrounding soil by means of a shovel. The excess of soil was removed by hand and the soil within the root system mass, i.e., the rhizosoil (RS), was collected. The above ground plant part was separated into the main stem, the accompanying tillers, and the corresponding heads, and the fluctuation dynamics of iron, sulfate and organic sulfur were monitored in each plant part during the crops' development.

2.2. The Nature of the FBS⁰ Fertilizer Granules

2.2. The Nature of the FBS^0 Fertilizer Granules

According to the patent [5], the fertilizer granules are mixed initially with elemental sulfur in the form of dust at a percentage of 2% to 4% (w/w, S^0: fertilizer's granules). The mixture passes through a shower of fine droplets consisting of a 1:1 mixture of molasses and glycerol, which acts as the binding system, in such a way that the whole surface is exposed. The binder is added at a percentage of 0.4% to 1.2% (w/w, binder: fertilizer granules). Then, the mixture is led to a mixer, where the sticky S^0 dust is evenly attached onto the sticky fertilizer granules, thus forming the final product FBS^0.

2.3. Determinations of Soil Parameters

Determinations of pH, organic matter, $CaCO_3$, NO_3^-, P-Olsen, exchangeable potassium, Mn-DTPA (DTPA: diethylenetriaminepentaacetic acid), Cu-DTPA, Zn-DTPA, and humic substances content in the rhizosoil were performed according to the procedures described by Jones (1999) [16].

2.4. Preparation of Dry Mass Digests

Samples of the separated plant parts (main stems, tillers, heads) were oven-dried at 80 °C and ground to pass a 40-mesh screen using an analytical mill (IKA, model A10). Prior to iron analysis, samples were digested with hot H_2SO_4 and repeated additions of 30% H_2O_2 until the digestion was complete [17].

2.5. Rhizosoil Fractionation Scheme

Rhizosoil samples were fractionated with the modified version of the BCR-three step sequential extraction procedure, as described by Pueyo et al. (2008) [18].

Exchangeable and weak acid soluble fraction (1st fractionation step; 1st fraction): 1 g soil sample was extracted with 40 mL of 0.11 mol L^{-1} acetic acid solution by shaking in a mechanical, end-over-end shaker at 30 rpm at room temperature (20 °C) for 16 h. The extract was separated by centrifugation at 3000 rpm for 20 min, collected in polyethylene bottles and stored at 4 °C until analysis. The residue was washed by shaking for 15 min with 20 mL of doubly distilled water and then centrifuged, discarding the supernatant.

Reducible fraction (2nd fractionation step; 2nd fraction): 40 mL of 0.5 mol L^{-1} hydroxylammonium chloride solution was added to the residue from the 1st step, and the mixture was shaken at 3000 rpm at 22 °C for 16 h. The acidification of this reagent was by the addition of a 2.5% (v/v) HNO_3 solution (prepared by weighing from a suitable concentrated solution). The extract was separated and the residue was washed as in the first step.

Oxidizable fraction (3rd fractionation step; 3rd fraction): 10 mL of 8.8 mol L^{-1} hydrogen peroxide solution was carefully added to the residue from the 2nd step. The mixture was digested for 1 h at 22 °C and for 1 h at 85 °C, and the volume was reduced to less than 3 mL. A second aliquot of 10 mL of H_2O_2 was added, the mixture was digested for 1 h at 85 °C, and the volume was reduced to about 1 mL. The residue was extracted with 50 mL of 1 mol L^{-1} ammonium acetate solution, adjusted to pH 2.0, at 3000 rpm and 22 °C for 16 h. The extract was separated and the residue was washed as in previous steps.

Residual fraction (4th fractionation step; 4th fraction): the residue from the 3rd step was digested with aqua regia. In this case, the amount of acid used to attack 1 g of sample was reduced to keep the same volume/mass ratio: 7.0 mL of HCl (37%) and 2.3 mL of HNO_3 (70%) were added.

In the diluted dry mass (DM) digests or rhizosoil fraction extracts, Fe was determined by atomic absorption spectrophotometry (GBC, Model Avanta spectrophotometer, GBC Scientific Equipment PTY LTD, Dandenong, Victoria, Australia) [17].

2.6. Sulfate and Total Sulfur Determination in the DM of the Plant Parts

Sulfate concentration was determined by extracting the oven-dried samples with 2% (*v*/*v*) acetic acid aqueous solution and by analyzing by a turbidimetric method [19,20]. Total sulfur content was determined after dry ashing at 600 °C [21]. The ash was dissolved in 2% (*v*/*v*) acetic acid aqueous solution, filtered through Whatman No. 42 paper, and total sulfate was determined turbidimetrically [19,20]. The content of the organic sulfur (Sorg) was calculated by subtracting the total sulfate content from that of the total sulfur.

2.7. Statistical Analysis

The comparisons between the corresponding FBS^0-crop and F-crop values in each case were submitted to *t*-test variance analysis with two-tailed distribution and two-sample equal variance, at $p \leq 5\%$. Where the differences between means of FBS^0-crop and F-crop samples were statistically significant, the percentage of the relative change is provided in the text.

3. Results

3.1. Dynamics of the Iron Fractions in Rhizosoil during the Crops' Development

No iron was found as iron attached to carbonates of the rhizosoil (i.e., the exchangeable and weak acid soluble fraction; fr1) throughout the experiment. At sowing, the total extracted iron from the rhizosoil amounted to 8.3 mg g^{-1} RS, allocated by 11% as iron adsorbed on iron-manganese oxides (i.e., the reducible fraction; fr2), 16% as iron bound or adsorbed to the organic matter of RS (i.e., the oxidizable fraction; fr3), and 73% as iron associated with the clay particles (i.e., the residual fraction; fr4). In the F-crop the total extracted Fe started to lower at day 61, presented the lowest value at day 125 and then increased progressively, reaching the initial value at the end of the cultivation period. In the FBS^0-crop, the lowest value appeared already on day 61 and then it increased and stabilized at a value of 6 mg g^{-1} RS, i.e., 72.3% of the initial, after day 105 onward (Figure 2A). Analyzing the percentage contribution of each fraction to the total extracted iron a different profile appeared in each treatment. In the F-crop (Figure 2B) the fluctuations of fr3 were the opposite of those of fr4. More specifically, fr4 decreased between days 105–167, then it increased till the end of the cultivation, whilst fr3 followed exactly the opposite course. The contribution of fr2 was a minor one and it could be considered as rather stable. In the FBS^0-crop (Figure 2C), the percentage contributions of fr4 and fr3 fluctuated in an exactly opposite fashion, and the fluctuation pattern was different from that of the F-crop. In fact it appears that the percentage contributions of both Fe fractions oscillated around the initial values of 73% and 16% respectively. Again the contribution of fr2 was a minor one, it was at the same percentage (ca. 12%) of that of the F-crop and could be considered as stable, too. Regardless of the pattern, fluctuations of fr3 were exactly the opposite of those of fr4 in both treatments, suggesting iron exchange between the pool of iron bound or adsorbed to the organic matter and the pool of iron associated with the clay particles. Analyzing the fractions dynamics it seems that day 125 was a turning point in the various fluctuation patterns.

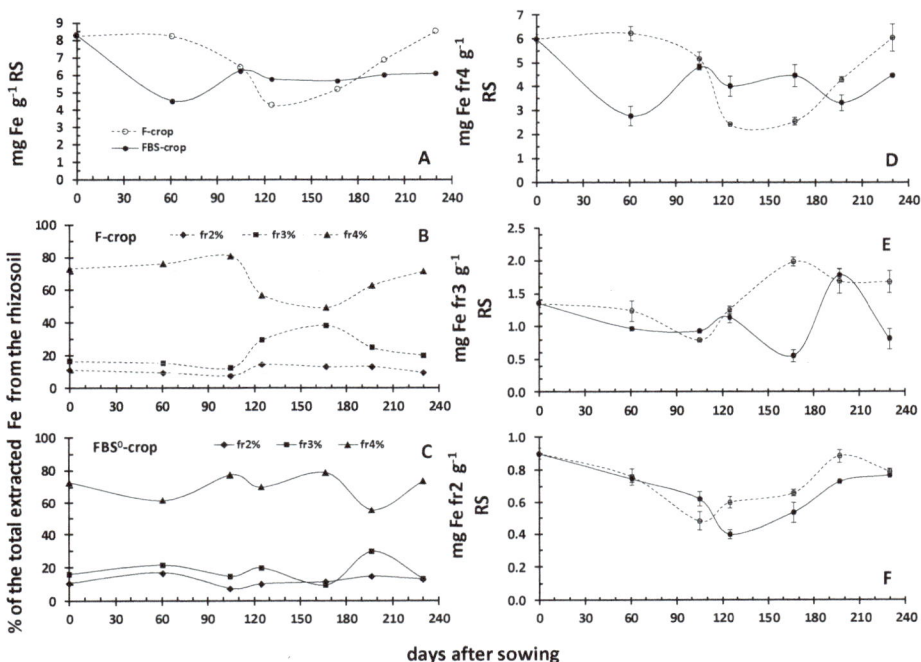

Figure 2. The dynamics of iron pools in the rhizosoil during the crop's development. The total extracted iron from the rhizosoil (RS; (**A**)), the percentage contribution of each fraction in the F-crop (**B**) and in the FBS^0-crop (**C**), along with the iron contents of 4th (**D**), 3rd (**E**), and the 2nd (**F**) fraction.

Was there a role of humic substances (HS) in the fluctuations of fr3? The third fraction contains iron that is closely related to OM and HS are a functional component of the rhizosoil's OM. The overall picture of the HS content's fluctuation of the rhizosoil revealed an oscillation around the value of 4.4 mg HS g^{-1} RS [6]. The incorporated S^0 did not affect the oscillation pattern. However, up to day 125 there was a tendency for lower HS content, and a tendency for higher HS content afterwards [6]. The fr3-to-HS ratio presented the same fluctuation pattern around 0.3 mg fr3 per mg HS up to day 125, and it differentiated thereafter (Figure 3A). In the FBS^0-crop it followed the oscillation pattern of the HS content in the rhizosoil around the value of 0.5 mg fr3 per mg HS, whilst in the F-crop it followed the pattern exactly in reverse around the value of 0.2 mg fr3 per mg HS. Analyzing the patterns in relation to organic matter dynamics (Figure 3B,C), again day 125 was a turning point. Before day 125, in the FBS^0-crop there was a tendency for less HS and more Fe per unit mass of organic matter, which reversed thereafter. In relation to the initial conditions, at the end of the crop: (i) The HS content per unit of RS was that of the initial one. (ii) The fr3:HS mass ratio increased (+48%) in the F-crop, whilst it decreased (−29%) in the FBS^0-crop. (iii) The HS:OM mass ratio decreased in both the F-crop (−8.5%) and the FBS^0-crop (−23%). (iv) The fr3:OM mass ratio increased (+43%) in the F-crop and decreased (−43%) in the FBS^0-crop.

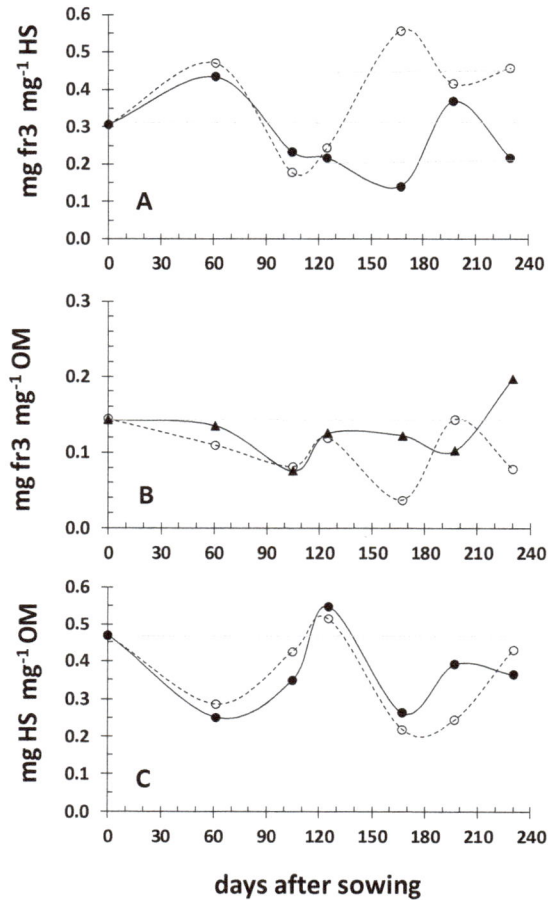

Figure 3. The dynamics of the 3rd fraction content in relation to the contents of the organic matter (OM), and the humic substances (HS) in the rhizosoil during the development of the F-crop (open cycles; dashed line) and the FBS^0-crop (full circles; solid line). (**A**) fr3-to-HS, (**B**) fr3-to-OM, and (**C**) HS-to-OM mass ratios.

3.2. Dynamics of Iron and Sulfur Nutrition in the Crop's above Ground Plant Parts during the Crops' Development, in Relation to the Dynamics of Rhizosoil Iron Pools

At the initiation of the tillering stage (day 61) both crops presented the highest Fe concentration in the DM of the main stems. In the FBS^0-crop in relation to F-crop, iron concentration was 3300 ppm, i.e., higher by 74% (Figure 4A), which resulted in higher accumulation of Fe in the above ground part (+42%, Figure 5B), because, in addition, dry mass per plant was higher (+18%, Figure 5A). Both sulfate concentration and accumulation remained unchanged (Figures 4B and 5C). Organic sulfur concentration was lower (−24%, Figure 4C) and its accumulation lower (−37%, Figure 5D), too. Total extracted Fe from the rhizosoil presented its lowest value (54.5% of the initial iron content at sowing, Figure 2D). All fractions contributed to this loss by 16% (fr2), 28% (fr3), and 54% (fr4) (Figure 2C).

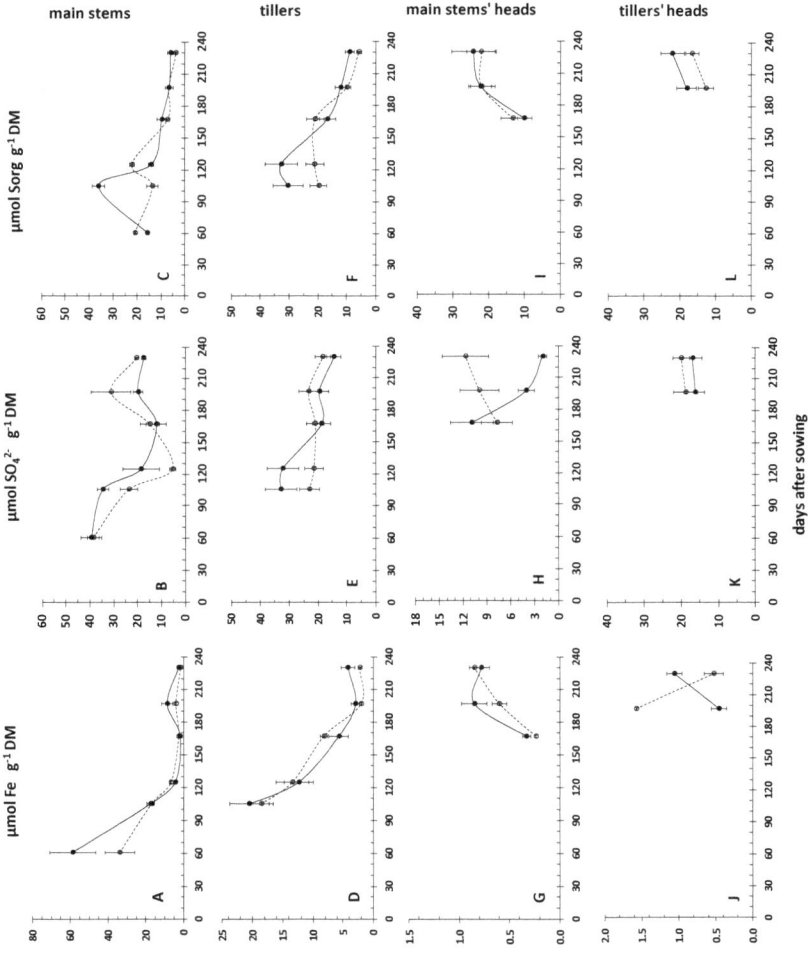

Figure 4. Dynamics or iron (Fe: (**A,D,G,J**)), sulfate (SO_4^{2-}: (**B,E,H,K**)), and organic sulfur (Sorg: (**C,F,I,L**)) concentrations in the main stems, the tillers, the main stems' heads, and the tillers' heads, respectively, during the development of the F-crop (open cycles; dashed line) and the FBS^0-crop (full circles; solid line).

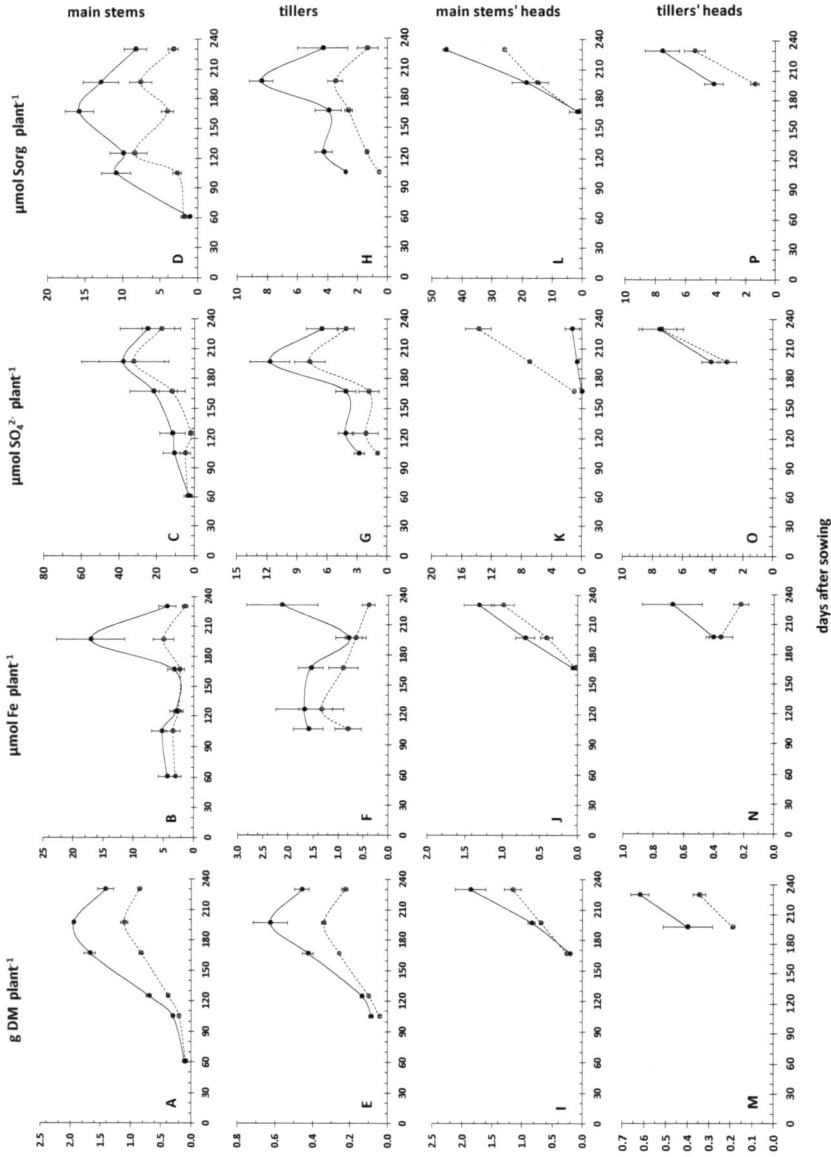

Figure 5. Dynamics of dry mass (DM: (**A,E,I,M**)), iron (Fe: (**B,F,J,N**)), sulfate (SO_4^{2-}: (**C,G,K,O**)), and organic sulfur (Sorg: (**D,H,L,P**)) accumulations per plant, in the main stems, the tillers, the main stems' heads, and the tillers' heads, respectively, during the development of the F-crop (open cycles; dashed line) and the FBS^0-crop (full circles; solid line).

At the initiation of the main stem elongation (day 105), the plants of both crops carried seven leaves in the main stems and five leaves in the tillers. Iron concentration of the main stem (Figure 4A) was found significantly reduced in both treatments in such a way that iron accumulation was the same with that of day 61, indicating a dilution effect. Tillers carried the same iron concentration with the main stem (Figure 4D). In both crops the total extracted iron from the rhizosoil was 78% of that at sowing (Figure 2A). In the F-crop, 32% of the plants carried no tillers, i.e., there were main stems only, whilst 5% of them carried three or more tillers (Table 2). In the FBS^0-crop, 44% of the plants carried no tillers and 16% 3+ tillers. Plants accumulated more dry mass in the main stems (+51.9%, Figure 5A) and in the tillers (+102.3%, Figure 5E). Relative to day 61, the total extracted iron from the rhizosoil increased (Figure 2A). The increase was due to increase of fr4 (+73%, Figure 2D), whilst fr3 remained unchanged (Figure 2E) and fr2 decreased further (−37%, Figure 2F). Sulfate concentration was higher in both the main stem (+42%, Figure 4B) and the tillers (+39%, Figure 4E), with a corresponding increase in sulfate accumulation per plant in both the main stem (+120%, Figure 5C) and the tillers (+180%, Figure 5G). Organic S concentration was higher in both the main stem (+157%, Figure 4C) and the tillers (+50%, Figure 4F). The same held true for organic S accumulation per plant in both the main stem (+66.7%, Figure 5D) and the tillers (+367%, Figure 5H).

Table 2. Tillering capacity and dynamics of the studied durum wheat variety, as affected by the applied fertilization schemes.

number	Days after Sowing			
	F-crop			
of tillers	105	125	167	197
0	32	60	63	57
1	38	22	19	21
2	25	11	15	20
3+	5	7	3	2
	FBS^0-crop			
	105	125	167	197
0	44	50	82	85
1	15	20	14	13
2	25	20	2	1
3+	16	10	2	1

F-crop: the crop that was subject to conventional fertilization (F-) treatment according to the local agricultural practices. FBS^0-crop: the crop which received the corresponding FBS^0-treatment.

At the stage of entering fast stem elongation (day 125), the plants of both crops carried eight leaves in the main stems and six leaves in the tillers. In the F-crop, 60% of the plants carried no tillers, and 7% three or more tillers. Between days 105 and 125 the main stem's length increased at a rate of 0.18 cm per day. Per plant, dry mass was allocated by 18% in the tillers (Figure 6A). In the FBS^0-crop, 50% of the plant carried no tillers, and 10% 3+ tillers. Between days 105 and 125 the main stem's length increased with a rate of 0.31 cm per day which resulted in higher main stems (+34.5%). Per plant, dry mass was allocated by 22% in the tillers (Figure 6B). In relation to F-crop, plants were characterized by higher accumulation of dry mass in the main stems (+84.7%, Figure 5A) and in the tillers (+35.6%, Figure 5E). Iron concentration was less (−33.5%, Figure 4A) in the main stem and the same in the tillers (Figure 4D), which resulted in a decrease in accumulated iron per main stem (Figure 5B) due to further dilution and in no change in the accumulated iron per tiller (Figure 5F). Sulfate concentration was higher in both the main stem (+270%, Figure 4B) and the tillers (+45%, Figure 4E), which resulted in higher sulfate accumulation, too (+500%, Figure 5C; +50%, Figure 5G, respectively). Fr2 presented its lowest value (Figure 2F) and the other two fractions minor changes.

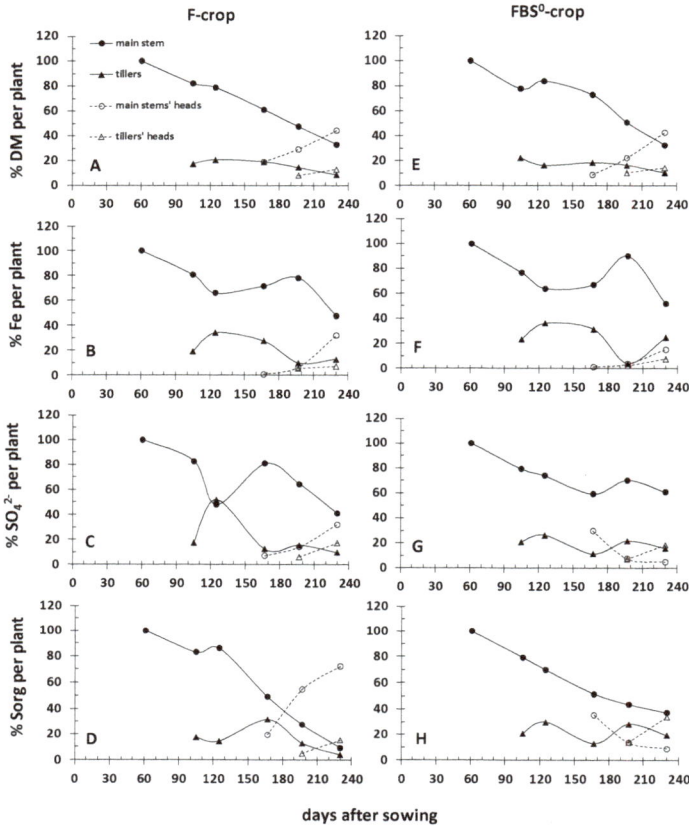

Figure 6. Comparison of the percentage contribution of the accumulated dry mass (DM: (**A,E**)), iron (Fe: (**B,F**)), sulfate ($SO_4{}^{2-}$: (**C,G**)), and organic sulfur (Sorg: (**D,H**)) per plant within the main stem, the tillers and their heads, in the F-crop and the FBS^0-crop during development, respectively.

At the stage of the complete emergence of head on the main stem (day 167), the plants of both crops carried eleven leaves and heads in the main stems. Stem elongation started to retard, whilst the boot was just visibly swollen in the tillers. Prior to this stage and between days 146 and 167 two topdressing and two herbicide applications took place. Between days 125 and 167, in the F-crop the lengths of the main stems increased with a rate of 0.61 cm per day. Of the plants 63% carried no tillers, and 3% three or more tillers. Tillers were carrying nine or ten leaves. In the FBS^0-crop the main stems lengths increased with a rate of 0.65 cm per day, which resulted in statistically higher main stems (+16.3%). Tillers were carrying five or six leaves. The 82% of the plants carried no tillers, and 2% three or more tillers. Plants had accumulated more dry mass in the main stems (+103.2%, Figure 5A), in the tillers (+63.1%, Figure 5E), and in the main stems' heads (+36.7%, Figure 5I). In the tillers, iron concentration decreased to half (Figure 4D) relative to day 125 and it was less by 30.6% relative to F-crop. These changes kept the accumulated amount of iron higher in both the main stems (+41.2%, Figure 5B) and the tillers (+73.4%, Figure 5F). In the heads of the main stems, iron concentration was higher (+37.5%, Figure 4G), which resulted in more accumulated iron per head (+121.4%, Figure 5J). Fr4 was at the same concentration compared with day 125 (Figure 2D), fr3 presented its lowest value (Figure 2E), whilst fr2 kept increasing (Figure 2F) at the same rate as in F-crop. Sulfate concentration continued to lower in both the main stem (Figure 4B) and the tillers (Figure 4E), which resulted in statistically the same amount per plant in the main stem (Figure 5C) but higher in the tillers (Figure 5F). Organic S concentrations decreased in both the main stems (Figure 4C) and the tillers (Figure 4F), whilst organic S accumulation presented its highest value in the main stem (Figure 5D) and remained stable in the tillers (Figure 5H).

At the stage of complete flowering and the initiation of milk development (day 197), the plants carried 11 leaves and heads in the main stems of both crops, stem elongation ceased, and the tillers carried heads in both crops. In the F-crop, 57% of the plants carried no tillers, and 2% three or more tillers. In the FBS^0-crop, 85% of the plants carried no tillers, and 1% three or more tillers. Main stems were higher by 17.4%. Plants presented more accumulated dry mass in the main stems (+75.2%, Figure 5A), in the tillers (+83.9%, Figure 5E), in the main stems' heads (+22.4%, Figure 5I), and in the tillers' heads (+109.1%, Figure 5M). The accumulated DM reached its peak in both the main stem and the tillers. In the FBS^0-crop, iron concentration was higher in main stems (+111.3%, Figure 4A) and their heads (+41.7%, Figure 4G), as well as in the tillers (+9%, Figure 4D), and less in the tillers' heads (−70.7%, Figure 4J). This resulted in a strong accumulation of iron in the main stem (+238%, Figure 5B), and in their heads (+70%, Figure 5J). However, in the tillers a strong dilution effect took place (Figure 5F), whilst in the tillers' heads no change was observed (Figure 5N). Fr3 was twice that of day 167 (Figure 2E), whilst fr2 increased (Figure 2F), too. Sulfate concentrations were higher in the main stems (+63%, Figure 4B), and lower in the tillers (−17.4%, Figure 4E), the main stems' heads (−59%, Figure 4H) and the tillers' heads (−15.8%, Figure 4K), which translated into the same accumulations of sulfate in the main stems (Figure 5C) and the tillers' heads (Figure 5O), higher accumulation in the tillers (+50%, Figure 5G), and a lower one in the main stems' heads (−900%, Figure 5K). As regards the concentrations of organic sulfur, they were the same in the main stem (Figure 4C), the tillers (Figure 4F), and the main stem's heads (Figure 4I), but higher in the tillers' heads (+50%, Figure 4L). The corresponding accumulations were higher in the main stems (+71.1%, Figure 5D), the tillers (+140%, Figure 5H), and the tillers' heads (+185.7%, Figure 5P), whilst the same in the main stem's heads (Figure 5L).

At the stage of dough development (day 231), the accumulated DM decreased in both the main stem (Figure 5A) and the tillers (Figure 5E) of both crops due to loss of lower leaves. In contrast, heads still accumulated DM in both the main stems (Figure 5I) and the tillers (Figure 5M) of both crops. In the FBS^0-treated crop the same pattern was observed: plants with more accumulated dry mass in the main stems (+64.5%), in the tillers (+100.1%), in the main stems' heads (+61.4%), and in the tillers' heads (+80.1%). In the FBS^0-crop, relative to day 197 the iron concentration of the main stems decreased (Figure 4A), it was stabilized in the tillers (Figure 4D) and the main stems' heads (Figure 4G), whilst it increased in the tillers' heads (Figure 4J). As regards iron accumulation it decreased in the main stems (Figure 5B) and increased in the tillers (Figure 5F), tillers' heads (Figure 5N) and main stems' heads (Figure 5J). Sulfate and organic S concentrations did not change (Figure 4), with the exception of sulfate concentration in the main stem's heads (−82.9%, Figure 4H).

3.3. FBS^0 Fortified the Crop's Iron and Sulfur Nutrition and Increased the Yield

The used durum wheat cultivar created up to three tillers; plants with four or more tillers were less than 1%. After day 167, in the F-crop the ratio of plants with main stems only stabilized to 60% (Table 2), whilst the rest of the plants with one or two tillers shared the same percentage contribution (20%). In the FBS^0-crop the ratio of plants with main stems only stabilized to 85%, whilst the rest of the plants carried only one tiller stabilized to 13%. FBS^0 decreased the tillers' production.

Moreover, the aforementioned dynamics clearly show that the corresponding fluctuations of the rhizosoil's iron did benefit the above ground plant parts in terms of dry mass, iron, sulfate, and organic sulfur nutrition. Prior to harvesting in the FBS^0-crop, per plant all plant parts had accumulated more dry mass, i.e., they were heavier (main stems: +55%, tillers: +100%, main stems' heads: +69%, tillers' heads: +82%). In accordance, the obtained commercial yield of the FBS^0-crop was higher by 27.3%.

More iron and organic sulfur accumulated in these plant parts, too. Iron fortification was observed in the main stems (+193%), in the tillers (+437%), in the main stems' heads (+31%), and in the tillers' heads (+250%). Organic sulfur fortification was found in the main stems (+191%), in the tillers (+231%), in the main stems' heads (+75%), and in the tillers' heads (+38%). As regards sulfate, it was higher in the main stems (+47%) and in the tillers (+58%). In the main stems' heads it was less (−91%), and unchanged in the tillers' heads.

The F-crop started with high iron and sulfate concentrations in the main stem. Then, Fe concentration decreased in parallel with sulfate concentration, whilst organic S concentration remained stable. Again, day 125 was a turning point. Thereafter, Fe and organic S concentrations remained low, whilst sulfate concentration increased significantly. In the tillers Fe concentration decreased, sulfate concentration remained stable, whilst organic S concentration started decreasing later on.

Therefore, the use of FBS^0 exerted some fortification effects. Initially iron concentration was even higher (early fortification effect), quickly decreasing thereafter, sulfate concentration remained high and organic S concentration increased. This fortification effect was observed to a lesser extent in the F-crop at day 125. After day 125, all iron, sulfate, and organic S concentrations remained rather stable, fluctuating around a mean value with a ratio of 1:4:2 in the main stem and 1:5:2 in the tillers. A late fortification effect was observed in the iron concentration of the main stems and their heads after the stage of complete flowering within the examined parameters.

The percentage partitioning of iron, sulfate, and organic sulfur were altered (Figure 6). Among these alterations, the most prominent was that of sulfate at day 125 (Figure 6C), where more sulfate was partitioned to the tillers. On that day, the sulfate concentration in the main stems presented a characteristically low value (Figure 4B)

4. Discussion

The used FBS^0 fertilizer in this work, apart from 2% S^0, also contained sulfate as the accompanying anion of ammonium and potassium cations. Therefore, it constitutes a third category in addition to those distinguished by Jansen et al. (1986) [1], i.e., it contains both directly and non-directly available S for plant uptake. Furthermore, both molasses and glycerol are water soluble and can be used by soil microorganisms along with the elemental sulfur, thus sustaining microbial action around the granules.

In the rhizosoil, apart from the crop's root systems, there are at least two more iron consumers, i.e., microbes and the root systems of the various weeds. In the case of the FBS^0-crop, S^0 increased the total amount of bacterial populations [22]. Therefore, apart from the phytosiderophores (PS) released by the durum wheat crop, more siderophore production, obviously of microbial origin, contributes to iron mobilization in the rhizosoil. On the other hand, both stategies of iron uptake are in action due to mixed root systems in the rhizosoil [14]. Being a graminaceous species, wheat utilizes Strategy II for iron uptake by the root system. It releases the phytosiderophore 2'-deoxymugineic acid (DMA) from roots, a chelating compound which is able to produce stable complexes with iron. Then the Fe-DMA complex is transported through the cell membrane [21]. There are only a few studies that have looked at PS concentrations in soil solution and release from soil-grown plants. It has been demonstrated [23] that carbon and energy investment into Fe acquisition under natural growth conditions is significantly smaller than previously derived from zero Fe-hydroponic studies. It has been reported that during the investigated period (21–47 days after germination), PS release initially exceeded Fe plant uptake 10-fold, but it significantly declined ca five weeks after germination [23]. PS released by strategy II plants are highly susceptible to microbial decomposition. However, to date very little is known about the fate of PS in soil. As regards the mineralization dynamics in the wheat rhizosphere and bulk soil of alkaline soils, between 40% and 65% of the DMA was either respired or incorporated into soil microbial biomass after 24 h, with the largest part of total incorporated DMA being recovered in gram negative bacteria [24]. Considering root growth dynamics, and taking into account that PS are mainly exuded from root tips, the significantly slower mineralization of DMA in bulk soil is of high ecological importance to enhance the Fe scavenging efficiency of PS released into the soil [24]. The adsorption and desorption of phytosiderophores by the soil solid phase has also been studied [25].

Another fact was that the FBS^0-crop hosted less weeds, which definitely merits further investigation [6]. This most probably explains the reported statistically lower organic matter concentration in the FBS^0-crop's rhizosoil at day 197, twenty days after the second herbicide application [6]. The significant increase in the organic matter of the F-crop's rhizosoil at that day seems to originate from the dead root system of the eliminated weeds. After the herbicide application, the dead root system of weeds was added to the already existing organic matter and the conventional F-crop showed much higher organic matter. This probably explains the replenishment of F-crop's Fe content and the observed deficit of FBS^0-crop's Fe content in the rhizosoil at the end of the cultivation period.

The dynamics of HS content in the rhizosoil is another point which requires a further study. Humic substances are heterogeneous high-molecular-weight organic materials which are ubiquitous in the soil. In the present study, the rhizosoil environment was well oxygenated, i.e., not reduced. The interactions between the carbon, nitrogen, sulfur, and iron biogeochemical cycles in dynamic redox environments have been illustrated [26]. The findings of this work suggest that HS are involved in the affected processes in the rhizosoil and this may be due to the presence of elemental sulfur itself or to the various intermediates of the S biogeochemical cycle, as described in [26]. The aforementioned dynamics clearly depict that the addition of elemental sulfur as an ingredient of the fertilizer along with the initial fertilization differentiated the behavior of the rhizosoil's examined parameters. The rhizosoil's pH did not decrease [6]; instead the system behaved as buffer. Furthermore, the added FBS^0 in a follow up field experiment increased the total microbial population, as well as promoting the activity of the sulfatase producing microbial populations in the rhizosoil of the studied commercial wheat crop grown on calcareous soil [22]. The amount of S^0 provided with the fertilizer significantly enhanced the mobilization of the labile pool of sulfate esters in the organic matter of the soil. Therefore, apart from the sulfate added by the fertilizer at sowing, there was sulfate input, too, due to sulfatase activity, which suggests a sustained input of sulfate during crop development.

The developmental events during wheat growth and development are complex [27] and the developmental sequence of the generic winter wheat shoot apex for optimal conditions has been discussed in detail [28]. Furthermore, the characterization of the wheat leaf metabolome (including sulfate) during grain filling and under varied N-supply has been reported [29]. On the other hand, the interactions between iron nutrition and sulfur nutrition have received increasing attention. The finding that methionine is the sole percursor of mugineic acid [30] and the demonstration that the methionine cycle (or Yang cycle) in roots is one of the methionine sources [31] led Astolfi et al. [21,32,33] to further investigate the interactions between Fe nutrition and S nutrition, concluding that "the interaction can occur determining rapid adjustments of sulfate uptake and assimilation which conceivably lead to a re-distribution of reduced S pool. Fe availability might represent a signal able to modulate production and utilization of thiols at root level, where phytosiderophores are effectively synthesized" [32]. The iron nutrition of both crops was adequate. According to Mills and Jones [34], the sufficiency range for iron concentration in the top two leaves just before heading ranges between 10 ppm and 300 ppm for winter wheat collected from production fields. This held true for S nutrition, too, based on the detailed data of Reuter and Robinson plant analysis interpretation manual and the references therein for wheat [35].

5. Conclusions

The incorporation of 2% elemental sulfur as an ingredient of the applied fertilizer at sowing affected the iron fractions of the rhizosoil towards iron mobilization, thus providing more iron to the crop, which apart from the iron nutrition aspect fortified the crop's sulfur nutrition, too. Summarizing the prominent facts, no iron was found as iron attached to carbonates of the rhizosoil. Fluctuations of the iron pool bound or adsorbed to the organic matter were exactly the opposite to those of the iron pool associated with the clay particles in both treatments, suggesting iron exchange between the two pools. Replenishment of the F-crop's Fe content and a deficit in the FBS^0-crop's Fe content in the rhizosoil were found at the end of the cultivation period. The initiation of the fast stem elongation stage (day 125) marked a turning point. Before day 125, the use of FBS^0 increased the iron concentration in the main stems and this was an early fortification effect, followed by an increase in the organic S concentration. After day 125, the FBS^0-crop consisted of plants with higher main stems and less tillers. A late fortification effect was observed in the iron concentration of the main stems and their heads after the stage of complete flowering. Prior to harvesting in the FBS^0-crop, all the plant's parts were heavier, more iron and organic sulfur were accumulated in these plant parts, and the obtained commercial yield of the FBS^0-crop was higher by 27.3%.

Acknowledgments: The authors are grateful to the greek fertilizer company Sulphur Hellas S.A. for providing the fertilizers and the consumables for chemical analyses, the farmer Panagiotis (Valvidas) Diamandis for applying the FBS^0 product in his commercial crop, the agronomists Dimitris Petrakos, Harris Mavrogiannis, and Filippa Maniou for their help during sampling, and the three anonymous reviewers for their comments. The soil and plant samples of this study were collected by consent of the land owner.

Author Contributions: D.L.B. and S.N.C. conceived and designed the experiment, elaborated the research questions, analyzed the data, and wrote the article. M.M., G.I.S., and P.P.S. carried out the chemical analyses. D.L.B. carried out the field work with the contribution of P.P.S.

Conflicts of Interest: The authors declare no conflict of interest. The funding sponsor had no role in the design of the study; in the collection, analyses, or interpretation of data; in the writing of the manuscript, nor in the decision to publish the results.

Abbreviations

RS	Rhizosoil
OM	Organic matter
HS	Humic substances
DM	Dry mass
fr	Fraction
F	Granular conventional fertilizer
S^0	Elemental sulfur
B	Binder

References

1. Janzen, H.H.; Bettany, J.R. Release of available sulfur from fertilizers. *Can. J. Soil Sci.* **1986**, *66*, 91–103. [CrossRef]
2. Germida, J.J.; Jansen, H.H. Factors affecting the oxidation of elemental sulfur in soils. *Fertil. Res.* **1993**, *35*, 101–114. [CrossRef]
3. Somani, L.L.; Totawat, K.L. Mined and industrial waste products capable of generating gypsum in soil. In *Handbook of Soil Conditioners*; Wallace, A., Terry, R.E., Eds.; Marcel Dekker, Inc.: New York, NY, USA, 1998; pp. 257–291.
4. Messick, D. Agricultural Demand for Sulphur—The Challenges, The Future, In TFI-FITR's Outlook and Technology Conference, Georgia, USA, 19 November 2014. Available online: http://www.firt.org/sites/default/files/TFI%20FIRT%20Outlook%20-%20Agricultural%20Demand%20for%20Sulphur%20-%20TSI.pdf (accessed on 28 October 2017).
5. Benardos, D. Method for Coating Fertilizer Beads with Elemental Sulfur. U.S. Patent WO2017077350A, 11 May 2017.
6. Chorianopoulou, S.N.; Saridis, G.I.; Sigalas, P.P.; Margetis, M.; Benardos, D.; Mavrogiannis, H.; Bouranis, D.L. The Application of S^0-coated Fertilizer to Durum Wheat Crop. In *Sulfur Metabolism in Higher Plants—Fundamental, Environmental and Agricultural Aspects; Proceedings of the International Plant Sulfur Workshop*; De Kok, L.J., Rennenberg, H., Hawkesford, M.J., Eds.; Springer International Publishing: Cham, Switzerland, 2017; pp. 115–121.

7. Lindsay, W.L.; Schwab, A.P. The chemistry of iron in soils and its availability to plants. *J. Plant Nutr.* **1982**, *5*, 821–840. [CrossRef]
8. Lucena, J.J. Effects of bicarbonate, nitrate and other environmental factors on iron chlorosis. A review. *J. Plant Nutr.* **2000**, *23*, 1591–1606. [CrossRef]
9. Lindsay, W.L. Iron oxide solubilisation by organic matter and its effect on iron availability. In *Iron Nutrition and Interaction in Plants*; Chen, Y., Hadar, Y., Eds.; Kluwer: Dordrecht, The Netherlands, 1991; pp. 29–36.
10. Römheld, V.; Marschner, H. Mobilization of iron in the rizosphere of different plant species. In *Advances in Plant Nutrition*; Tinker, B., Laüchli, A., Eds.; Praeger Scientific: New York, NY, USA, 1986; pp. 155–204.
11. Rodríguez-Lucena, P.; Tomasi, N.; Pinton, R.; Hernández-Apaolaza, L.; Lucena, J.J.; Cesco, S. Evaluation of ^{59}Fe-lignosulfonates complexes as Fe-sources for plants. *Plant Soil* **2009**, *325*, 53–63. [CrossRef]
12. Lindsay, W.L. Soil and plant relationships associated with iron deficiency with emphasis on nutrient interactions. *J. Plant Nutr.* **1984**, *7*, 489–500. [CrossRef]
13. Tyler, G. Mineral nutrient limitations of calcifuge plants in phosphate sufficient limestone soil. *Ann. Bot.* **1996**, *77*, 649–656. [CrossRef]
14. Zohlen, A.; Tyler, G. Immobilization of tissue iron on calcareous soil: Differences between calcicole and calcifuge plants. *Oikos* **2000**, *89*, 95–106. [CrossRef]
15. Broadley, M.; Brown, P.; Cakmak, I.; Rengel, Z.; Zhao, F. Function of Nutrients: Micronutrients. In *Marschner's Mineral Nutrition of Higher Plants*; Marschner, P., Ed.; Academic Press: Cambridge, MA, USA, 2012; pp. 191–200.
16. Jones, J.B., Jr. *Soil Analysis Handbook of Reference Methods*; CRC Press: Boca Raton, FL, USA, 1999; ISBN 9780849302053.
17. Chorianopoulou, S.N.; Nikologiannis, S.; Gasparatos, D.; Bouranis, D.L. Relationships between iron, sulfur, nitrogen and phosphorus in lawns grown on a calcareous soil irrigated by slightly saline water. *Fresenius Environ. Bull.* **2017**, *26*, 1240–1246.
18. Pueyo, M.; Mateu, J.; Rigol, A.; Vidal, M.; Lopez-Sanchez, J.F.; Rauret, G. Use of the modified BCR three-step sequential extraction procedure for the study of trace element dynamics in contaminated soils. *Environ. Pollut.* **2008**, *152*, 330–341. [CrossRef] [PubMed]
19. Sörbo, B. Sulfate: Turbidimetric and nephelometric methods. In *Methods in Enzymology: Sulfur and Sulfur Amino Acids*; Jakoby, W.B., Griffith, O.W., Eds.; Academic Press, Inc.: New York, NY, USA, 1987.
20. Miller, R.O. Extractable chloride, nitrate, orthophosphate, potassium, and sulfate sulfur in plant tissue: 2% acetic acid extraction. In *Handbook of Reference Methods for Plant Analysis*; Kalra, Y.P., Ed.; CRC Press LLC: Boca Raton, FL, USA, 1998; pp. 115–118.
21. Astolfi, S.; Zuchi, S.; Passera, C.; Cesco, S. Does the sulfur assimilation pathway play a role in the response to Fe deficiency in maize (*Zea mays* L.) plants? *J. Plant Nutr.* **2003**, *26*, 2111–2121. [CrossRef]
22. Chorianopoulou, S.N.; Venieraki, A.; Maniou, F.; Mavrogiannis, H.; Benardos, D.; Katinakis, P.; Bouranis, D.L. Sulfogrow®: A new type of plant biostimulant? In Proceedings of the 13th International Conference on Protection and Restoration of the Environment, Mykonos, Greece, 3–8 July 2016; pp. 733–738.
23. Oburger, E.; Gruber, B.; Schindlegger, Y.; Schenkeveld, W.D.C.; Hann, S.; Kraemer, S.M.; Wenzel, W.W.; Puschenreiter, M. Root exudation of phytosiderophores from soil-grown wheat. *New Phytol.* **2014**, *203*, 1161–1174. [CrossRef] [PubMed]
24. Oburger, E.; Gruber, B.; Wanek, W.; Watzinger, A.; Stanetty, C.; Schindlegger, Y.; Hann, S.; Schenkeveld, W.D.C.; Kraemer, S.M.; Puschenreiter, M. Microbial decomposition of 13C- labeled phytosiderophores in the rhizosphere of wheat: Mineralization dynamics and key microbial groups involved. *Soil Biol. Biochem.* **2016**, *98*, 196–207. [CrossRef]
25. Walter, M.; Oburger, E.; Schindlegger, Y.; Hann, S.; Puschenreiter, M.; Kraemer, S.M.; Schenkeveld, W.D.C. Retention of phytosiderophores by the soil solid phase—Adsorption and desorption. *Plant Soil* **2016**, *404*, 85–97. [CrossRef] [PubMed]
26. Li, Y.; Yu, S.; Strong, J.; Wang, H. Are the biogeochemical cycles of carbon, nitrogen, sulfur, and phosphorus driven by the "FeIII–FeII redox wheel" in dynamic redox environments? *J. Soils Sediments* **2012**, *12*, 683–693. [CrossRef]
27. White, J.; Edwards, J. *Wheat Growth and Development*; Procrop; New South Wales, Department of Primary Industries: Orange, NSW, Australia, 2007.

28. McMaster, G.S. Phytomers, phyllochrons, phenology and temperate cereal development. *J. Agric. Sci.* **2005**, *143*, 137–150. [CrossRef]

29. Heyneke, E.; Watanabe, M.; Erban, A.; Duan, G.; Buchner, P.; Walther, D.; Kopka, J.; Hawkesford, M.; Hoefgen, R. Characterization of the wheat leaf metabolome during grain filling and under varied N-supply. *Front. Plant Sci.* **2017**, *8*, 2048. [CrossRef] [PubMed]

30. Mori, S.; Nishizawa, N. Methionine as a dominant precursor of phytosiderophores in Graminea plants. *Plant Cell Physiol.* **1987**, *28*, 1081–1092.

31. Ma, J.F.; Shinada, T.; Matsuda, C.; Kyosuke, N.J. Biosynthesis of phytosiderophores, mugineic acid, associated with methionine cycle. *Biol. Chem.* **1995**, *270*, 16549–16554. [CrossRef]

32. Astolfi, S.; Zuchi, S.; Cesco, S.; Varanini, Z.; Pinton, R. Influence of iron nutrition on sulphur uptake and metabolism in maize (*Zea mays* L.) roots. *Soil Sci. Plant Nutr.* **2004**, *50*, 1079–1083. [CrossRef]

33. Astolfi, S.; Zuchi, S.; Hubberten, M.-H.; Pinton, R.; Hoefgen, R. Supply of sulphur to S-deficient young barley seedlings restores their capability to cope with iron shortage. *J. Exp. Bot.* **2010**, *61*, 799–806. [CrossRef] [PubMed]

34. Mills, H.A.; Jones, J.B., Jr. *Plant Analysis Handbook II*; MicroMacroPublishing, Inc.: Athens, GA, USA, 1996.

35. Reuter, D.J.; Robinson, J.B. *Plant Analysis, an Interpretation Manual*, 2nd ed.; CSIRO Publishing: Clayton, Australia, 1997; pp. 244–245.

agriculture

MDPI

Article

Effect of Magnesium on Gas Exchange and Photosynthetic Efficiency of Coffee Plants Grown under Different Light Levels

**Kaio Gonçalves de Lima Dias [1,*], Paulo Tácito Gontijo Guimarães [2],
Antônio Eduardo Furtini Neto [3], Helbert Rezende Oliveira de Silveira [2] and
Julian Junio de Jesus Lacerda [4]**

[1]   Departamento de Ciência do Solo/DCS, Universidade Federal de Lavras/UFLA, Cx. P. 3037,
     37200-000 Lavras-MG, Brazil
[2]   Empresa de Pesquisa Agropecuária de Minas Gerais/EPAMIG, Cx. P. 176, 37200-000 Lavras-MG, Brazil;
     paulotgg@epamig.ufla.br (P.T.G.G.); herosrezende@yahoo.com.br (H.R.O.d.S.)
[3]   Instituto Tecnológico Vale Desenvolvimento Sustentável, 66055-090 Belém-PA, Brazil; furtinineto@gmail.com
[4]   Campus Professora Cinobelina Elvas, Universidade Federal do Piauí, Rodovia Municipal Bom Jesus Viana,
     km 1, 64900-000 Bom Jesus-PI, Brazil; julianlacerda@gmail.com
*   Correspondence: kaiogld@gmail.com; Tel.: +55-35-991713076

Received: 12 July 2017; Accepted: 14 September 2017; Published: 30 September 2017

Abstract: The aim of the present study was to investigate the effects of magnesium on the gas exchange and photosynthetic efficiency of Coffee seedlings grown in nutrient solution under different light levels. The experiment was conducted under controlled conditions in growth chambers and nutrient solution at the Department of Plant Pathology of the Federal University of Lavras. The treatments consisted of five different Mg concentrations (0, 48, 96, 192 and 384 mg·L^{-1}) and four light levels (80, 160, 240 and 320 µmol photon m^{-2}·s^{-1}). Both the Mg concentration and light levels affected gas exchange in the coffee plants. Photosynthesis increased linearly with the increasing light, indicating that the light levels tested were low for this crop. The highest CO$_2$ assimilation rate, lowest transpiration, and highest water use efficiency were observed with 250 mg·Mg·L^{-1}, indicating that this concentration was the optimal Mg supply for the tested light levels.

Keywords: coffee plant nutrition; photoinhibition; photoprotection; leaf scald

1. Introduction

Coffee was originally an understory plant, however, it is now mostly grown under full sunshine conditions in Brazil [1]. This crop presents the lowest net CO$_2$ assimilation rates reported for C$_3$ woody plants grown in tropical climates [2]. The low photosynthetic capacity of coffee plants is a physiological trait characteristic of shade-adapted plants grown under full sunshine conditions [3].

Coffee leaves are saturated at relatively low light levels (between 300 and 700 µmol·m^{-2}·s^{-1}) due to the induction of strong stomatal control of photosynthesis [4]. On a clear day (without clouds), the photon flux may reach approximately 2000 µmol·m^{-2}·s^{-1} during the afternoon [2,3]. Therefore, irradiance levels higher than the coffee photosynthesis saturation point are common.

When leaves are exposed to more light than they can use, the photosystem II (PSII) reaction center is inactivated and frequently damaged. Chlorophylls in their excited state may react with molecular oxygen due to the absorption of excess light, resulting in the production of reactive oxygen species (ROS), which damage the photosynthetic apparatus [5,6]. This stress resulting from excess light is known as photoinhibition and in more serious cases may result in photooxidation, with visible damage to leaf tissues [7]. Photooxidation is most likely responsible for leaf scald symptoms in coffee plants, which are increasingly more common, especially in the frontal face of planting rows (facing the afternoon sun) [8].

The expansion of coffee plantations towards Cerrado areas and climate changes such as long periods of dry weather with heat waves and increased irradiance peaks [9] have worsened this problem. Crop shading is not a good option in large-scale cultivation. Although this approach decreases photoinhibition, it also typically decreases coffee plant productivity [1] due to a lower CO_2 assimilation rate, and causes greater stimulation of vegetative growth with negative effects on floral bud emission, a reduced number of nodes, and lower development of flowers per node [10]. Additionally, shading restricts mechanization and increases production costs.

In chloroplasts, the light triggers activation of ribulose-1,5-bisphosphate (RuBP) carboxylase, the main enzyme responsible for the photosynthesis process. On the other hand, magnesium (Mg) deficiency negatively affects many fundamental physiological and biochemical processes that are required for plant growth and development. Suitable Mg concentrations increase the activity of RuBP carboxylase and also of other stromal enzymes. Recent studies showed that Mg-deficient plants were more susceptible to photooxidation damage, indicating that plants growing under high light conditions have higher Mg requirements [11,12]. Thus, Mg availability in the environment could lead to better photosynthetic capacity, especially at high light levels.

The aim of the present study was to investigate the effects of Mg on the gas exchange and photosynthetic efficiency of *Coffea arabica* L. seedlings grown in nutrient solution under different light levels.

2. Materials and Methods

The experiments were conducted under controlled conditions in growth chambers at the Department of Plant Pathology (Departamento de Fitopatologia) of the Federal University of Lavras (Universidade Federal de Lavras—UFLA). The plants were grown in nutrient solution. Five different Mg concentrations and four light levels were tested. The Mg concentrations tested were 0, 48 (the Mg concentration in Hoagland solution; [13]), 96, 192 and 384 mg·Mg·L^{-1}. The light levels tested were 80, 160, 240 and 320 μmol·m^{-2}·s^{-1}. The lowest light level was chosen to resemble the low lighting conditions experienced by coffee plants grown in understories under shaded conditions or under a high planting density. The highest light level (320 μmol·m^{-2}·s^{-1}) simulated the light level in which coffee leaves should be saturated [4]. Two intermediate light levels (160 and 240 μmol·m^{-2}·s^{-1}) were also tested to establish a gradient of light incidence on plants and enable the fitting of regression equations.

Three-liter pots were used in these experiments. A randomized block experimental design was applied with a 5×4 factorial scheme, with 6 replicates and one plant per experimental unit in a total of 120 plots.

Seedlings of the coffee cultivar Mundo Novo IAC 379/19 were used. The seedlings had 4 pairs of true leaves and were grown in soil not subjected to liming at the Experimental Farm of the Agricultural Research Company of Minas Gerais (Fazenda Experimental da Empresa de Pesquisa Agropecuária de Minas Gerais (EPAMIG)) in Machado. The seedlings were removed from the soil and placed in trays containing deionized water for 10 days until new roots emerged. Then, the seedlings were transferred to 3 L pots containing half-strength Hoagland and Arnon nutrient solution [13] from which Mg was omitted. The seedlings remained in the pots for 15 days with constant aeration.

Following this period, the seedlings were transferred to full-strength Hoagland solution with one of the following Mg concentrations: 0, 48, 96, 192 or 384 mg·Mg·L^{-1}. The nutrient solution was constantly aerated. The solution volume was refilled daily with deionized water, and the pH was corrected to 5.0–5.5 using 0.1 mol·L^{-1} HCl or 0.1 mol·L^{-1} NaOH. When Mg depletion reached 70% of the initial concentration, all solutions were exchanged for corresponding solutions to maintain approximately constant Mg availability during the experimental period.

Light was provided by daylight tubular fluorescent lamps (Osram 20 W). Different light levels were achieved by varying the distance between the plants and the light source using different shelf heights. The plants were grown under a 12 hour light: 12 hour dark photoperiod. The light levels at the different plant heights were measured using a quantum sensor (Licor LI-190SA; Li-Cor Biosciences, Inc., Lincoln, NE, USA).

Ninety days following the beginning of the treatments, ecophysiological measurements were performed in fully expanded leaves using an infrared gas analyzer (IRGA) (LI-6400XT Portable Photosynthesis System, LI-COR, Lincoln, NE, USA). The following parameters were directly measured: delta CO_2 and delta H_2O, leaf temperature, light intensity in the chamber and gas flux; while the following were indirectly assessed, based on the algorithms, by the software of the system: CO_2 internal concentration (Ci; $\mu mol \cdot m^{-2} \cdot s^{-1}$), transpiration (E; mmol H_2O $m^{-2} \cdot s^{-1}$), stomatal conductance (Gs; mol H_2O $m^{-2} \cdot s^{-1}$), vapor pressure deficit (VPD; kPa), and CO_2 assimilation rate (photosynthesis, A; $\mu mol \cdot m^{-2} \cdot s^{-1}$), water use efficiency (WUE; A/E; μmol CO_2 mol^{-1} H_2O) and instantaneous carboxylation efficiency (A/Ci) [14,15].

Measurements were performed one hour after the onset of illumination in the growth chamber. The measurements were performed within a closed chamber (Blue + Red LED LI-6400-02B, LI-COR, Lincoln, NE, USA) using an artificial source of photosynthetically active radiation (PAR) at the same light intensity under which the plants were grown. The CO_2 assimilation rate in the chamber was measured using the environment CO_2 concentration (453.1 $\pm$ 40 μmol CO_2 mol^{-1}).

After photosynthetic evaluation, the leaves were washed and packaged separately in paper bags and oven dried at 60 ° C until reaching constant weight. The coffee leaves were ground for chemical analysis.

A variance analysis using the F test was applied to test for significant differences between treatments. When significant differences were found, the effect of Mg concentrations and light level were analyzed using regression analysis. Non-significant interactions were not shown, thus in those cases, each factor was analyzed using the mean of another. All analyses were performed using the Sisvar software [16], and graphs were built using the SigmaPlot 11.0 software souced by Systat Software Inc. Chicago, USA. The maximum and minimum points for the quadratic equations were calculated by equaling the first derivative to zero.

3. Results and Discussion

3.1. Stomatal Conductance and Leaf Temperature

A significant interaction between the Mg concentration and light level was observed for stomatal conductance (Gs) (Figure 1A). The Gs decreased with the increasing Mg concentration at all light levels, with a tendency to stabilize from 192 mg·Mg·L^{-1}. This result was related to the decrease in potassium (K) availability with the increasing Mg concentration (Figure 2). K plays an important role in stomatal conductance. Its accumulation and release by stomatal guard cells leads to changes in cell turgor, resulting in stomatal opening and closing [6,17].

The reduction in leaf K contents as a function of the increase in Mg doses is due to the antagonistic effect among these nutrients. In general, increasing the amount absorbed from one cation can result in the reduction of the absorption of another cation [18]. This inhibition between these nutrients is competitive, that is, there is competition by the same site of the "carriers" in the membrane [19]. The antagonistic relationship between Mg and K was observed in experiments with several cultures [20–22].

Another factor that decreased stomatal conductance was the vapor pressure deficit (VPD) (Figure 1B). The difference between vapor pressure inside the leaves and in the air induces stomatal movement; this difference depends on the total leaf transpiration rate and water potential gradient between guard cells and other epidermal cells [23]. High VPD values may cause stomatal closing to prevent excessive water loss through transpiration [6]. Marenco et al. [24] observed a pronounced decrease in Gs and photosynthesis with the increasing VPD.

Chaves et al. [25] studied coffee plants under field conditions in 2007 and observed low Gs values (approximately 0.06 mol H_2O $m^{-2} \cdot s^{-1}$) with high VPDs. Therefore, VPD and Gs were negatively correlated.

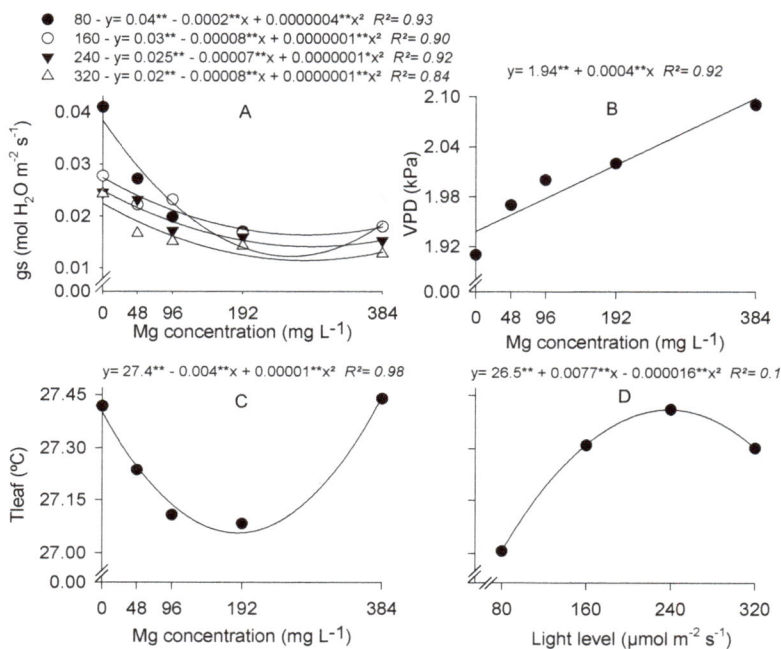

Figure 1. (**A**) Stomatal conductance (Gs; mol H_2O $m^{-2} \cdot s^{-1}$), (**B**) vapor pressure deficit (VPD; kPa), and (**C,D**) leaf temperature (Tleaf; °C) in coffee seedlings grown with different Mg concentrations and under different light levels. (*) Significant according to the *t* test at $p < 0.05$. (**) Significant according to the *t* test at $p < 0.01$.

The average Gs for coffee plants is 0.108 mol H_2O $m^{-2} \cdot s^{-1}$ [26]. The low Gs values observed in the present study (even for the control treatment 0 mg·Mg·L^{-1}, which presented a lower VPD) were probably due to the low light levels.

The Gs value was highest with the lowest light level tested (80 µmol·m$^{-2} \cdot s^{-1}$), which was related to the lower leaf temperatures observed for this light level (Figure 1D). This finding was especially true for the treatment lacking Mg (0 mg·Mg·L^{-1}) that did not have competition between Mg and K. So the K uptake had no negative effect. Leaf temperatures higher than the air temperature may cause stomatal closing and decrease the Gs [27].

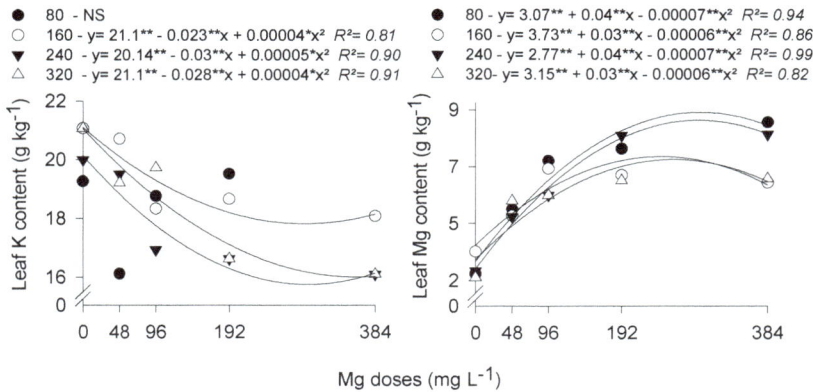

Figure 2. Contents of Mg and K in leaves of coffee seedlings grown with different Mg concentrations and under different light levels. (*) Significant according to the *t* test at $p < 0.05$. (**) Significant according to the *t* test at $p < 0.01$.

3.2. Gas Exchange

No significant interactions were observed between the Mg concentration and light level for the CO_2 internal concentration (Ci), transpiration (E) and photosynthesis (A) (Figure 3).

The Ci decreased linearly with the increasing Mg supply; its variation depending on the light level was best fitted by a positive quadratic equation (Figure 3A, B). The decrease in Ci with the increasing Mg supply was a result of the improved CO_2 use, due to the higher efficiency of the photosynthetic apparatus (Figure 3E). CO_2 concentrations tend to be lower with higher photosynthetic rates and Ci has a negative linear correlation with the photosynthetic rate [28]. Mg binding increases the affinity of Rubisco for CO_2 and doubles its maximum reaction velocity [11].

The lower Ci observed with the intermediate light levels might be related to the higher leaf temperatures observed for these light levels (Figure 1D). Increased leaf temperatures in coffee plants may cause a gradual increase in photorespiration and the internal CO_2 concentration [1].

The variation observed in the transpiration (E) with the increasing Mg supply was best fitted by a positive quadratic equation (Figure 3C). The leaf transpiration rate is primarily determined by the light level, VDP, and Gs [29]. The decrease in transpiration observed with 250 mg·Mg·L^{-1} down to 0.324 mmol H_2O m^{-2}·s^{-1} might have been related to the Gs, which presented variation with the increasing Mg, best fitted by a positive quadratic equation (Figure 1A). This result was in accordance with Assad et al. (2004), who observed a decrease in transpiration due to stomatal closing as a result of the increasing VPD. In a field study, Gutiérrez and Meinzer [30] attributed the decrease in transpiration of coffee plants to the stomatal closing induced by high temperatures and VPDs.

E decreased with increasing light levels starting from 125 μmol·m^{-2}·s^{-1} (Figure 3D). Coffee plants tend to decrease their transpiration and increase their photosynthetic capacity at high light levels by increasing their specific leaf mass (ratio between the leaf mass and area) [31]. In contrast, decreasing transpiration with decreasing irradiance has been observed in shaded coffee plants [4,32]. However, the light levels in the field are higher than those tested in the present study.

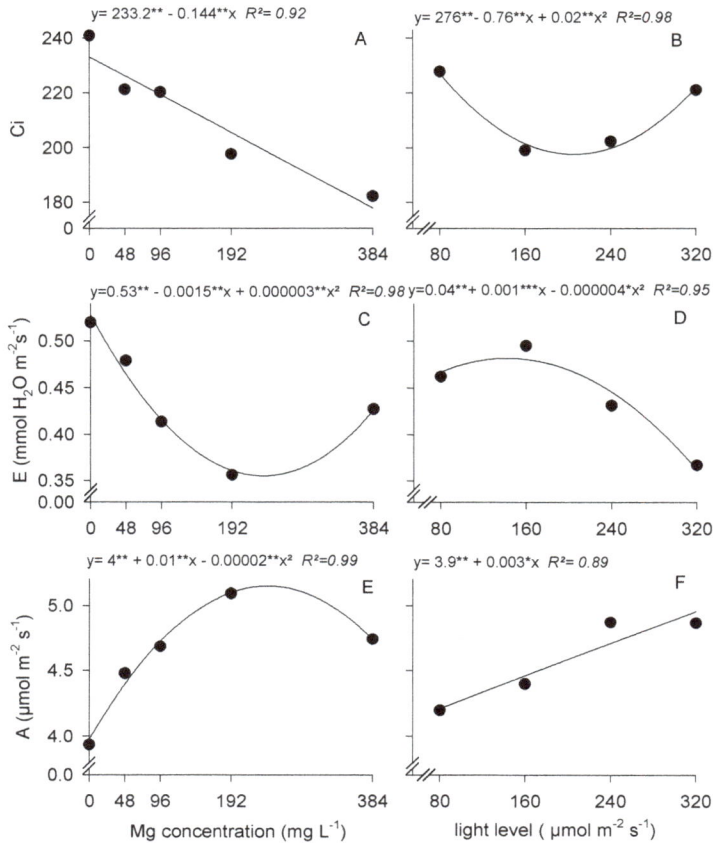

Figure 3. (**A,B**) CO_2 internal concentration (Ci; $\mu mol \cdot mol^{-1}$), (**C,D**) transpiration (E; mmol H_2O $m^{-2} \cdot s^{-1}$), and (**E,F**) CO_2 assimilation rate (photosynthesis, A; $\mu mol \cdot m^{-2} \cdot s^{-1}$) in coffee seedlings grown with different Mg concentrations and under different light levels. (*) Significant according to the *t* test at $p < 0.05$. (**) Significant according to the *t* test at $p < 0.01$.

Similar to the VPD, the high temperatures that usually accompany high light levels have a high impact on C assimilation and transpiration [1]. Additionally, the increase in VPD under higher temperatures results in stomatal closing and decreased transpiration [33], thereby preventing the evaporative cooling of leaves and the maintenance of leaf temperatures at higher light levels (Figure 3D).

The variation in photosynthesis (A) with the increasing Mg supply was best fitted by a negative quadratic equation, with a maximum value of 5.25 $\mu mol \cdot m^{-2} \cdot s^{-1}$ observed with 250 mg $Mg \cdot L^{-1}$. Photosynthesis exhibited a positive linear correlation with the light level (Figure 3E,F).

The increase in photosynthesis with the increasing Mg supply is related to several key roles of Mg in plant functions, including the regulation of photophosphorylation (adenosine triphosphate formation in chloroplasts), CO_2 photosynthetic fixation, protein synthesis, chlorophyll formation, phloem loading, the partitioning and utilization of photoassimilates, the generation of reactive oxygen species (ROSs), photooxidation in leaf tissues, and enzyme activation. Mg is the nutrient that activates most enzymes in plants (i.e. ATPases and Rubisco) [11]. Mg deficiency causes carbohydrate accumulation in leaves [11,17,34], which may affect the photosynthetic metabolism and decrease the use of absorbed light for photosynthesis.

The decrease in photosynthesis from 250 mg Mg·L^{-1} may be related to the imbalance caused by excess Mg, especially due to the decreased K uptake (Figure 2). K is extremely important for the activation of the carboxylase function of Rubisco [35], which has been proposed to explain the increased photosynthetic rates observed with an adequate K supply [36,37].

The photosynthesis rates in C$_3$ plants generally vary between 10 and 20 µmol·m^{-2}·s^{-1} [6]. However, due to the limiting CO$_2$ supply coffee plants present low net CO$_2$ assimilation rates (between 4 and 11 µmol·m^{-2}·s^{-1}) [1]. Therefore, the net CO$_2$ assimilation rates of 4 to 6 µmol·m^{-2}·s^{-1} observed in the present study can be considered normal for coffee plants, even with the low tested light levels and the low observed Gs (Figure 1A).

Chaves et al. [20] observed photosynthesis rates of approximately 2.5 µmol·m^{-2}·s^{-1} in the field. According to these authors, coffee plant leaves present plasticity and are able to adapt to environments with different light levels. However, higher light than the level needed for photosynthesis may results in an energy imbalance, often resulting in photoinhibition. Photoinhibition is a complex set of molecular processes that leads to the inhibition of photosynthesis due to excess light [25].

Prolonged exposure of plants or organelles to excess light may result in photodestruction of photosynthetic pigments as a result of light- and oxygen-dependent bleaching. This process is usually called photooxidation and may lead to cell or organism death. In most cases, photooxidation is a secondary phenomenon that occurs following a distinct lag phase. During this lag phase, there is a decline in photosynthesis that is dependent on the light intensity and exposure time (photoinhibition) and a lack of change in the composition of the pigment reserves [7]. Photoinhibition is not a consequence of pigment destruction; instead, pigment bleaching occurs following a certain degree of photoinhibition. Thus, these processes represent two different phenomena [7].

Photooxidation is most likely responsible for leaf scald symptoms in coffee plants. ROS are produced during photoinhibition and can result in oxidative stress (photooxidation) if the plant antioxidant complex is not capable of removing the generated ROS [38].

Leaf scald symptoms have been observed in several coffee producing regions. Photoinhibition preceding photooxidation should be even more common because the photon flux from early morning until midday generally varies between 800 and 1200 µmol·m^{-2}·s^{-1}, and may reach 2000 µmol·m^{-2}·s^{-1} during the afternoon [2,3] and coffee leaves are saturated by relatively low irradiances between 300 and 700 µmol·m^{-2}·s^{-1} [1]. Photorespiration is an effective mechanism for protection against photoinhibition and Mg plays a direct role in photorespiration. During photorespiration, excess energy stored as ATP and NADPH during the photochemical phase of photosynthesis is dissipated [6]. Energy transference from chlorophylls to carotenoids formed during the xanthophyll cycle leads to energy dissipation as heat at the PSII light-harvesting complex [39].

Although the photosynthesis rate increased with the increasing light level (Figure 3F), an increase in the superoxide dismutase (SOD) and ascorbate peroxidase (APX) activities was also observed [17]. Prior to the appearance of the visual symptoms of Mg deficiency, enzyme activities from the plant antioxidant complex delay the photooxidative damages caused by ROS and the inactivation of photosynthetic enzymes. Photosynthesis is only decreased during the more advanced phases of Mg deficiency [40].

3.3. Water Use Efficiency and Instantaneous Carboxylation Efficiency

The variation in the water use efficiency (WUE)—the ratio between photosynthesis and transpiration—with the increasing Mg supply was best fitted by a negative quadratic equation, with a maximum value observed with 245 mg·Mg·L^{-1} (Figure 4A). The WUE increased due to higher photosynthetic efficiency and decreased transpiration with the increasing Mg supply. The lower WUEs observed with the higher Mg concentrations were related to the imbalance caused by excess Mg, which had a negative effect on the photosynthetic rates.

The WUE increased linearly with the increasing light level (Figure 4B) due to the increased photosynthetic rate (Figure 3F) and decreased transpiration (Figure 3D). In field experiments with shaded coffee plants, the WUE was higher with 0% and 50% shading [32].

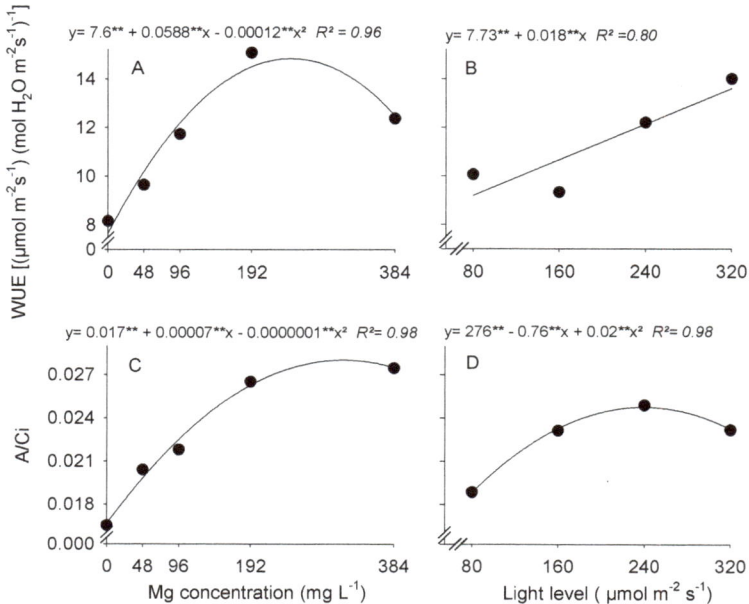

Figure 4. (**A**,**B**) Water use efficiency (WUE) and (**C**,**D**) instantaneous carboxylation efficiency (A/Ci) in coffee seedlings grown with different Mg concentrations and under different light levels. (*) Significant according to the *t* test at $p < 0.05$. (**) Significant according to the *t* test at $p < 0.01$.

The instantaneous carboxylation efficiency (A/Ci) calculated as the ratio between photosynthesis and the CO_2 internal concentration are closely related to the intracellular CO_2 concentration and CO_2 assimilation rate [41]. The variations in the A/Ci with the increasing Mg supply and light levels were best fitted by negative quadratic equations (Figure 4A, B). An increased A/Ci with an increasing Mg supply and light level is related to an increase in the photosynthesis rate (Figure 3E, F) and a decrease in the internal C concentration (Figure 3A,B).

The highest CO_2 assimilation rate, lowest transpiration, and highest water use efficiency were observed with 250 mg·Mg·L^{-1}, indicating that this concentration was the optimal Mg supply for the tested light levels. The critical Mg supply for coffee plants most likely varies with the light level. For example, coffee plants grown in the west region of Bahia may need more Mg than those grown in the south of Minas or in regions with lower light levels.

4. Conclusions

Both the Mg supply and the light level affected gas exchange in coffee plants. The positive linear correlation between photosynthesis and the light level showed that the tested light levels were low for coffee plants. The highest CO_2 assimilation rate, lowest transpiration, and highest water use efficiency were observed with approximately 250 mg Mg·L^{-1}, indicating that this concentration represented the optimal Mg supply for the tested light levels.

Acknowledgments: We acknowledgements the collaborators in particular to UFLA, EPAMIG and the Coffee Research Consortium.

Author Contributions: All the authors assisted in the development of the research and in the discussion of the results.

Conflicts of Interest: The authors declare no conflicts of interest.

References

1. Damatta, F.M.; Ronchi, C.P.; Maestri, M.; Barros, R.S. Ecophysiology of coffee growth and production. *Braz. J. Plant Physiol.* **2007**, *19*, 485–510. [CrossRef]
2. Larcher, W. *Ecofisiologia Vegetal [Plant Ecophysiology]*; Rima: São Carlos, Brazil, 2000.
3. Ronquim, J.C. Assimilação de Carbono e Fluorescência da Clorofila do Cafeeiro (Coffea arábica L.) sob Condições Contrastantes de Irradiância, Temperatura e Disponibilidade de CO_2 [Carbon Assimilation and Chlorophyll Fluorescence of Coffee Trees (*Coffea arábica* L.) under Contrasting Conditions of Irradiance, Temperature and CO_2 Availability]. Ph.D. Thesis, Federal University of São Carlos, São Carlos, Brazil, 2007.
4. Damatta, F.M. Ecophysiological constraints on the production of shaded and unshaded coffee: A review. *Field Crops Res.* **2004**, *86*, 99–114. [CrossRef]
5. Silveira, J.A.; Silva, S.L.F.; Silva, E.N.; Viégas, R.A. Mecanismos biomoleculares envolvidos com a resistência ao estresse salino em plantas [Biomolecular mechanisms involved in plant saline stress resistance]. In *Manejo da Salinidade na Agricultura: Estudos Básicos e Aplicados [Salinity Management in Agriculture: Basic and Applied Studies]*; Gheyi, H., Dias, N.S., Lacerda, C.F., Eds.; INCTSal: Fortaleza, Brazil, 2010; pp. 161–180.
6. Taiz, L.; Zeiger, E. *Fisiologia Vegetal*, 5th ed.; Artmed: Porto Alegre, Brazil, 2013.
7. Araújo, S.A.C.; Deminicis, B.B. Fotoinibição da fotossíntese [Photosynthesis Inhibition]. *Rev. Bras. Biociênc.* **2009**, *7*, 463–472.
8. Oliveira, K.M.G.; Carvalho, L.G.; Lima, L.A.; Gomes, R.C.C. Modelagem para estimativa da orientação de linhas de plantio de cafeeiros [Modeling for estimating the orientation of planting rows of coffee trees]. *Eng. Agrícola* **2012**, *32*, 293–305. [CrossRef]
9. Salati, E.; Santos, A.A.; Nobre, C. As Mudanças Climáticas Globais e seus Efeitos nos Ecossistemas Brasileiros [Global Climate Change and Its Effects on Brazilian Ecosystems]. 2012. Available online: www.comciencia. br/reportagens/clima/clima14.htm (accessed on 13 January 2015).
10. Morais, H.; Caramori, P.H.; Ribeiro, A.M.A.; Gomes, J.C.; Koguishi, M.S. Microclimatic characterization and productivity of coffee plants grown under shade of pigeon pea in Southern Brazil. *Pesqui. Agropecu. Bras.* **2006**, *41*, 763–770. [CrossRef]
11. Cakmak, I.; Yazici, A.M. Magnesium: A forgotten element in crop production. *Better Crops Plant Food* **2010**, *94*, 23–25.
12. Cakmak, I.; Kirkby, E.A. Role of magnesium in carbon partitioning and alleviating photooxidative damage. *Physiol. Plant.* **2008**, *133*, 692–704. [CrossRef] [PubMed]
13. Hoagland, D.R.; Arnon, D.I. *The Water-Culture Method for Growing Plants without Soil*; Agricultural Experiment Station: Berkley, CA, USA, 1950.
14. Jaimez, R.E.; Rada, F.; Garcia-Núñez, C.; Azócar, A. Seasonal variations in leaf gas exchange of platain cv. Hartón (Musa AAB) under different soil water conditions in a humid tropical region. *Sci. Hortic.* **2005**, *104*, 79–89. [CrossRef]
15. Melo, A.S.; Silva Júnior, C.D.; Fernandes, P.D.; Sobral, L.; Brito, M.E.B.; Dantas, J.D.M. Alterações das características fisiológicas da bananeira sob condições de fertirrigação [Alteration of the physiologic characteristics in banana under fertigation conditions]. *Ciênc. Rural* **2009**, *39*, 733–741. [CrossRef]
16. Ferreira, D.F. SISVAR: A computer statistical analysis system. *Ciênc. Agrotecnol.* **2011**, *35*, 1039–1042. [CrossRef]
17. Dias, K.G.L. Nutrição, Bioquímica e Fisiologia de Cafeeiros Supridos com Magnésio [Nutrition, Biochemistry and Physiology in Coffee Plants Supplied with Mg]. Ph.D. Thesis, Federal University of Lavras, Lavras-MG, Brazil, 2015.
18. Marschner, H. *Mineral Nutrition of Higher Plants*, 3rd ed.; Academic Press: London, UK, 2012.
19. Ding, Y.; Xu, G. Low magnesium with high potassium supply changes sugar partitioning and root growth pattern prior to visible magnesium deficiency in leaves of Rice (*Oryza sativa* L.). *Am. J. Plant Sci.* **2011**, *2*, 601–608. [CrossRef]

20. Wallau, R.L.R.; Soares, A.P.; Camargos, S.L. Concentração e acúmulo de macronutrientes em mudas de mogno cultivadas em solução nutritiva. *Rev. Ciênc. Agroambient.* **2008**, *6*, 1–12.
21. Marques, D.J.; Broetto, F.; Silva, E.C.; Carvalho, J.G. Dinâmica de cátions na raiz e folhas de berinjela cultivada sobre doses crescentes de potássio oriundas de duas fontes. *Idesia (Arica)* **2011**, *29*, 69–77. [CrossRef]
22. Veigas, I.J.M.; Sousa, G.O.; Silva, A.F.; Carvalho, J.G.; Lima, M. M. Composição mineral e sintomas visuais de deficiências de nutrientes em plantas de pimenta-longa (Piper hispidinervum C. DC.). *Acta Amaônica* **2013**, *43*, 43–50. [CrossRef]
23. Yong, J.W.H.; Wong, S.C.; Farquhar, G.D. Stomatal responses to changes in vapour pressure difference between leaf and air. *Plant Cell Environ.* **1997**, *20*, 1213–1216. [CrossRef]
24. Marenco, R.A.; Siebke, K.; Farquhar, G.D.; Ball, M.C. Hydraulically based stomatal oscillations and stomatal patchiness in Gossypium hirsutum. *Funct. Plant Biol.* **2006**, *33*, 1103–1113. [CrossRef]
25. Chaves, A.R.M.; Martins, S.C.V.; Batista, K.D.; Celin, E.F.; DaMatta, F.M. Varying leaf-to-fruit ratios affect branch growth and dieback, with little to no effect on photosynthesis, carbohydrate or mineral pools, in different canopy positions of field-grown coffee trees. *Environ. Exp. Bot.* **2012**, *77*, 207–218. [CrossRef]
26. Cavatte, P.C.; Mrtins, S.V.C.; Wolfgramm, R.; DaMatta, F.M. Physiological responses of two coffee (*Coffea canephora*) genotypes to soil water deficit. In *Droughts: Causes, Effects and Predictions*; Sanchez, J.M., Ed.; Nova Science Publishers: New York, NY, USA, 2008; pp. 306–330.
27. Dias, D.P.; Marenco, R.A. Fotossíntese e fotoinibição em mogno e acariquara em função da luminosidade e temperatura foliar [Photosynthesis and photoinhibition in mahogany and acariquara as a function of irradiance and leaf temperature]. *Pesqui. Agropecu. Bras.* **2007**, *42*, 305–311. [CrossRef]
28. Concenço, G.; Ferreira, E.A.; Silva, A.A.; Ferreira, F.A.; Galon, L.; Reis, M.R.; D'Antonino, L.; Vargas, L.; Silva, L.V.B.D. Fotossíntese de biótipos de azevém sob condição de competição [Photosynthesis of ryegrass biotypes under different competition levels]. *Planta Daninha* **2008**, *26*, 595–600.
29. Mariano, K.R.; Barreto, L.S.; Silva, A.H.B.; Neiva, G.K.P.; Ribeiro, A.J.; Amorim, S.M.C. Fotossíntese e tolerância protoplasmática foliar em Myracrodruonurundeuva fr. All. submetida ao déficit hídrico [Photosynthesis and leaf protoplasmatic tolerance in Myracrodruonurundeuva fr. All. under water deficit]. *Caatinga* **2009**, *22*, 72–77.
30. Gutierrez, M.V.; Meinzer, F.C. Carbon isotope discrimination and photosynthetic gas exchange in coffee hedgerows during canopy development. *Aust. J. Plant Physiol.* **1994**, *21*, 207–219. [CrossRef]
31. Lee, D.W.; Baskaran, K.; Mansor, M.; Mohamad, H.; Yap, S.K. Irradiance and spectral quality affect Asian tropical rain forest tree seedling development. *Ecology* **1996**, *77*, 568–580. [CrossRef]
32. Baliza, D.P.; Cunha, R.L.; Castro, A.M.; Barbosa, J.P.R.A.D.; Pires, M.F.; Gomes, R.A. Trocas gasosas e características estruturais adaptativas de cafeeiros cultivados em diferentes níveis de radiação [Gas exchange and adaptative structural characteristics of coffee plants grown in different levels of radiation]. *Coffee Sci.* **2012**, *7*, 250–258.
33. Assad, E.D.; Pinto, H.S.; Zullo, J.R.J.; Ávila, A.M.H. Impacto das mudanças climáticas no zoneamento agroclimático do café no Brasil [Climate changes impact the agroclimatic zonning of coffee in Brazil]. *Pesqui. Agropec. Bras.* **2004**, *39*, 1057–1064. [CrossRef]
34. Silva, D.M.; Brandão, I.R.; Alves, J.D.; Santos, M.O.; Souza, K.R.D.; Silveira, H.R.O. Physiological and biochemical impacts of magnesium-deficiency in two cultivars of coffee. *Plant Soil* **2014**, *128*, 133–150. [CrossRef]
35. Prado, R.M. *Nutrição de Plantas [Plant Nutrition]*; Editora da Unesp: São Paulo, Brazil, 2008.
36. Jia, Y.; Yang, X.; Islam, E.; Feng, Y. Effects of potassium deficiency on chloroplast ultrastructure and chlorophyll fluorescence in inefficient and efficient genotypes of rice. *J. Plant Nutr.* **2008**, *31*, 2105–2118. [CrossRef]
37. Catuchi, T.A.; Vitolo, H.F.; Bertolli, S.C.; Souza, G.M. Tolerance to water deficiency between two soybean cultivars: Transgenic versus conventional. *Ciênc. Rural* **2011**, *31*, 373–378. [CrossRef]
38. Logan, B.A.; Kornyeyev, D.; Hardison, J.; Holaday, A.S. The role of antioxidant enzymes in photoprotection. *Photosynth. Res.* **2006**, *88*, 119–132. [CrossRef] [PubMed]
39. Pandey, V.; Dixit, V.; Shyam, R. Chromium effect on ROS generation and detoxification in pea (*Pisun sativum*) leaf chloroplasts. *Protoplasma* **2010**, *236*, 85–95. [CrossRef] [PubMed]

40. Kaiser, M.W. The effect of hydrogen peroxide on CO_2 fixation of isolated chloroplast. *Biochim. Biophys. Acta* **1976**, *440*, 476–482. [CrossRef]

41. Machado, E.C.; Schmidt, P.T.; Medina, C.L.; Ribeiro, R.V. Respostas da fotossíntese de três espécies de citros a fatores ambientais [Photosynthetic responses of three citrus species to environmental factors]. *Pesqui. Agropec. Bras.* **2005**, *40*, 1161–1170. [CrossRef]

agriculture

MDPI

Review

Phosphorus Transport in Arabidopsis and Wheat: Emerging Strategies to Improve P Pool in Seeds

Mushtak Kisko [1], Vishnu Shukla [2], Mandeep Kaur [2], Nadia Bouain [1], Nanthana Chaiwong [1], Benoit Lacombe [1], Ajay Kumar Pandey [2] and Hatem Rouached [1,*,†]

[1]   BPMP, Univ Montpellier, CNRS, INRA, SupAgro, 34060 Montpellier, France; mushtakkisko@gmail.com (M.K.); nadia.bouain@gmail.com (N.B.); nantana.c189@gmail.com (N.C.); benoit.lacombe@supagro.fr (B.L.)
[2]   Department of Biotechnology, National Agri-Food Biotechnology Institute, Sector 81, S.A.S. Nagar (Mohali), Punjab 140306, India; shuklavrk@gmail.com (V.S.); mandeep@nabi.res.in (M.K.); pandeyak1974@gmail.com (A.K.P.)
*   Correspondence: hatem.rouached@inra.fr; Tel.: +33-(0)-4-99-61-31-54; Fax: +33-(0)-4-67-52-57-37
†   Present Address: Department of Plant Biology, Carnegie Institution for Science, Stanford, CA 94305, USA.

Received: 21 November 2017; Accepted: 10 February 2018; Published: 14 February 2018

Abstract: Phosphorus (P) is an essential macronutrient for plants to complete their life cycle. P taken up from the soil by the roots is transported to the rest of the plant and ultimately stored in seeds. This stored P is used during germination to sustain the nutritional demands of the growing seedling in the absence of a developed root system. Nevertheless, P deficiency, an increasing global issue, greatly decreases the vigour of afflicted seeds. To combat P deficiency, current crop production methods rely on heavy P fertilizer application, an unsustainable practice in light of a speculated decrease in worldwide P stocks. Therefore, the overall goal in optimizing P usage for agricultural purposes is both to decrease our dependency on P fertilizers and enhance the P-use efficiency in plants. Achieving this goal requires a robust understanding of how plants regulate inorganic phosphate (Pi) transport, during vegetative growth as well as the reproductive stages of development. In this short review, we present the current knowledge on Pi transport in the model plant *Arabidopsis thaliana* and apply the information towards the economically important cereal crop wheat. We highlight the importance of developing our knowledge on the regulation of these plants' P transport systems and P accumulation in seeds due to its involvement in maintaining their vigour and nutritional quality. We additionally discuss further discoveries in the subjects this review discusses substantiate this importance in their practical applications for practical food security and geopolitical applications.

Keywords: phosphate; seeds; Arabidopsis; wheat

1. Introduction

Phosphorus (P) is an essential macronutrient for plant growth and production [1,2]. P deficiency is a generally widespread stressor occurring in natural and agricultural environments. Consequently, global agricultural crop production has been severely affected [2,3]. It is estimated that crop yield on 30–40% of the world's total arable land is limited by inorganic phosphate (Pi) bioavailability [4,5]. Many causes could explain the low availability of Pi to plants, such as the Pi (HPO_4^{2-}) interaction with soil cations such as zinc (Zn^{2+}) or iron (Fe^{2+}), which form an insoluble complex [6–8]. In addition, and most importantly, global Pi reserves are rapidly decreasing due to an increase in its demand [2,9–11]. Analysis of data collected over 14 years revealed that the usage of global P fertilizer considerably increased at a rate of about 357,000 t/annum (i.e., an annual increase of 2.4%) [12,13]. Experts agree that the world is facing serious P crisis [14,15] and that the global P reserve is not distributed uniformly [5]. Taken together, these issues constitute compelling evidence justifying a direct link between Pi availability and the overwhelming world food security in coming few years. Developing a better understanding of how plants regulate Pi uptake for transport to different organs and seeds will help design new strategies to increase crop yield and simultaneously reduce P input.

Approximately 75% of Pi taken up by plant roots for use by vegetative tissues and vital storage in developing seeds is stored in the form of phytic acid (PA) [3]. Although P fertilizer supply has increased the yield of cereal grains such as wheat, a recent analysis showed that about 37% of wheat areas worldwide have experienced yield stagnation [16], highlighting the importance of precise management of P fertilizer application to achieve high wheat yield and quality. Achieving this objective necessitates a better understanding of how crops regulate P homeostasis. During the last decades, our knowledge on the molecular regulation of P transport and P redistribution in different plant organs during vegetative phase has progressed mainly in model plants such as *Arabidopsis thaliana* (for review, [17]). However, limited progress was achieved in crops such as wheat, a major dietary source of calories and protein for humans. In this short review, we present the recent progress of our understanding on Pi transport in wheat through information obtained for *A. thaliana*. We next highlight the importance of the knowledge generated on the molecular mechanisms which regulate the Pi transport and its accumulation in grains.

2. Phosphate Uptake and Transport in Wheat and Arabidopsis

In plants, Pi deficiency affects growth that manifests itself at a phenotypic level by affecting the aboveground and underground biomass. The effects of Pi availability on wheat growth is presented in Figure 1.

Figure 1. Changes in wheat growth under different phosphate conditions. (**A**) Wheat were grown hydroponically for three weeks in presence (+P) or absence (−P) of inorganic phosphate (KH$_2$PO$_4$). (**B**) shoot and (**C**) root of wheat grown either in presence (0.5 mM) or in absence of inorganic phosphate for three weeks.

Pi is acquired by root systems, which undergo a significant architectural change in response to Pi deficiency (for review, [18]) that is associated with increasing Pi uptake capacity through an upregulation of regulators and transporters involved in P-homeostasis [19]. To improve P use efficiency in crop plants, it is therefore prerequisite to understand P distribution within the plant.

Numerous Pi transporters have been identified in wheat [20–22] Gradual elucidation of these components and their roles have been effectively achieved through experiments either increasing P remobilization from senescing tissue or reducing the partition of P to developing grains [23]. However, initial identification of these transporters were generated from data collected for those of the model plant Arabidopsis [24]. In this dicot model plant, Pi transporters generally belong to a gene family referred to as phosphate transporter (*PHT*) [25]. The PHT family is divided into five groups (*PHT1*, *PHT2*, *PHT3*, *PHT4*, and *PHT5*) differentiated primarily by their sub-cellular localization and functional properties. Plasma membrane-bound proteins belonging to the *PHT1* subfamily are primarily responsible for Pi uptake in Arabidopsis [25]. In contrast, PHT2 proteins are localized in chloroplasts, while PHT3/MPT proteins are mostly mitochondrial membrane transporters and *PHT4* proteins are Golgi-apparatus located transporters. A vacuolar Pi transporter was identified in Arabidopsis and designated as PHT5/VPT/SPX-MFS proteins [26].

Pi is also transported outside the root cells for distribution between different plant organs. The Phosphate 1 (*PHO1*) gene family contains 11 Pi exporter proteins mostly involved in the translocation of Pi from roots to shoots [27,28]. Among these molecular players, the mutation of both *PHT1;1* and *PHT1;4* or *PHO1* and *PHO1;H1* causes the most severe phenotype that is characterized by a decreased Pi accumulation in Arabidopsis [19,28], demonstrating the importance of these Pi transporters and exporters in Pi uptake and Pi translocation to shoots. For more information, readers are referred to these reviews [25,29].

The identification of Arabidopsis proteins involved in Pi distribution accelerated the discovery of wheat Pi transporters by exploring regions of the wheat genome with similar sequence to the genomic sequences of these Arabidopsis Pi transporters [20]. Validation and characterization of identified wheat Pi transporter candidates have been attained through evaluation of their genetic function either in Arabidopsis or by means of complementation with yeast mutants defective in Pi transport [30,31]. A cumulative list of members of the crop origin Pi-transporters is summarized in Table 1.

Table 1. Inventory for the list of genes and regulators those are involved in sensing, uptake and signaling during Pi limiting conditions in model plant Arabidopsis and *Triticum aestivum*.

Components of Phosphate Transport	*Arabidopsis thaliana*	*Triticum aestivum*
Sensing	AtSIZ1 ([32])	NA*
	AtSPX1, AtSPX2, AtSPX3 & AtSPX4 ([33,34])	TaSPX1 ([35])
	AtIPK1 ([36])	TaIPK1 ([37])
Uptake & Transport	AtPHT1;1–AtPHT1;9 ([38])	TaPHT1.1-TaPHT1.13 ([20–22,39])
	AtPHT2;1 ([40])	TaPHT2.1 ([41])
	AtPHT3;1–AtPHT3.3 ([42,43])	TaPHT3.1–TaPHT3.3 ([20,44])
	AtPHT4;1–AtPHT4;6 ([45])	TaPHT4.1–TaPHT4.6 ([20])
	AtPHT5;1–AtPHT5;3 ([46])	NA
	AtPHO1 ([27])	TaPHO1 ([47])
Signaling	AtPHR1 ([48])	TaPHR1 ([49])
	AtPHL1 ([50])	NA
	AtPHF1 ([51])	NA
	AtPHO1 ([52])	NA
	AtPHO2 ([53])	TaPHO2 ([47])
	AtZAT6 ([54])	NA
	AtARP6 ([55])	NA
	AtH2A.Z ([55])	NA
	AtWRKY6 ([56])	NA
	AtWRKY42 ([57])	NA
	AtWRKY45 ([58])	NA
	AtWRKY75 ([59])	NA
	AtNLA ([60])	NA
	AtIPS1 ([61])	TaIPS1 ([44])
	AtmiR399 ([62])	Tae-miR399-A1 ([47])
	AtMiR827 ([63])	NA
	AtALIX ([64])	NA

* NA: Not applicable.

The wheat genome contains several TaPHT members that could be divided into four subfamilies, PHT1 (*TaPHT1.1–1.13*), PHT2 (*TaPHT2.1*), PHT3 (*TaPHT3.1–3.3*), and PHT4 (*TaPHT4.1–4.6*). Their transcripts demonstrate enhanced expression in Pi limited roots and shoots ([20]). The complexity in discerning a total number of TaPHT1 members within the wheat genome has been recently discussed ([21]) while experiments characterizing the role of specific members involved in P uptake within two wheat genotypes (*KN9204* and *SJZ8*) have been performed. Differential expression pattern of high-affinity *TaPHTs* was observed in different wheat varieties under varying Pi regimes ([39]). Interestingly, the expression of *TaPHT1.1, 1.2, 1.9,* and *1.10* at flowering positively correlated with P uptake after stem elongation in different wheat varieties under field conditions supplemented with different P rates ([21]). Under Pi deprivation, Pi uptake increases and involves a high-affinity PHT1 member *TaPT2* ([65]). Down-regulation of *TaPHT2.1* was able to induce a pronounced decrease in Pi accumulation in both sufficient and Pi-deficient wheat, suggesting its association with other PHTs involved in Pi uptake and translocation within plants ([41]). This reinforces the impact the intracellular Pi transport mechanism has upon regulating the plant Pi uptake efficiency. Still, in contrast to the data known about Arabidopsis, little is known about the regulation of wheat Pi transporters at the protein levels ([66]). It is worth to note that apart from its role in Pi uptake, *TaPHT2.1* was functionally characterized as an important P signalling component involved in Pi translocation from cytosol to chloroplast in Pi-stressed leaves ([41]). Whether any Pi transporter in wheat could play an additional transceptor role remains an open question. Membrane proteins that fulfil a dual nutrient transport function, such as the *PHO1* ([67]) or nitrogen transporter *NRT1* ([68]), serve as extant examples. Other than *PHT2* members and some reports showing differential expression of *TaPHT3* and four transcript abundance under Pi-depleted roots and shoots ([20,44]), there is no detail on their precise biological role in wheat tissues/cell. In addition, recent expression profiles of wheat *PHT1* sub-family genes during hydroponic and field-grown plant tissues were correlated with the presence of cis-acting promoter elements ([22]). These studies showed growing interests in the crop Pi-transporters among researchers.

3. Phosphate Sensing and Signalling in Arabidopsis and Wheat

How plants sense and signal Pi deficiency has been a long-standing question. In Arabidopsis, decades of research eventually identified the Pi starvation signalling pathway, defined as *SPX1-PHR1-miR399-PHO2-PHT1/PHO1* (for review: [29,69]). Key genes encoding for SPX proteins was initially reported to be upregulated in response to P deficiency via transcriptome analysis ([70]). *SPX* genes perform diverse functions in plant tolerance to Pi starvation ([33]) and play an important role in sensing P concentration in the cytosol through its ability to bind the inositol polyphosphate signalling molecule ([71]). *SPX1* interacts with the transcription factor Phosphate Response 1 (*PHR1*) in presence of Pi and dissociates under Pi deficiency ([71]). *PHR1* regulates many Pi-related genes ([50]), such as the *miRNA399* that eventually targets Phosphate 2 (PHO2) transcripts. The reduction of PHO2 protein abundance leads to the accumulation of PHT1 and PHO1 proteins ([72]) and consequently an increase of the plant capacity to uptake Pi and translocate Pi to shoots ([73]). Noteworthy, the proper functioning of this signalling pathway requires the contribution of many others genes such the *SUMO E3 ligase SIZ1* ([32]), *PHOSPHATE TRANSPORTER TRAFFIC FACILITATOR1* (*PHF1*), ([51]) and *NITROGEN LIMITATION ADAPTATION* (NLA), ([60]). *SIZ1* is involved in the regulation of *PHR1* by sumoylation ([32]). *PHF1* is required for the trafficking of Pi transporters to the plasma membrane [51]. *NLA* is proposed to function at the plasma membrane to direct the degradation of *PHT1s* ([60]), a fine-tuning process required for Pi uptake capacity of plants.

In wheat, transcript profiles of genes involved in Pi starvation response with organ-specific Pi allocation patterns were studied in roots and shoots of Chinese 80–55 (P-efficient cultivar) and Machete (less-efficient cultivar) under Pi deficiency [44]. This report revealed the distinct modes for allocation of Pi and organic P compounds between the source and sink tissues that modulate the adaptation under varying Pi condition. The P-allocation patterns in the multiple plant organs correlated with the transcript expression patterns, suggestive of molecular signatures for improved phosphorus use efficiency (PUE) during limited Pi supply. Few genes involved in Pi starvation signalling responses have been reported for hexaploid wheat [35], such as an ortholog of the Arabidopsis transcription factor PHR1 characterized for its function in regulating Pi-signalling and plant growth in wheat [49]. Under both Pi-sufficient and deficient conditions, over-expression of the *TaPHR1-A1* homolog moderately up-regulated the expression levels of *TaPHR1* throughout the plant, resulting in a moderate increase of leaf Pi concentration and thus avoiding resultant toxicity ([49]). Pi uptake was positively favoured by *TaPHR1-A1* over-expression by increasing root tip number, lateral root length, and *TaPHTs* expression (*TaPHT1.2* in roots and *TaPHT1.6* in shoots). Utilizing bimolecular fluorescence complementation assays, it has been confirmed that wheat PHR1 forms a homodimer and confers transcriptional activation of a putative downstream target Pi-transporter *TaPHT1.2* [49].

The presence of Arabidopsis PHO2 orthologs in hexaploid wheat has also been proposed [35]. Detailed analysis of respective mutant lines for three *TaPHO2* genes from homologous group1 (*A1*, *B1*, and *D1*) showed remarkably different effects on P uptake, distribution, and plant growth [47]. The overall expression of *TaPHO2* in wheat was severely reduced in a *tapho2-d1* mutant, leading to high total shoot P under limited Pi conditions, but also showed inhibited growth and yield [47]. This resembled the phenotype observed in a *pho2* mutant of both monocots (e.g., rice) and dicots (e.g., Arabidopsis) [74,75]. Interestingly, *tapho2-a1* knockout mutant plants showed reduced *TaPHO2* expression that leads to only a moderate increase of total P and Pi levels in leaf under both sufficient and deficient P conditions [47]. Unlike the *tapho2-d1* mutant, *tapho2-a1* mutants demonstrated a moderate increase in P levels and accumulation alongside improved plant growth and grain yield [47]. In light of these interesting data, the involvement of *TaPHO2-D1* in Pi homeostasis to maintain plant growth rather than a simple Pi starvation signalling pathway has been proposed [47]. The Pi starvation signalling pathway *PHR1-IPS1-miR399-UBC24/PHO2-PHT1/PHO1* seems to be conserved and functional in numerous plant species. Manipulating the components of this pathway could be an important strategy for improving Pi nutrition in crops.

These genes and molecular mechanisms involved in Pi stress response are specifically induced during Pi deficiency and not under any other modes of stress known to alter Pi homeostasis. These observations indicate the existence of additional unknown genes and pathways regulating the Pi content in plants [8,52]. For instance, it is now well established that Pi content in plants is altered when plants are challenged by zinc limitation ($-$Zn) [6,76–79]. Intriguingly, under single $-$Zn stress, an excess of Pi supply causes loss of wheat biomass in comparison with plants grown under –P-Zn simultaneous stress [78]. Nevertheless, despite its fundamental importance, very little is known about the regulatory network established during Zn deficiency to control Pi homeostasis [52]. Studying Zn/Pi homeostasis interactions will lead us to uncover new genes and pathways controlling plant Pi homeostasis. This knowledge will be an additional resource for the improving Pi usage through perturbing Zn deficiency signalling pathways.

These experiments have all begun to clarify the components and systems that regulate P/Pi sensing and signalling within plants. The emergence of genome editing tools holds promise for further studies perturbing specific Pi-related genes/pathways, or genes involved in modulating the Pi content, such as genes involved in Zn deficiency signalling, to ultimately improve Pi nutrition in crops [80].

4. Phosphate in Seeds

The accumulation of nutrients in the seed is important for seed vigour and germination [81]. The amount of micronutrients in seeds has declined systematically since the beginning of the green revolution in the 1960s [82,83]. In contrast, phytic acid (PA), the organic form of P in seeds, has increased following global changes (e.g., elevated atmospheric CO_2) in various plant species including wheat [84]. Since PA is considered an anti-nutrient, increasing Pi content in grain while decreasing PA has become a trait of great interest [85].

In general, while Pi uptake and its intracellular and long-distance transport in plants has been extensively studied, the Pi transport in seeds has received little attention [86]. Current knowledge on the role of seed-specific PHTs is largely lacking [87]. In the seeds, nutrients reach the embryo via various pathways and at different developmental stages. Transfer of nutrients from the maternal seed coat to the filial endosperm and embryo is required for seed production and quality. In the case of Pi, the transfer of this element from the seed coat to the embryo requires Pi exporters. Recently, the *PHO1* gene was shown to be expressed in the chalazal seed coat in Arabidopsis, suggesting a role in the transfer of P from the seed coat to the embryo in developing seeds [88]. Consistently, Pi transfer from the seed coat to the embryo is perturbed in the *pho1* mutant. This observation and experimental data state the ignition point for a deep investigation of Pi transport in seeds, which may help to start understanding the mechanism regulating P accumulation.

In wheat, the mature grain may contain up to 90% of the total shoot P, with 20–90% of this being translocated from other tissues (for details, see [89]). PA accounts for up to 1–2% of the total weight [90,91]. P and PA concentrations in the grain increases as P application increases [92]. Increased PA concentration greatly decreases the bioavailability of nutritional minerals in wheat grain, such as Zn [92]. Therefore, reduction of PA in cereal grains is considered an important trait that is generated either through breeding or biotechnological approaches. Reducing PA in grains may provide a dual gain with less grain P loss and more micronutrient retention [93,94]. Generation of low PA crops may be achieved by targeting PA biosynthesis genes or transport [29,95–98]. As an alternative strategy to achieve low PA grains, roles for other families of transporters are also emerging. For instance, knockdown of the rice Pi transporter *OsPHT1.8* resulted in lowered PA accumulation in the embryo and mature grains [94]. Subsequently, they showed that rice *PHT1.8* performs a novel biological function during crosstalk between Pi and auxin signalling. This was one of the recent reports that provided clues for the link between auxin and -Pi responses [99]. Sulphate transporters have also implicated in grain PA and P content regulation. Map-based cloning and complementation tools resulted in the identification of rice sulphate transporters referred as *OsSULTR3;3*, which are involved in compositional changes of Pi and PA in developing grains [100]. Subsequently, another sulphate transporter family gene named the *SULTR-like phosphorus distribution transporter (SPDT)* has demonstrated involvement in the intervascular transfer of P, especially at the nodes by unloading P from xylem toward the phloem [101]. Therefore, such studies have suggested that node-localized transporters could affect the preferential accumulation of P in grains [101].

Nevertheless, the relative dearth of information has led to few varieties of wheat being studied in regards to grain total P and PA. It is still unclear whether transportation of P to the grain occurs directly from phloem or via xylem through recycling from roots, and to what extent translocation of P between plant organs is altered at different P-regimes [102]. In wheat, only two significant transgenic studies that include over-expression of *TaPHR1-A1* and knockout of *TaPHO2-A1* were able to achieve enhanced P uptake and grain yield under low Pi condition [47,49]. It is reasonable to speculate that manipulating these Pi-related genes, among others yet to be discovered, will allow intentional modulation of Pi loading in grains [20,88].

Although the current knowledge of *PHT* expression and that of other regulators in seeds is in its early stage across plant species, some rice transporters demonstrate function in seed filling with Pi. A recent study investigating suppression of rice *OsPHT1;8* suggested its role in P redistribution and allocation of Pi in both embryo and endosperm seed tissue [94]. Hence, it will be important to implement such functional strategies for addressing transport and accumulation of Pi into grains [20].

5. Conclusions

It has been reported that plants use only 20%–30% of the Pi fertilizers applied to soil [103]. The significant remaining Pi is lost and can leach into aquatic ecosystems, instigating ecological issues such as eutrophication. Thus, it is clear that the excessive use of Pi fertilizers is not only an unsustainable and costly practice, but also ecologically unfriendly. Therefore, research on Pi nutrition in plants should lead to changes in agricultural practices that would be both economically and environmentally beneficial.

How societally pertinent crop plants such as wheat maintain P homeostasis and respond to changes of Pi concentration remain poorly understood. Improving wheat Pi nutrition will require a full understanding of the physiology and molecular regulation of P remobilization from vegetative tissues to grains. In the future, it will be interesting to identify the complete list of genes that are involved in Pi transport between different wheat grain tissues, Pi acquisition, and Pi mobilization in embryo development. Although some uptake Pi transporters were discovered through classical molecular approaches, their regulatory mechanisms at the transcriptional and posttranscriptional levels remain obscure. This is particularly challenging in wheat because of the current unavailability of the complete genomic sequence. For the ones identified, the availability of sequenced mutant populations [104] alongside current genome editing tools like Clustered regularly interspaced short palindromic repeats-Cas9 technology will constitute an invaluable resource for their functional validation. In addition, a proper combination of omics approaches (such as RNA-seq), empowered with system biology tools, will help to construct regulatory pathways regulating Pi accumulation in wheat during its different developmental stages. Gaining this knowledge is vital to create crop varieties with improved P-use efficiency and modulate the Pi accumulation in grain.

Acknowledgments: The author thanks Benjamin Jin (Carnegie Institution for Science, Stanford, CA, USA) for critically reading the manuscript and providing valuable suggestions. This work was funded by the "Institut National de la Recherche Agronomique,Montpellier, France" INRA and by the Région Languedoc-Roussillon: Chercheur d'Avenir 2015, Projet cofinancé par le Fonds Européen de Développement Régional to HR. Work in AKP lab was supported by NABI-Core grant. Authors thanks The Editors for the invitation to contribute to this special issue "Plant Nutrient Dynamics in Stressfuls Environments".

Author Contributions: M.K., N.B., N.C., B.L., and H.R. reviewed knowledge on the regulation of Pi in Arabidopsis. V.S., M.K., A.K.P. reviewed knowledge on the regulation of Pi in wheat. H.R. prepared figure. A.K. prepared table. A.K.P., H.R. approved final version of the review.

Conflicts of Interest: The authors declare no conflict of interest.

References

1. Dessibourg, O. Arsenic-based bacteria point to new life forms. *New Sci.* **2010**, *18*, 2.
2. Heuer, S.; Gaxiola, R.; Schilling, R.; Herrera-Estrella, L.; López-Arredondo, D.; Wissuwa, M.; Delhaize, E.; Rouached, H. Improving phosphorus use efficiency: A complex trait with emerging opportunities. *Plant J.* **2017**, *90*, 868–885. [CrossRef] [PubMed]
3. Lott, J.N.; Ockenden, I.; Raboy, V.; Batten, G.D. Phytic acid and phosphorus in crop seeds and fruits: A global estimate. *Seed Sci. Res.* **2000**, *10*, 11–33.
4. Runge-Metzger, A. Closing the cycle: Obstacles to efficient P management for improved global food security. *Scope-Sci. Comm. Probl. Environ. Int. Counc. Sci. Unions* **1995**, *54*, 27–42.
5. MacDonald, G.K.; Bennett, E.M.; Potter, P.A.; Ramankutty, N. Agronomic phosphorus imbalances across the world's croplands. *Proc. Natl. Acad. Sci. USA* **2011**, *108*, 3086–3091. [CrossRef] [PubMed]
6. Bouain, N.; Shahzad, Z.; Rouached, A.; Khan, G.A.; Berthomieu, P.; Abdelly, C.; Poirier, Y.; Rouached, H. Phosphate and zinc transport and signalling in plants: Toward a better understanding of their homeostasis interaction. *J. Exp. Bot.* **2014**, *65*, 5725–5741. [CrossRef] [PubMed]
7. Mongon, J.; Chaiwong, N.; Bouain, N.; Prom-u-Thai, C.; Secco, D.; Rouached, H. Phosphorus and Iron Deficiencies Influences Rice Shoot Growth in an Oxygen Dependent Manner: Insight from Upland and Lowland Rice. *Int. J. Mol. Sci.* **2017**, *18*, 607. [CrossRef] [PubMed]
8. Rouached, H.; Rhee, S.Y. System-level understanding of plant mineral nutrition in the big data era. *Curr. Opin. Syst. Biol.* **2017**, *4*, 71–77. [CrossRef]
9. Van Kauwenbergh, S.J. *World Phosphate Rock Reserves and Resources*; IFDC: Muscle Shoals, AL, USA, 2010.
10. Van Vuuren, D.P.; Bouwman, A.F.; Beusen, A.H. Phosphorus demand for the 1970–2100 period: A scenario analysis of resource depletion. *Glob. Environ. Chang.* **2010**, *20*, 428–439. [CrossRef]
11. Vance, C.P.; Uhde-Stone, C.; Allan, D.L. Phosphorus acquisition and use: Critical adaptations by plants for securing a nonrenewable resource. *New Phytol.* **2003**, *157*, 423–447. [CrossRef]

12. Lott, J.N.; Kolasa, J.; Batten, G.D.; Campbell, L.C. The critical role of phosphorus in world production of cereal grains and legume seeds. *Food Secur.* **2011**, *3*, 451–462. [CrossRef]

13. Walan, P.; Davidsson, S.; Johansson, S.; Höök, M. Phosphate rock production and depletion: Regional disaggregated modeling and global implications. *Resour. Conserv. Recycl.* **2014**, *93*, 178–187. [CrossRef]

14. Abelson, P.H. A potential phosphate crisis. *Science* **1999**, *283*, 2015. [CrossRef] [PubMed]

15. Cordell, D.; Drangert, J.-O.; White, S. The story of phosphorus: Global food security and food for thought. *Glob. Environ. Chang.* **2009**, *19*, 292–305. [CrossRef]

16. Ray, D.K.; Ramankutty, N.; Mueller, N.D.; West, P.C.; Foley, J.A. Recent patterns of crop yield growth and stagnation. *Nat. Commun.* **2012**, *3*, 1293. [CrossRef] [PubMed]

17. Rouached, H.; Arpat, A.B.; Poirier, Y. Regulation of phosphate starvation responses in plants: Signaling players and cross-talks. *Mol. Plant* **2010**, *3*, 288–299. [CrossRef] [PubMed]

18. Bouain, N.; Doumas, P.; Rouached, H. Recent advances in understanding the molecular mechanisms regulating the root system response to phosphate deficiency in Arabidopsis. *Curr. Genom.* **2016**, *17*, 308–314. [CrossRef] [PubMed]

19. Shin, H.; Shin, H.S.; Dewbre, G.R.; Harrison, M.J. Phosphate transport in Arabidopsis: Pht1; 1 and Pht1; 4 play a major role in phosphate acquisition from both low-and high-phosphate environments. *Plant J.* **2004**, *39*, 629–642. [CrossRef] [PubMed]

20. Shukla, V.; Kaur, M.; Aggarwal, S.; Bhati, K.K.; Kaur, J.; Mantri, S.; Pandey, A.K. Tissue specific transcript profiling of wheat phosphate transporter genes and its association with phosphate allocation in grains. *Sci. Rep.* **2016**, *6*. [CrossRef] [PubMed]

21. Teng, W.; Zhao, Y.-Y.; Zhao, X.-Q.; He, X.; Ma, W.-Y.; Deng, Y.; Chen, X.-P.; Tong, Y.-P. Genome-wide Identification, Characterization, and Expression Analysis of PHT1 Phosphate Transporters in Wheat. *Front. Plant Sci.* **2017**, *8*. [CrossRef] [PubMed]

22. Grün, A.; Buchner, P.; Broadley, M.R.; Hawkesford, M.J. Identification and expression profiling of Pht1 phosphate transporters in wheat in controlled environments and in the field. *Plant Biol.* **2017**. [CrossRef] [PubMed]

23. Veneklaas, E.J.; Lambers, H.; Bragg, J.; Finnegan, P.M.; Lovelock, C.E.; Plaxton, W.C.; Price, C.A.; Scheible, W.R.; Shane, M.W.; White, P.J.; et al. Opportunities for improving phosphorus-use efficiency in crop plants. *New Phytol.* **2012**, *195*, 306–320. [CrossRef] [PubMed]

24. Koornneef, M.; Meinke, D. The development of Arabidopsis as a model plant. *Plant J.* **2010**, *61*, 909–921. [CrossRef] [PubMed]

25. Nussaume, L.; Kanno, S.; Javot, H.; Marin, E.; Pochon, N.; Ayadi, A.; Nakanishi, T.M.; Thibaud, M.-C. Phosphate import in plants: Focus on the PHT1 transporters. *Front. Plant Sci.* **2011**, *2*, 83. [CrossRef] [PubMed]

26. Liu, J.; Yang, L.; Luan, M.; Wang, Y.; Zhang, C.; Zhang, B.; Shi, J.; Zhao, F.-G.; Lan, W.; Luan, S. A vacuolar phosphate transporter essential for phosphate homeostasis in Arabidopsis. *Proc. Natl. Acad. Sci. USA* **2015**, *112*, E6571–E6578. [CrossRef] [PubMed]

27. Hamburger, D.; Rezzonico, E.; Petétot, J.M.-C.; Somerville, C.; Poirier, Y. Identification and characterization of the Arabidopsis *PHO1* gene involved in phosphate loading to the xylem. *Plant Cell* **2002**, *14*, 889–902. [CrossRef] [PubMed]

28. Stefanovic, A.; Ribot, C.; Rouached, H.; Wang, Y.; Chong, J.; Belbahri, L.; Delessert, S.; Poirier, Y. Members of the *PHO1* gene family show limited functional redundancy in phosphate transfer to the shoot, and are regulated by phosphate deficiency via distinct pathways. *Plant J.* **2007**, *50*, 982–994. [CrossRef] [PubMed]

29. Secco, D.; Bouain, N.; Rouached, A.; Prom-U-Thai, C.; Hanin, M.; Pandey, A.K.; Rouached, H. Phosphate, phytate and phytases in plants: From fundamental knowledge gained in Arabidopsis to potential biotechnological applications in wheat. *Crit. Rev. Biotechnol.* **2017**, *37*, 898–910. [CrossRef] [PubMed]

30. Hassler, S.; Lemke, L.; Jung, B.; Möhlmann, T.; Krüger, F.; Schumacher, K.; Espen, L.; Martinoia, E.; Neuhaus, H.E. Lack of the Golgi phosphate transporter PHT4; 6 causes strong developmental defects, constitutively activated disease resistance mechanisms and altered intracellular phosphate compartmentation in Arabidopsis. *Plant J.* **2012**, *72*, 732–744. [CrossRef] [PubMed]

31. Jia, F.; Wan, X.; Zhu, W.; Sun, D.; Zheng, C.; Liu, P.; Huang, J. Overexpression of mitochondrial phosphate transporter 3 severely hampers plant development through regulating mitochondrial function in Arabidopsis. *PLoS ONE* **2015**, *10*, e0129717. [CrossRef] [PubMed]

32. Miura, K.; Rus, A.; Sharkhuu, A.; Yokoi, S.; Karthikeyan, A.S.; Raghothama, K.G.; Baek, D.; Koo, Y.D.; Jin, J.B.; Bressan, R.A. The Arabidopsis SUMO E3 ligase SIZ1 controls phosphate deficiency responses. *Proc. Natl. Acad. Sci. USA* **2005**, *102*, 7760–7765. [CrossRef] [PubMed]

33. Duan, K.; Yi, K.; Dang, L.; Huang, H.; Wu, W.; Wu, P. Characterization of a sub-family of Arabidopsis genes with the SPX domain reveals their diverse functions in plant tolerance to phosphorus starvation. *Plant J.* **2008**, *54*, 965–975. [CrossRef] [PubMed]

34. Puga, M.I.; Mateos, I.; Charukesi, R.; Wang, Z.; Franco-Zorrilla, J.M.; de Lorenzo, L.; Irigoyen, M.L.; Masiero, S.; Bustos, R.; Rodríguez, J. SPX1 is a phosphate-dependent inhibitor of PHOSPHATE STARVATION RESPONSE 1 in Arabidopsis. *Proc. Natl. Acad. Sci. USA* **2014**, *111*, 14947–14952. [CrossRef] [PubMed]

35. Oono, Y.; Kobayashi, F.; Kawahara, Y.; Yazawa, T.; Handa, H.; Itoh, T.; Matsumoto, T. Characterisation of the wheat (*Triticum aestivum* L.) transcriptome by de novo assembly for the discovery of phosphate starvation-responsive genes: Gene expression in Pi-stressed wheat. *BMC Genom.* **2013**, *14*, 77. [CrossRef] [PubMed]

36. Kuo, H.F.; Chang, T.Y.; Chiang, S.F.; Wang, W.D.; Charng, Y.Y.; Chiou, T.J. Arabidopsis inositol pentakisphosphate 2-kinase, AtIPK1, is required for growth and modulates phosphate homeostasis at the transcriptional level. *Plant J.* **2014**, *80*, 503–515. [CrossRef] [PubMed]

37. Bhati, K.K.; Aggarwal, S.; Sharma, S.; Mantri, S.; Singh, S.P.; Bhalla, S.; Kaur, J.; Tiwari, S.; Roy, J.K.; Tuli, R. Differential expression of structural genes for the late phase of phytic acid biosynthesis in developing seeds of wheat (*Triticum aestivum* L.). *Plant Sci.* **2014**, *224*, 74–85. [CrossRef] [PubMed]

38. Rausch, C.; Bucher, M. Molecular mechanisms of phosphate transport in plants. *Planta* **2002**, *216*, 23–37. [CrossRef] [PubMed]

39. Davies, T.; Ying, J.; Xu, Q.; Li, Z.; Li, J.; Gordon-Weeks, R. Expression analysis of putative high-affinity phosphate transporters in Chinese winter wheats. *Plant Cell Environ.* **2002**, *25*, 1325–1339. [CrossRef]

40. Daram, P.; Brunner, S.; Rausch, C.; Steiner, C.; Amrhein, N.; Bucher, M. Pht2; 1 encodes a low-affinity phosphate transporter from Arabidopsis. *Plant Cell* **1999**, *11*, 2153–2166. [CrossRef] [PubMed]

41. Guo, C.; Zhao, X.; Liu, X.; Zhang, L.; Gu, J.; Li, X.; Lu, W.; Xiao, K. Function of wheat phosphate transporter gene TaPHT2; 1 in Pi translocation and plant growth regulation under replete and limited Pi supply conditions. *Planta* **2013**, *237*, 1163–1178. [CrossRef] [PubMed]

42. Poirier, Y.; Bucher, M. Phosphate transport and homeostasis in Arabidopsis. *Arabidopsis Book* **2002**, e0024. [CrossRef] [PubMed]

43. Zhu, W.; Miao, Q.; Sun, D.; Yang, G.; Wu, C.; Huang, J.; Zheng, C. The mitochondrial phosphate transporters modulate plant responses to salt stress via affecting ATP and gibberellin metabolism in Arabidopsis thaliana. *PLoS ONE* **2012**, *7*, e43530. [CrossRef] [PubMed]

44. Aziz, T.; Finnegan, P.M.; Lambers, H.; Jost, R. Organ-specific phosphorus-allocation patterns and transcript profiles linked to phosphorus efficiency in two contrasting wheat genotypes. *Plant Cell Environ.* **2014**, *37*, 943–960. [CrossRef] [PubMed]

45. Guo, B.; Jin, Y.; Wussler, C.; Blancaflor, E.; Motes, C.; Versaw, W.K. Functional analysis of the Arabidopsis PHT4 family of intracellular phosphate transporters. *New Phytol.* **2008**, *177*, 889–898. [CrossRef] [PubMed]

46. Liu, T.-Y.; Huang, T.-K.; Yang, S.-Y.; Hong, Y.-T.; Huang, S.-M.; Wang, F.-N.; Chiang, S.-F.; Tsai, S.-Y.; Lu, W.-C.; Chiou, T.-J. Identification of plant vacuolar transporters mediating phosphate storage. *Nat. Commun.* **2016**, *7*, 111.95. [CrossRef] [PubMed]

47. Ouyang, X.; Hong, X.; Zhao, X.; Zhang, W.; He, X.; Ma, W.; Teng, W.; Tong, Y. Knock out of the PHOSPHATE 2 gene TaPHO2-A1 improves phosphorus uptake and grain yield under low phosphorus conditions in common wheat. *Sci. Rep.* **2016**, *6*, 29850. [CrossRef] [PubMed]

48. Rubio, V.; Linhares, F.; Solano, R.; Martín, A.C.; Iglesias, J.; Leyva, A.; Paz-Ares, J. A conserved MYB transcription factor involved in phosphate starvation signaling both in vascular plants and in unicellular algae. *Genes Dev.* **2001**, *15*, 2122–2133. [CrossRef] [PubMed]

49. Wang, J.; Sun, J.; Miao, J.; Guo, J.; Shi, Z.; He, M.; Chen, Y.; Zhao, X.; Li, B.; Han, F. A phosphate starvation response regulator Ta-PHR1 is involved in phosphate signalling and increases grain yield in wheat. *Ann. Bot.* **2013**, *111*, 1139–1153. [CrossRef] [PubMed]

50. Bustos, R.; Castrillo, G.; Linhares, F.; Puga, M.I.; Rubio, V.; Pérez-Pérez, J.; Solano, R.; Leyva, A.; Paz-Ares, J. A central regulatory system largely controls transcriptional activation and repression responses to phosphate starvation in Arabidopsis. *PLoS Genet.* **2010**, *6*, e1001102. [CrossRef] [PubMed]

51. González, E.; Solano, R.; Rubio, V.; Leyva, A.; Paz-Ares, J. PHOSPHATE TRANSPORTER TRAFFIC FACILITATOR1 is a plant-specific SEC12-related protein that enables the endoplasmic reticulum exit of a high-affinity phosphate transporter in Arabidopsis. *Plant Cell* **2005**, *17*, 3500–3512. [CrossRef] [PubMed]
52. Rouached, H.; Stefanovic, A.; Secco, D.; Arpat, A.B.; Gout, E.; Bligny, R.; Poirier, Y. Uncoupling phosphate deficiency from its major effects on growth and transcriptome via PHO1 expression in Arabidopsis. *Plant J.* **2011**, *65*, 557–570. [CrossRef] [PubMed]
53. Bari, R.; Pant, B.D.; Stitt, M.; Scheible, W.-R. PHO2, microRNA399, and PHR1 define a phosphate-signaling pathway in plants. *Plant Physiol.* **2006**, *141*, 988–999. [CrossRef] [PubMed]
54. Devaiah, B.N.; Nagarajan, V.K.; Raghothama, K.G. Phosphate homeostasis and root development in Arabidopsis are synchronized by the zinc finger transcription factor ZAT6. *Plant Physiol.* **2007**, *145*, 147–159. [CrossRef] [PubMed]
55. Smith, A.P.; Jain, A.; Deal, R.B.; Nagarajan, V.K.; Poling, M.D.; Raghothama, K.G.; Meagher, R.B. Histone H2A. Z regulates the expression of several classes of phosphate starvation response genes but not as a transcriptional activator. *Plant Physiol.* **2010**, *152*, 217–225. [CrossRef] [PubMed]
56. Chen, Y.-F.; Li, L.-Q.; Xu, Q.; Kong, Y.-H.; Wang, H.; Wu, W.-H. The WRKY6 transcription factor modulates PHOSPHATE1 expression in response to low Pi stress in Arabidopsis. *Plant Cell* **2009**, *21*, 3554–3566. [CrossRef] [PubMed]
57. Su, T.; Xu, Q.; Zhang, F.-C.; Chen, Y.; Li, L.-Q.; Wu, W.-H.; Chen, Y.-F. WRKY42 modulates phosphate homeostasis through regulating phosphate translocation and acquisition in Arabidopsis. *Plant Physiol.* **2015**, *167*, 1579–1591. [CrossRef] [PubMed]
58. Wang, H.; Xu, Q.; Kong, Y.-H.; Chen, Y.; Duan, J.-Y.; Wu, W.-H.; Chen, Y.-F. Arabidopsis WRKY45 transcription factor activates PHOSPHATE TRANSPORTER1; 1 expression in response to phosphate starvation. *Plant Physiol.* **2014**, *164*, 2020–2029. [CrossRef] [PubMed]
59. Devaiah, B.N.; Karthikeyan, A.S.; Raghothama, K.G. WRKY75 transcription factor is a modulator of phosphate acquisition and root development in Arabidopsis. *Plant Physiol.* **2007**, *143*, 1789–1801. [CrossRef] [PubMed]
60. Lin, W.-Y.; Huang, T.-K.; Chiou, T.-J. Nitrogen Limitation Adaptation, a target of microRNA827, mediates degradation of plasma membrane–localized phosphate transporters to maintain phosphate homeostasis in Arabidopsis. *Plant Cell* **2013**, *25*, 4061–4074. [CrossRef] [PubMed]
61. Wu, P.; Ma, L.; Hou, X.; Wang, M.; Wu, Y.; Liu, F.; Deng, X.W. Phosphate starvation triggers distinct alterations of genome expression in Arabidopsis roots and leaves. *Plant Physiol.* **2003**, *132*, 1260–1271. [CrossRef] [PubMed]
62. Huang, T.-K.; Han, C.-L.; Lin, S.-I.; Chen, Y.-J.; Tsai, Y.-C.; Chen, Y.-R.; Chen, J.-W.; Lin, W.-Y.; Chen, P.-M.; Liu, T.-Y. Identification of downstream components of ubiquitin-conjugating enzyme PHOSPHATE2 by quantitative membrane proteomics in Arabidopsis roots. *Plant Cell* **2013**, *25*, 4044–4060. [CrossRef] [PubMed]
63. Kant, S.; Peng, M.; Rothstein, S.J. Genetic regulation by NLA and microRNA827 for maintaining nitrate-dependent phosphate homeostasis in Arabidopsis. *PLoS Genet.* **2011**, *7*, e1002021. [CrossRef] [PubMed]
64. Cardona-López, X.; Cuyas, L.; Marín, E.; Rajulu, C.; Irigoyen, M.L.; Gil, E.; Puga, M.I.; Bligny, R.; Nussaume, L.; Geldner, N. ESCRT-III-associated protein ALIX mediates high-affinity phosphate transporter trafficking to maintain phosphate homeostasis in Arabidopsis. *Plant Cell* **2015**, *27*, 2560–2581. [CrossRef] [PubMed]
65. Guo, C.; Guo, L.; Li, X.; Gu, J.; Zhao, M.; Duan, W.; Ma, C.; Lu, W.; Xiao, K. TaPT2, a high-affinity phosphate transporter gene in wheat (*Triticum aestivum* L.), is crucial in plant Pi uptake under phosphorus deprivation. *Acta Physiol. Plant.* **2014**, *36*, 1373–1384. [CrossRef]
66. Bayle, V.; Arrighi, J.-F.; Creff, A.; Nespoulous, C.; Vialaret, J.; Rossignol, M.; Gonzalez, E.; Paz-Ares, J.; Nussaume, L. Arabidopsis thaliana high-affinity phosphate transporters exhibit multiple levels of posttranslational regulation. *Plant Cell Online* **2011**, *23*, 1523–1535. [CrossRef] [PubMed]
67. Wege, S.; Khan, G.A.; Jung, J.-Y.; Vogiatzaki, E.; Pradervand, S.; Aller, I.; Meyer, A.J.; Poirier, Y. The EXS domain of PHO1 participates in the response of shoots to phosphate deficiency via a root-to-shoot signal. *Plant Physiol.* **2016**, *170*, 385–400. [CrossRef] [PubMed]

68. Krouk, G.; Lacombe, B.; Bielach, A.; Perrine-Walker, F.; Malinska, K.; Mounier, E.; Hoyerova, K.; Tillard, P.; Leon, S.; Ljung, K. Nitrate-regulated auxin transport by NRT1. 1 defines a mechanism for nutrient sensing in plants. *Dev. Cell* **2010**, *18*, 927–937. [CrossRef] [PubMed]

69. Lin, W.-Y.; Lin, S.-I.; Chiou, T.-J. Molecular regulators of phosphate homeostasis in plants. *J. Exp. Bot.* **2009**, *60*, 1427–1438. [CrossRef] [PubMed]

70. Misson, J.; Thibaud, M.-C.; Bechtold, N.; Raghothama, K.; Nussaume, L. Transcriptional regulation and functional properties of Arabidopsis Pht1; 4, a high affinity transporter contributing greatly to phosphate uptake in phosphate deprived plants. *Plant Mol. Biol.* **2004**, *55*, 727–741. [CrossRef] [PubMed]

71. Wild, R.; Gerasimaite, R.; Jung, J.-Y.; Truffault, V.; Pavlovic, I.; Schmidt, A.; Saiardi, A.; Jessen, H.J.; Poirier, Y.; Hothorn, M. Control of eukaryotic phosphate homeostasis by inositol polyphosphate sensor domains. *Science* **2016**, *352*, 986–990. [CrossRef] [PubMed]

72. Liu, T.-Y.; Huang, T.-K.; Tseng, C.-Y.; Lai, Y.-S.; Lin, S.-I.; Lin, W.-Y.; Chen, J.-W.; Chiou, T.-J. PHO2-dependent degradation of PHO1 modulates phosphate homeostasis in Arabidopsis. *Plant Cell* **2012**, *24*, 2168–2218. [CrossRef] [PubMed]

73. Briat, J.-F.; Rouached, H.; Tissot, N.; Gaymard, F.; Dubos, C. Integration of P, S, Fe, and Zn nutrition signals in Arabidopsis thaliana: Potential involvement of PHOSPHATE STARVATION RESPONSE 1 (PHR1). *Front. Plant Sci.* **2015**, *6*, 290. [CrossRef] [PubMed]

74. Aung, K.; Lin, S.-I.; Wu, C.-C.; Huang, Y.-T.; Su, C.-l.; Chiou, T.-J. PHO2, a phosphate overaccumulator, is caused by a nonsense mutation in a microRNA399 target gene. *Plant Physiol.* **2006**, *141*, 1000–1011. [CrossRef] [PubMed]

75. Hu, B.; Zhu, C.; Li, F.; Tang, J.; Wang, Y.; Lin, A.; Liu, L.; Che, R.; Chu, C. LEAF TIP NECROSIS1 plays a pivotal role in the regulation of multiple phosphate starvation responses in rice. *Plant Physiol.* **2011**, *156*, 1101–1115. [CrossRef] [PubMed]

76. Khan, G.A.; Bouraine, S.; Wege, S.; Li, Y.; de Carbonnel, M.; Berthomieu, P.; Poirier, Y.; Rouached, H. Coordination between zinc and phosphate homeostasis involves the transcription factor PHR1, the phosphate exporter PHO1, and its homologue PHO1; H3 in Arabidopsis. *J. Exp. Bot.* **2014**, *65*, 871–884. [CrossRef] [PubMed]

77. Kisko, M.; Bouain, N.; Rouached, A.; Choudhary, S.P.; Rouached, H. Molecular mechanisms of phosphate and zinc signalling crosstalk in plants: Phosphate and zinc loading into root xylem in Arabidopsis. *Environ. Exp. Bot.* **2015**, *114*, 57–64. [CrossRef]

78. Ova, E.A.; Kutman, U.B.; Ozturk, L.; Cakmak, I. High phosphorus supply reduced zinc concentration of wheat in native soil but not in autoclaved soil or nutrient solution. *Plant Soil* **2015**, *393*, 147–162. [CrossRef]

79. Pal, S.; Kisko, M.; Dubos, C.; Lacombe, B.; Berthomieu, P.; Krouk, G.; Rouached, H. TransDetect identifies a new regulatory module controlling phosphate accumulation. *Plant Physiol.* **2017**, *175*, 916–926. [CrossRef] [PubMed]

80. Rouached, H. Recent developments in plant zinc homeostasis and the path toward improved biofortification and phytoremediation programs. *Plant Signal. Behav.* **2013**, *8*, e22681. [CrossRef] [PubMed]

81. White, P.J. Veneklaas Erik, J. Nature and nurture: The importance of seed phosphorus content. *Plant Soil* **2012**, *357*, 1–8. [CrossRef]

82. Fan, Y.; Van den Dool, H. A global monthly land surface air temperature analysis for 1948–present. *J. Geophys. Res. Atmos.* **2008**, *113*. [CrossRef]

83. DeFries, R.; Fanzo, J.; Remans, R.; Palm, C.; Wood, S.; Anderman, T.L. Metrics for land-scarce agriculture. *Science* **2015**, *349*, 238–240. [CrossRef] [PubMed]

84. Dietterich, L.H.; Zanobetti, A.; Kloog, I.; Huybers, P.; Leakey, A.D.; Bloom, A.J.; Carlisle, E.; Fernando, N.; Fitzgerald, G.; Hasegawa, T. Impacts of elevated atmospheric CO_2 on nutrient content of important food crops. *Sci. Data* **2015**, *2*, 150063. [CrossRef] [PubMed]

85. Belgaroui, N.; Zaidi, I.; Farhat, A.; Chouayekh, H.; Bouain, N.; Chay, S.; Curie, C.; Mari, S.; Masmoudi, K.; Davidian, J.-C. Over-expression of the bacterial phytase US417 in Arabidopsis reduces the concentration of phytic acid and reveals its involvement in the regulation of sulfate and phosphate homeostasis and signaling. *Plant Cell Physiol.* **2014**, *55*, 1912–1924. [CrossRef] [PubMed]

86. Wang, F.; Rose, T.; Jeong, K.; Kretzschmar, T.; Wissuwa, M. The knowns and unknowns of phosphorus loading into grains, and implications for phosphorus efficiency in cropping systems. *J. Exp. Bot.* **2015**, *67*, 1221–1229. [CrossRef] [PubMed]

87. Zhang, F.; Sun, Y.; Pei, W.; Jain, A.; Sun, R.; Cao, Y.; Wu, X.; Jiang, T.; Zhang, L.; Fan, X. Involvement of OsPht1; 4 in phosphate acquisition and mobilization facilitates embryo development in rice. *Plant J.* **2015**, *82*, 556–569. [CrossRef] [PubMed]

88. Vogiatzaki, E.; Baroux, C.; Jung, J.-Y.; Poirier, Y. PHO1 exports phosphate from the chalazal seed coat to the embryo in developing Arabidopsis seeds. *Curr. Biol.* **2017**, *27*, 2893–2900. [CrossRef] [PubMed]

89. Batten, G.; Wardlaw, I. Senescence and grain development in wheat plants grown with contrasting phosphorus regimes. *Aust. J. Plant Physiol.* **1987**, *14*, 253–265. [CrossRef]

90. Raboy, V. Approaches and challenges to engineering seed phytate and total phosphorus. *Plant Sci.* **2009**, *177*, 281–296. [CrossRef]

91. Sparvoli, F.; Cominelli, E. Seed biofortification and phytic acid reduction: A conflict of interest for the plant? *Plants* **2015**, *4*, 728–755. [CrossRef] [PubMed]

92. Zhang, Y.-Q.; Sun, Y.-X.; Ye, Y.-L.; Karim, M.R.; Xue, Y.-F.; Yan, P.; Meng, Q.-F.; Cui, Z.-L.; Cakmak, I.; Zhang, F.-S. Zinc biofortification of wheat through fertilizer applications in different locations of China. *Field Crops Res.* **2012**, *125*, 1–7. [CrossRef]

93. Raboy, V.; Cichy, K.; Peterson, K.; Reichman, S.; Sompong, U.; Srinives, P.; Saneoka, H. Barley (*Hordeum vulgare* L.) low phytic acid 1–1: An endosperm-specific, filial determinant of seed total phosphorus. *J. Hered.* **2014**, *105*, 656–665. [CrossRef] [PubMed]

94. Li, Y.; Zhang, J.; Zhang, X.; Fan, H.; Gu, M.; Qu, H.; Xu, G. Phosphate transporter OsPht1; 8 in rice plays an important role in phosphorus redistribution from source to sink organs and allocation between embryo and endosperm of seeds. *Plant Sci.* **2015**, *230*, 23–32. [CrossRef] [PubMed]

95. Shi, J.; Wang, H.; Schellin, K.; Li, B.; Faller, M.; Stoop, J.M.; Meeley, R.B.; Ertl, D.S.; Ranch, J.P.; Glassman, K. Embryo-specific silencing of a transporter reduces phytic acid content of maize and soybean seeds. *Nat. Biotechnol.* **2007**, *25*, 930–937. [CrossRef] [PubMed]

96. Bhati, K.K.; Alok, A.; Kumar, A.; Kaur, J.; Tiwari, S.; Pandey, A.K. Silencing of ABCC13 transporter in wheat reveals its involvement in grain development, phytic acid accumulation and lateral root formation. *J. Exp. Bot.* **2016**, *67*, 4379–4389. [CrossRef] [PubMed]

97. Ali, N.; Paul, S.; Gayen, D.; Sarkar, S.N.; Datta, K.; Datta, S.K. Development of low phytate rice by RNAi mediated seed-specific silencing of inositol 1, 3, 4, 5, 6-pentakisphosphate 2-kinase gene (IPK1). *PLoS ONE* **2013**, *8*, e68161. [CrossRef] [PubMed]

98. Ali, N.; Paul, S.; Gayen, D.; Sarkar, S.N.; Datta, S.K.; Datta, K. RNAi mediated down regulation of myo-inositol-3-phosphate synthase to generate low phytate rice. *Rice* **2013**, *6*, 12. [CrossRef] [PubMed]

99. Li, Z.; Zhang, X.; Zhao, Y.; Li, Y.; Zhang, G.; Peng, Z.; Zhang, J. Enhancing auxin accumulation in maize root tips improves root growth and dwarfs plant height. *Plant Biotechnol. J.* **2018**, *16*, 86–99. [CrossRef] [PubMed]

100. Zhao, H.; Frank, T.; Tan, Y.; Zhou, C.; Jabnoune, M.; Arpat, A.B.; Cui, H.; Huang, J.; He, Z.; Poirier, Y. Disruption of OsSULTR3; 3 reduces phytate and phosphorus concentrations and alters the metabolite profile in rice grains. *New Phytol.* **2016**, *211*, 926–939. [CrossRef] [PubMed]

101. Yamaji, N.; Takemoto, Y.; Miyaji, T.; Mitani-Ueno, N.; Yoshida, K.T.; Ma, J.F. Reducing phosphorus accumulation in rice grains with an impaired transporter in the node. *Nature* **2017**, *541*, 92–95. [CrossRef] [PubMed]

102. Peng, Z.; Li, C. Transport and partitioning of phosphorus in wheat as affected by P withdrawal during flag-leaf expansion. *Plant Soil* **2005**, *268*, 1–11. [CrossRef]

103. Syers, K.; Johnston, A.E.; Curtin, D. *Efficiency of Soil and Fertilizer Phosphorus Use: Reconciling Changing Concepts of Soils Phosphorus Behaviour with Agronomic Information*; FAO Fertilizer and Plant Nutrition Bulletin 18; Food and Agriculture Organization of the United Nations: Rome, Italy, 2008.

104. Krasileva, K.V.; Vasquez-Gross, H.A.; Howell, T.; Bailey, P.; Paraiso, F.; Clissold, L.; Simmonds, J.; Ramirez-Gonzalez, R.H.; Wang, X.; Borrill, P. Uncovering hidden variation in polyploid wheat. *Proc. Natl. Acad. Sci. USA* **2017**, *114*, E913–E921. [CrossRef] [PubMed]

agriculture

MDPI

Article

Limits to the Biofortification of Leafy Brassicas with Zinc

Philip J. White [1,2,*], **Paula Pongrac** [1,†], **Claire C. Sneddon** [1,‡], **Jacqueline A. Thompson** [1] and **Gladys Wright** [1]

[1]   Ecological Science Group, The James Hutton Institute, Invergowrie, Dundee DD2 5DA, UK; paula.pongrac@gmail.com (P.P.); c.c.sneddon@dundee.ac.uk (C.C.S.); jacqueline.thompson@hutton.ac.uk (J.A.T.); gladys.wright@hutton.ac.uk (G.W.)
[2]   Distinguished Scientist Fellowship Program, King Saud University, Riyadh 11451, Saudi Arabia
*   Correspondence: philip.white@hutton.ac.uk; Tel.: +44-1382-560043
†   Current Address: Biotechnical Faculty, University of Ljubljana, SI-1000 Ljubljana, Slovenia.
‡   Current Address: Medical Research Institute, Ninewells Hospital and Medical School, University of Dundee, Dundee DD1 9SY, UK.

Received: 24 January 2018; Accepted: 22 February 2018; Published: 27 February 2018

Abstract: Many humans lack sufficient zinc (Zn) in their diet for their wellbeing and increasing Zn concentrations in edible produce (biofortification) can mitigate this. Recent efforts have focused on biofortifying staple crops. However, greater Zn concentrations can be achieved in leafy vegetables than in fruits, seeds, or tubers. Brassicas, such as cabbage and broccoli, are widely consumed and might provide an additional means to increase dietary Zn intake. Zinc concentrations in brassicas are limited primarily by Zn phytotoxicity. To assess the limits of Zn biofortification of brassicas, the Zn concentration in a peat:sand (v/v 75:25) medium was manipulated to examine the relationship between shoot Zn concentration and shoot dry weight (DW) and thereby determine the critical shoot Zn concentrations, defined as the shoot Zn concentration at which yield is reduced below 90%. The critical shoot Zn concentration was regarded as the commercial limit to Zn biofortification. Experiments were undertaken over six successive years. A linear relationship between Zn fertiliser application and shoot Zn concentration was observed at low application rates. Critical shoot Zn concentrations ranged from 0.074 to 1.201 mg Zn g^{-1} DW among cabbage genotypes studied in 2014, and between 0.117 and 1.666 mg Zn g^{-1} DW among broccoli genotypes studied in 2015–2017. It is concluded that if 5% of the dietary Zn intake of a population is currently delivered through brassicas, then the biofortification of brassicas from 0.057 to > 0.100 mg Zn g^{-1} DW through the application of Zn fertilisers could increase dietary Zn intake substantially.

Keywords: biofortification; *Brassica oleracea* L.; broccoli; cabbage; nutrition; toxicity; zinc

---

1. Introduction

It is estimated that over one-fifth of the world's population suffers from zinc (Zn) deficiency, which results in impaired development, ill health, and a reduction in gross domestic product [1–5]. One strategy to increase human dietary Zn intake is to increase Zn concentrations in edible produce. This strategy is termed biofortification and can be achieved through the use of Zn fertilisers on plant genotypes that have greater ability to acquire and accumulate Zn in their edible tissues [1,4–9]. Zinc might be applied to the soil as inorganic or organic fertilisers or to foliage as soluble salts [1,7,9–11]. Inorganic fertilisers are often preferred because of their consistent composition; foliar applications are most effective where the phytoavailability of Zn decreases rapidly when applied to the soil [1,5]. Recent biofortification efforts have focused largely on developing germplasm and agronomic strategies to increase Zn concentrations in staple crops including cereals, pulses, cassava and potatoes, and Zn concentrations approaching 0.02–0.10 mg g^{-1} dry weight (DW), depending upon the crop, have been achieved [1,6,9,12–21]. However, greater Zn concentrations can be achieved in leafy vegetables than in fruits, seeds or tubers because Zn transport in the phloem limits Zn accumulation in the latter tissues [13]. Zinc concentrations in leafy vegetables appear to be limited primarily by Zn phytotoxicity, suggesting that concentrations of 0.10–0.70 mg Zn g^{-1} DW shoot might be achieved without loss of yield [13]. Thus, Zn biofortification of leafy vegetables might also provide a means to increase Zn intake by human populations.

Leafy vegetables are a significant source of micronutrients for human populations, especially those with low incomes or with a vegetarian diet [3,4]. Brassicaceous vegetables, such as cabbage (*Brassica oleracea* var. *capitata*) and broccoli (*B. oleracea* L. var. *italica*), are among the most commonly consumed and economically important vegetables in the world [22,23]. The health benefits of brassicaceous vegetables are not only associated with their mineral composition but also their vitamin content and the presence of other organic compounds, particularly glucosinolates [22,24,25]. Although leafy vegetables currently contribute proportionally less Zn to human diets than animal products or cereals [3,4], their greater potential for Zn biofortification could be exploited to increase Zn intake and improve human health.

A large variation in shoot Zn concentration has been reported among the genotypes of *B. oleracea* [26–32]. For example, shoot Zn concentration among 36 cabbage genotypes grown together in a field in Himachal Pradesh, India, ranged from 0.002 to 0.005 mg Zn g^{-1} fresh weight [29]; significant differences in leaf Zn concentrations were observed among three cabbage genotypes grown together in the field in Pennsylvania, USA [26]; floret Zn concentrations of 10 broccoli genotypes grown together in Poznań, Poland, ranged from 0.042 to 0.066 mg Zn g^{-1} DW [30]; the average shoot Zn concentration of 22 kale (*B. oleracea* var. *acephala*) genotypes grown together in the field in New Hampshire, USA, over two years ranged from 0.033 to 0.060 mg Zn g^{-1} DW [27]; and leaf Zn concentrations of 6 kale genotypes grown in the field in KwaZulu-Natal Province, South Africa, ranged from 0.025 to 0.032 mg Zn g^{-1} DW [33]. However, although the heritability of shoot Zn concentration in *B. oleracea* is significant and several chromosomal quantitative trait loci (QTL) affecting shoot Zn concentration have been identified in this species, the heritability is often low and the QTL identified depends on the growth conditions [28,29,34]. The aim of the experiments reported here was, therefore, to determine the limit to Zn biofortification—using a Zn fertiliser—of two leafy brassicas, cabbage and broccoli, which contribute significantly to human diets worldwide. The Zn concentration in a peat:sand (*v*/*v* 75:25) potting medium was manipulated using zinc nitrate in order to examine the relationship between shoot Zn concentration and shoot dry biomass. This was to determine the critical shoot Zn concentration—defined as the shoot Zn concentration at which the yield was reduced below 90% [35]—of different genotypes. This value was regarded as the commercial limit to Zn biofortification. The manipulation of shoot Zn concentration through Zn applications to the substrate was necessary because the Zn concentrations of shoot tissues are difficult to determine unambiguously following the application of foliar Zn fertilisers. The Zn concentration in the substrate was manipulated using an inorganic Zn fertiliser rather than an organic amendment to avoid any potential effects of other components of organic amendments.

2. Materials and Methods

2.1. Plant Material

Five cabbage (*Brassica oleracea* L. var. *capitata*) genotypes were selected for study on the basis of their potentially large leaf zinc (Zn) concentrations based on data presented by [28]. The genotypes Bison, Cape Horn and Red Drumhead were grown in 2012, 2013 and 2014. The genotype Elisa was grown only in 2012. The genotype Tundra was grown only in 2013 and 2014. Seeds of Bison, Cape Horn, Red Drumhead and Tundra were obtained from Kings Seeds (Colchester, UK) and seeds of Elisa were obtained from Thompson & Morgan (Ipswich, UK). Four broccoli (*Brassica oleracea* L. var. *italica*) genotypes, Belstar, Chevalier, Marathon and Waltham 29, were studied in 2015, 2016 and 2017.

These genotypes were selected on the basis of their potentially contrasting leaf Zn concentrations based on data presented by [28]. Seeds of Belstar, Chevalier, Marathon and Waltham 29 were obtained from Van Meuwen (Spalding, UK), Kings Seeds Direct (Colchester, UK) and Unwins (Huntingdon, UK).

2.2. Growth Conditions

Seeds of cabbage were germinated in square Petri dishes (length × width × depth = 100 mm × 100 mm × 18 mm; Camlab, Cambridge, UK) on blue germination paper (Anchor Paper Company, St Paul, MN, USA) moistened with 10 mL deionised water. One week after placing the seeds in the Petri dishes, three to ten germinated seedlings of a genotype were transplanted to each pot containing 1 litre of the potting medium described in the next paragraph. Seedlings were thinned to a density of one seedling per pot two weeks after transplanting and grown for a further four weeks before harvesting. Ten seeds of broccoli were sown directly into pots containing 1 litre of potting medium. Broccoli seedlings were thinned to a density of one seedling per pot three weeks after sowing and grown for a further six weeks before harvesting. Thus, cabbage was harvested six weeks after transplanting pre-germinated seedlings to pots and broccoli was harvested nine weeks after sowing seeds into pots. Both cabbage and broccoli plants were at the true leaf/rosette stage of development, immediately prior to heading.

Plants were grown in an unheated glasshouse at The James Hutton Institute, Dundee, UK (latitude 56.4566° N, longitude 3.0708° W) in a potting medium, henceforth described as a "substrate", with Zn concentrations ranging from sufficient to phytotoxic. The standard substrate was made in bulk each year and comprised 75% peat (Sinclairs Professional Peat, Sinclair Pro, Ellesmere Port, UK) and 25% sand mixed with 0.225 g L^{-1} single superphosphate, 0.4 g L^{-1} ammonium nitrate, 0.75 g L^{-1} potassium nitrate, 2.25 g L^{-1} ground limestone, 2.25 g L^{-1} Magnesian limestone and 0.51 g L^{-1} of a trace element mixture containing 0.16% boron, 0.79% copper, 11.82% iron, 1.97% manganese, and 0.04% molybdenum by weight. The density of the substrate was 625 g L^{-1} when dry and 822 g L^{-1} when watered to holding capacity. The standard substrate contained 9.4 mg Zn L^{-1} and <0.1 mg L^{-1} water-extractable Zn (n = 6) prior to any further additions.

The standard substrate was thoroughly mixed and sieved before zinc nitrate was added to achieve the specified Zn concentrations. Zinc nitrate was employed rather than zinc sulphate to avoid any potential effect of increasing sulphate bioavailability on Zn accumulation. Six treatments were established in all experiments. In 2012, the treatments were the additions of 0, 0.075, 0.15, 1.5, 150 or 3000 mg Zn L^{-1} substrate. In 2013, the treatments were the additions of 0, 0.15, 150, 500, 1000 or 1500 mg Zn L^{-1} substrate. In 2014, 2015, 2016 and 2017 the treatments were the additions of 0, 1.5, 15, 150, 300 or 450 mg Zn L^{-1} substrate. The substrate was watered to holding capacity immediately before the experiments. When watered to holding capacity the pH of the solution in the substrate was ca. 6.8. Three replicate pots of each genotype x treatment were established in 2012 and 2013. Ten replicate pots of each genotype x treatment were established in 2014, 2015, 2016 and 2017. Pots were arranged in a blocked design with one replicate of each genotype x treatment combination in a random location within each block.

Surviving plants were harvested on 12 September 2012, 27 June 2013, 22 July 2014, 29 July 2015, 5 October 2016 and 25 July 2017. The accumulated temperature x time during plant growth in the glasshouse was 606.7 °C day in 2012, 525.3 °C day in 2013, 638.7 °C day in 2014, 820.1 °C day in 2015, 929.4 °C day in 2016 and 908.9 °C day in 2017. The accumulated solar radiation in the glasshouse during plant growth was 524.0 MJ m^{-2} in 2012, 702.5 MJ m^{-2} in 2013, 716.3 MJ m^{-2} in 2014, 1046.4 MJ m^{-2} in 2015, 702.4 MJ m^{-2} in 2016 and 1011.7 MJ m^{-2} in 2017.

2.3. Plant Analysis

Shoot fresh weight (FW) was determined at harvest and shoot dry weight (DW) was determined following drying to a constant weight in an oven at 70 °C. Zinc concentrations were determined following acid digestion of dried shoot material using inductively coupled plasma mass spectrometry (ICP-MS) as described by White et al. [36]. Accurately weighed subsamples (c. 50 mg DW) were digested in closed vessels using a microwave digester (MARS Xpress, CEM Microwave Technology, Buckingham, UK). Samples were first digested with 10 mL concentrated nitric acid (HNO_3) before 3 mL of 30% hydrogen peroxide (H_2O_2) was added to each vessel and digestion completed. Digested samples were diluted with milliQ (sterile, 18.2 $M\Omega$ cm) water before Zn concentrations were determined using ICP-MS (PerkinElmer ELAN, DRCe, Monza, Italy). Blank digestions were performed to determine background Zn concentrations and a tomato leaf standard (Reference 1573a; National Institute of Standards and Technology, NIST, Gaithersburg, MD, USA) was used as an analytical control.

2.4. Statistics

Data are expressed as the mean ± standard error of the mean (SE) for n observations. The relationships between shoot biomass (W) and the zinc concentration (Zn) in the substrate or shoot tissue were fitted to a sigmoidal function: W = a/(1 + EXP(b*((Zn) − c))) + d, where the minimum biomass equals d, the maximum biomass equals a + d, c is the (Zn) at the point of inflection and the slope of the relationship at the point of inflection is given by −ab/4.

3. Results

The relationships between substrate fertiliser Zn concentration, which is less than the actual substrate Zn concentration but is henceforth termed 'substrate Zn concentration' for brevity, and shoot fresh weight (FW), shoot dry weight (DW) and shoot Zn concentration of cabbage seedlings were studied in experiments performed in 2012, 2013 and 2014 (Table S1). The experiments performed in 2012 and 2013 narrowed the range of substrate Zn concentrations required to determine the 'critical' substrate and shoot Zn concentrations at which shoot DW was 90% of its maximal value [13]. The experiment performed in 2012 indicated that neither shoot FW nor shoot DW of cabbage genotypes was greatly reduced at substrate concentrations up to 150 mg L^{-1} and that shoot Zn concentration increased as the substrate Zn concentration was increased (Table S1). The genotype Elisa did not grow well in the substrate employed in these experiments and was replaced by Tundra in subsequent experiments. The experiment performed in 2013 indicated that the sensitivity of shoot FW and shoot DW to increasing substrate Zn concentration followed the sequence Red Drumhead > Bison > Cape Horn > Tundra (Table S1). The critical substrate Zn concentration at which shoot DW was 90% of its maximal value was < 0.15 mg L^{-1} for Red Drumhead, between 0.15 and 150 mg L^{-1} for Bison and between 150 and 500 mg L^{-1} for Cape Horn and Tundra. Shoot Zn concentration increased linearly with increasing substrate Zn concentration with gradients between 7.2 and 13.6 mg Zn kg^{-1} DW/ mg Zn L^{-1} (Bison = 9.9, Cape Horn = 8.4, Red Drumhead = 13.6, Tundra = 7.2). The critical shoot Zn concentrations were <0.18 mg g^{-1} DW for Red Drumhead, between 0.05 and 1.65 mg g^{-1} DW for Bison, between 1.36 and 4.27 mg g^{-1} DW for Cape Horn and between 1.05 and 3.73 mg g^{-1} DW for Tundra. Experiments in both 2012 and 2013 indicated considerable variation in the responses of individual plants of all genotypes grown under identical conditions to substrate Zn concentration, and that a large number of replicates would be required to obtain a more precise estimate of the critical substrate and shoot Zn concentrations for shoot DW accumulation.

In the experiments performed on cabbage in 2014, increasing substrate Zn concentration above about 100–150 mg L^{-1} reduced shoot DW of all cabbage genotypes studied (Figure 1A–D). However, estimates of critical substrate Zn concentrations were relatively imprecise. For Bison, the relationship between substrate Zn concentration and shoot DW could be fitted with a sigmoidal function, indicating a critical substrate Zn concentration of 96 mg fertiliser Zn g^{-1} DW substrate (Figure 1A, Table 1). Shoot DW of the Cape Horn genotype increased greatly with the addition of 0.15 mg L^{-1} Zn to the substrate, suggesting that the Zn in the substrate itself was insufficient to support maximal growth of this genotype in this experiment. The relationship between substrate Zn concentration above 0.15 mg L^{-1} and shoot DW could, however, be fitted by a sigmoidal function indicating a critical substrate Zn concentration of 100 mg L^{-1} (Figure 1B, Table 1). The relationship between substrate Zn concentration and shoot DW for Red Drumhead could not be fitted by a sigmoidal function but indicated a critical substrate Zn concentration of 5 mg L^{-1} (Figure 1C, Table 1). The genotype Tundra was relatively insensitive to the addition of <450 mg Zn L^{-1} and was estimated to have a critical substrate Zn concentration of 260 mg L^{-1} (Figure 1D, Table 1). The ranking of genotypes and estimates of critical values were generally consistent with data obtained in 2013.

Increasing the substrate Zn concentration increased shoot Zn concentration in all cabbage genotypes (Figure 1E–H). The relationship between substrate Zn concentration and shoot Zn concentration was linear in the genotypes Bison and Tundra, with a gradient of 5.80 and 4.62 L kg^{-1} DW, respectively, but asymptotic in Cape Horn and Red Drumhead, reaching a maximum of 1.70 and 1.91 mg Zn g^{-1} DW, respectively (Figure 1I–L, Table 2). From the experiments performed in 2014, the critical shoot Zn concentrations of Bison, Cape Horn, Red Drumhead and Tundra were estimated to be 0.79, 0.80, 0.074 and 1.20 mg Zn g^{-1} DW, respectively (Figure 1I–L, Table 1). The ranking of genotypes and the estimates of critical values were generally consistent with data obtained in 2013.

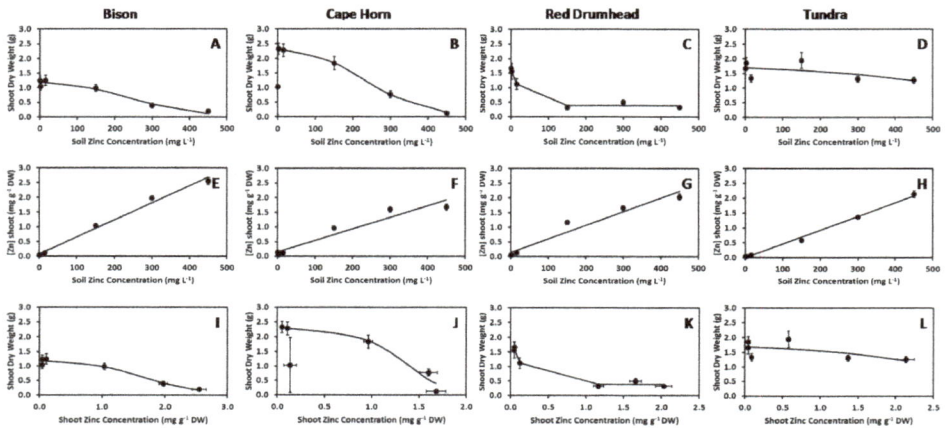

Figure 1. Response of cabbage genotypes to the addition of zinc (Zn) fertiliser to the substrate. *Top row*: Relationship between the addition of Zn fertiliser to substrate (substrate Zn concentration, ((Zn))) and shoot dry weight (DW) of (**A**) Bison, (**B**) Cape Horn, (**C**) Red Drumhead and (**D**) Tundra genotypes grown in 2014. *Middle row*: Relationship between substrate Zn concentration and shoot Zn concentration of (**E**) Bison, (**F**) Cape Horn, (**G**) Red Drumhead and (**H**) Tundra genotypes grown in 2014. Parameters for linear regression lines are given in Table 2. *Bottom row*: Relationship between shoot Zn concentration ((Zn)) and shoot dry weight of (**I**) Bison, (**J**) Cape Horn, (**K**) Red Drumhead and (**L**) Tundra genotypes. Regression lines are fitted to the equation DW = a/(1 + EXP(b*((Zn) − c))) + d, where the minimum biomass equals d, the maximum biomass equals a + d, c is the substrate (**A–D**) or shoot (**I–L**) Zn concentration at the point of inflection and the slope of the relationship at the point of inflection is given by −ab/4. For all lines d = 0 and a is the maximum shoot biomass. Parameters a, b and c were: (**A**) a = 1.26, b = 0.0114, c = 248 for Bison, (**B**) a = 2.41, b = 0.0130, c = 241 for Cape Horn, (**C**) not fitted for Red Drumhead, (**D**) a = 1.77, b = 0.0049, c = 633 for Tundra, (**I**) a = 1.21, b = 2.19, c = 1.68 for Bison, (**J**) a = 2.31, b = 4.18, c = 1.31 for Cape Horn, (**K**) not fitted for Red Drumhead, (**L**) a = 1.77, b = 1.05, c = 2.91 for Tundra. Data are shown as the mean and standard error of the mean (Table S1).

The relationships between substrate Zn concentration and shoot FW, shoot DW and shoot Zn concentration of broccoli seedlings were studied in experiments performed in 2015, 2016 and 2017 (Table S1; Figures 2–4). In all years there was considerable variation in the responses of individual plants of all genotypes grown under identical conditions to substrate Zn concentration (Table S1) and estimates of critical substrate and shoot Zn concentrations were relatively imprecise (Figures 2 and 4). Shoot DW was reduced by increasing substrate Zn concentration in all years in all genotypes (Figure 2).

However, the estimated critical substrate Zn concentration for shoot DW differed between years and, apparently, among genotypes (Table 1). The critical substrate Zn concentration was greater in 2017, ranging from 271 to 408 mg Zn L^{-1} among genotypes, than in 2016, ranging from 43 to 139 mg Zn L^{-1} among genotypes, and 2015, ranging from 107 to 121 mg Zn L^{-1} among genotypes (Table 1). In 2015, the critical substrate Zn concentration was similar for all genotypes but in both 2016 and 2017, the shoot DW of Belstar appeared less sensitive to substrate Zn concentration than the other cultivars, whereas the shoot DW of Waltham 29 was among the most sensitive genotypes to substrate Zn concentration (Table 1).

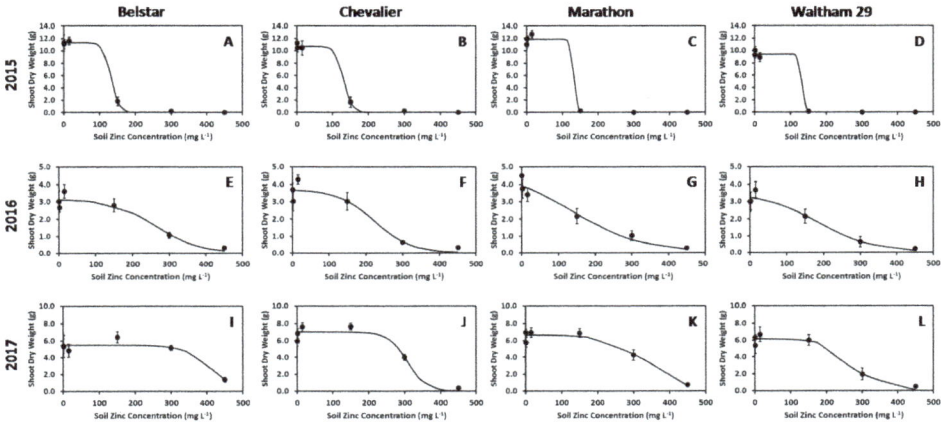

Figure 2. Relationship between substrate zinc concentration (Zn) and shoot dry weight (DW) of (**A,E,I**) Belstar, (**B,F,J**) Chevalier, (**C,G,K**) Marathon and (**D,H,L**) Waltham 29 broccoli genotypes grown in 2015 (**top row**), 2016 (**middle row**) and 2017 (**bottom row**). Regression lines are fitted to the equation DW = $a/(1 + EXP(b*((Zn) - c))) + d$, where the minimum biomass equals d, the maximum biomass equals a + d, c is the substrate Zn concentration at the point of inflection and the slope of the relationship at the point of inflection is given by $-ab/4$. For all lines d = 0 and a is the maximum shoot biomass. Parameters a, b and c were: (**A**) a = 11.3, b = 0.0893, c = 131 for Belstar in 2015, (**B**) a = 10.7, b = 0.0862, c = 130 for Chevalier in 2015, (**C**) a = 11.9, b = 0.2041, c = 131 for Marathon in 2015, (**D**) a = 9.40, b = 0.2181, c = 131 for Waltham 29 in 2015, (**E**) a = 3.16, b = 0.0164, c = 265 for Belstar in 2016, (**F**) a = 3.72, b = 0.0197, c = 225 for Chevalier in 2016, (**G**) a = 4.95, b = 0.01, c = 135 for Marathon in 2016, (**H**) a = 3.51, b = 0.0132, c = 189 for Waltham 29 in 2016, (**I**) a = 5.51, b = 0.0264, c = 408 for Belstar in 2017, (**J**) a = 7.03, b = 0.0380, c = 308 for Chevalier in 2017, (**K**) a = 6.68, b = 0.0193, c = 332 for Marathon in 2017, (**L**) a = 6.18, b = 0.0255, c = 271 for Waltham 29 in 2017. Data are shown as the mean and standard error of the mean (Table S1).

Shoot Zn concentration generally increased linearly with increasing substrate Zn concentration, although this relationship had an exponential tendency in Chevalier, Marathon and Waltham 29 in 2015 and tended towards an asymptotic maximum in Marathon in 2016 (Figure 3). The shoots of plants grown in 2015 had greater shoot Zn concentrations than those grown in 2016, which had greater shoot Zn concentrations than those grown in 2017 when grown in the same substrates (Table S1, Figure 3). Genotypes differed in their relationship between shoot Zn concentration and substrate Zn concentration (Figure 3). In general, the shoot Zn concentration in Belstar increased the least and the shoot Zn concentration in Waltham 29 the most, with increasing substrate Zn concentration across the three years of the study (Figure 3).

Table 1. Maximum shoot biomass and critical substrate zinc (Zn) and shoot Zn concentrations at which shoot dry biomass was reduced below 90%, estimated from the relationships shown in Figures 1 and 2 derived from the data expressed on a dry weight (DW) basis presented in Supplementary Table S1.

Year	Genotypes	Maximum Shoot Biomass		Critical Zn Concentration	
		Substrate Zn Regression (g DW)	Shoot Zn Regression (g DW)	Substrate (mg L^{-1})	Shoot (mg g^{-1} DW)
Cabbage					
2014	Bison	1.07	1.06	96	0.789
2014	Cape Horn	2.08	2.06	100	0.802
2014	Tundra	1.52	1.52	260	1.201
2014	Red Drumhead	1.48	1.45	5	0.074
Broccoli					
2015	Belstar	11.33	11.33	107	0.434
2015	Chevalier	10.72	10.72	105	0.499
2015	Marathon	11.87	11.87	121	0.514
2015	Waltham 29	9.40	9.64	121	0.117
2016	Belstar	3.12	3.12	139	1.018
2016	Chevalier	3.68	3.68	120	0.901
2016	Marathon	3.54	3.55	43	0.277
2016	Waltham 29	2.91	2.91	68	0.406
2017	Belstar	5.51	5.51	408	1.666
2017	Chevalier	7.03	7.03	308	1..424
2017	Marathon	6.68	6.70	332	1.527
2017	Waltham 29	6.18	6.24	271	1.195

Table 2. Gradients and intercepts for linear regressions between shoot zinc (Zn) concentrations and substrate Zn concentrations of four cabbage genotypes studied in 2014 and four broccoli genotypes studied in 2015, 2016 and 2017.

Year	Gradient (mg Zn kg^{-1} DW/mg Zn L^{-1}), Intercept (mg Zn kg^{-1} DW)			
	Bison	Cape Horn	Tundra	Red Drumhead
2014	5.802, 73.64	5.209, 85.92 (A)	4.620, 4.706	5.641, 83.76 (A)
	Belstar	Chevalier	Marathon	Waltham 29
2015	7.126, 70.16	5.170, 5.644 (E)	5.015, 12.83 (E)	6.853, 36.36 (E)
2016	5.654, 74.84	5.714, 80.16	5.894, 88.42	6.350, 12.32
2017	3.997, 49.46	4.837, 8.768	4.680, 7.523	4.452, 11.53

Data are regressions based on six substrate Zn concentrations, except for the cabbage genotypes Cape Horn and Red Drumhead in 2014 and the broccoli genotypes Chevalier, Marathon and Waltham 29 in 2015, which are regressions based on five substrate Zn concentrations $\leq$300 mg Zn L^{-1}. (A) Asymptotic, and (E) exponential regressions shown in Figures 1 and 3.

Shoot DW decreased with increasing shoot Zn concentration in all years in all genotypes (Figure 4). However, the estimated critical shoot Zn concentration for shoot DW accumulation differed between years and, apparently, among genotypes (Table 1). In general, shoot DW of Belstar appeared less sensitive to shoot Zn concentration than the other cultivars, particularly in 2016 and 2017, whilst Waltham 29 was among the most sensitive genotypes to shoot Zn concentration (Table 1). In 2015, the critical shoot Zn concentrations ranged between 0.117 mg g^{-1} DW for Waltham 29 and 0.514 mg g^{-1} DW for Marathon. In 2016, the critical shoot Zn concentrations ranged between 0.277 mg g^{-1} DW for Marathon and 1.018 mg g^{-1} DW for Belstar. In 2017, the critical shoot Zn concentrations ranged between 1.195 mg g^{-1} DW for Waltham 29 and 1.666 mg g^{-1} DW for Belstar.

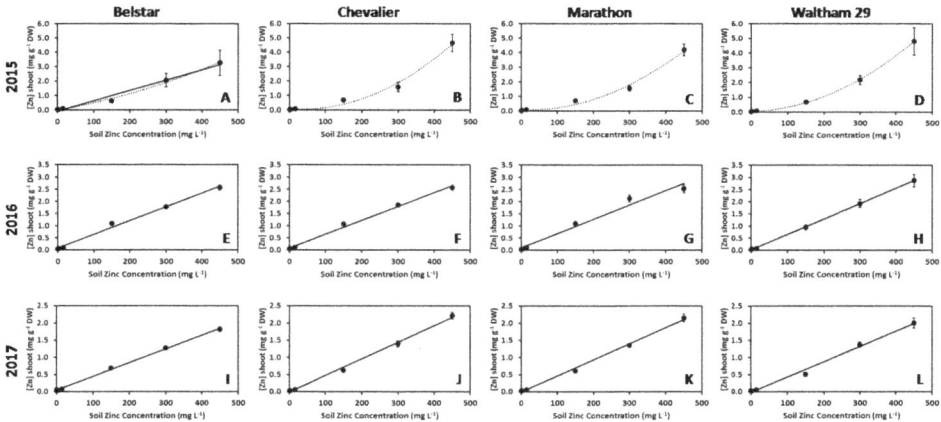

Figure 3. Relationship between substrate zinc (Zn) concentration and shoot Zn concentration of (**A,E,I**) Belstar, (**B,F,J**) Chevalier, (**C,G,K**) Marathon and (**D,H,L**) Waltham 29 broccoli genotypes grown in 2015 (**top row**), 2016 (**middle row**) and 2017 (**bottom row**). Parameters for linear regression lines are given in Table 2. Data are shown as the mean and standard error of the mean (Table S1).

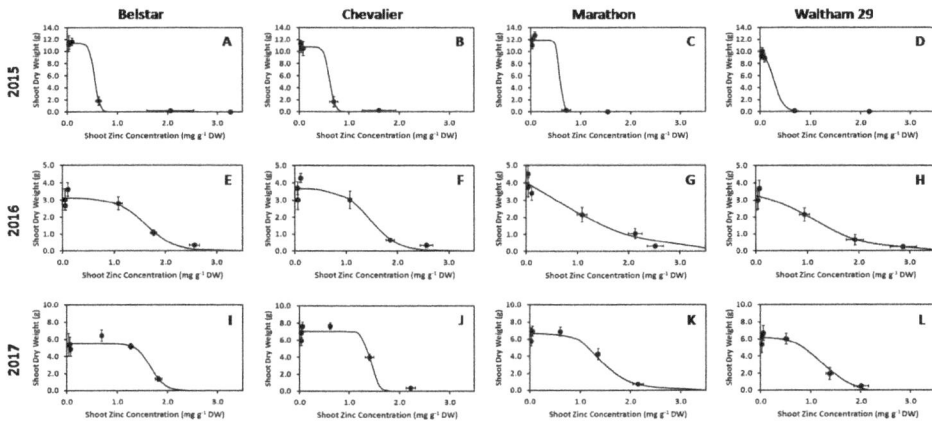

Figure 4. Relationship between shoot zinc concentration (Zn) and shoot dry weight (DW) of (**A,E,I**) Belstar, (**B,F,J**) Chevalier, (**C,G,K**) Marathon and (**D,H,L**) Waltham 29 broccoli genotypes grown in 2015 (**top row**), 2016 (**middle row**) and 2017 (**bottom row**). Regression lines are fitted to the equation $DW = a/(1 + EXP(b*((Zn) - c))) + d$, where the minimum biomass equals d, the maximum biomass equals $a + d$, c is the substrate Zn concentration at the point of inflection and the slope of the relationship at the point of inflection is given by $-ab/4$. For all lines $d = 0$ and a is the maximum shoot biomass. Parameters a, b and c were: (**A**) a = 11.3, b = 20.3, c = 0.542 for Belstar in 2015, (**B**) a = 10.7, b = 19.5, c = 0.612 for Chevalier in 2015, (**C**) a = 11.9, b = 30.8, c = 0.585 for Marathon in 2015, (**D**) a = 10.4, b = 10.5, c = 0.272 for Waltham 29 in 2015, (**E**) a = 3.13, b = 3.63, c = 1.61 for Belstar in 2016, (**F**) a = 3.69, b = 3.75, c = 1.47 for Chevalier in 2016, (**G**) a = 5.97, b = 1.19, c = 0.596 for Marathon in 2016, (**H**) a = 3.64, b = 1.89, c = 1.14 for Waltham 29 in 2016, (**I**) a = 5.51, b = 7.17, c = 1.66 for Belstar in 2017, (**J**) a = 7.02, b = 13.1, c = 1.42 for Chevalier in 2017, (**K**) a = 6.70, b = 3.66, c = 1.53 for Marathon in 2017, (**L**) a = 6.24, b = 4.05, c = 1.20 for Waltham 29 in 2017. Data are shown as the mean and standard error of the mean (Table S1).

There was no relationship between shoot DW of broccoli genotypes and either the critical substrate Zn concentration or the critical shoot Zn concentration (data not shown). However, although estimates of both the critical substrate Zn concentration and the critical shoot Zn concentration differed among experiments (Table 1), there was a linear relationship between the critical shoot Zn concentration and the critical substrate Zn concentration both among genotypes in 2016 and 2017 and among experiments performed in different years (Figure 5A). Data for cabbage genotypes obtained in 2014 also exhibited a similar relationship (Figure 5A). Thus, plants that are better able to tolerate the accumulation of Zn in their shoots can survive and grow in substrates with larger substrate Zn concentrations. There was also a negative relationship between the critical substrate Zn concentration and the initial linear rate of change of Zn uptake with increasing substrate Zn concentration in broccoli (Figure 5B), which might be explained because greater shoot Zn accumulation at any given substrate Zn concentration will result in a lower critical substrate Zn concentration should all other factors remain constant. However, there was also a negative relationship between the critical shoot Zn concentration and the initial linear rate of change of Zn uptake with increasing substrate Zn concentration in broccoli (Figure 5C). This observation might suggest that tolerance to increasing shoot Zn concentration is promoted by (unknown) substrate factors that change with increasing substrate Zn concentrations. Similar negative relationships between both the critical substrate and shoot Zn concentrations and the initial linear rate of change of Zn uptake with increasing substrate Zn concentration were also observed in cabbage (Figure 5B,C).

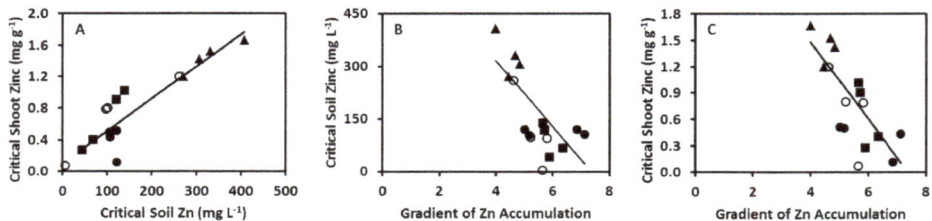

Figure 5. Relationship between (**A**) critical shoot zinc (Zn) concentrations and critical substrate Zn concentrations, (**B**) critical substrate Zn concentrations and the linear rates of change of Zn uptake with increasing substrate Zn concentration and (**C**) critical shoot Zn concentrations and the linear rate of change of Zn uptake with increasing substrate Zn concentration among four genotypes of broccoli studied in 2015 (closed circles), 2016 (squares) and 2017 (triangles), and four genotypes of cabbage studied in 2014 (open circles). The regression line in panels A ($y = 4.09x + 101$ mg kg^{-1}/mg L^{-1}, $R^2 = 0.840$, $n = 12$) and B ($y = -0.438x + 3.232$, $R^2 = 0.640$, $n = 12$) and C ($y = -93.96x + 693.2$, $R^2 = 0.585$, $n = 12$) are for broccoli only. Data are presented in Tables 1 and 2.

4. Discussion

Increasing the substrate Zn concentration above a critical value reduced the shoot DW of all cabbage and broccoli genotypes studied (Table S1, Figures 1 and 2). The critical substrate fertiliser Zn concentration, which is less than the total available Zn in the substrate but approximates the Zn readily available to the plant in the substrate, varied between years and differed among the genotypes, ranging from 5 to 260 mg Zn L^{-1} among cabbage genotypes studied in 2014 and between 43 and 408 mg Zn L^{-1} for the broccoli genotypes studied in 2015–2017 (Table 1). These values are much lower than statutory maximum annual Zn loading rates to soils in Europe and elsewhere [37].

The phytoavailability of soil Zn is affected by many factors [38]. These are often soil-specific. A major determinant of Zn phytoavailability in soils is pH but soil organic matter content, mineral and clay composition, porosity and moisture content are also influential. In addition, although brassicas are non-mycorrhizal, interactions with soil biota can also affect Zn phytoavailability in soils. It is, therefore, unwise to extrapolate from the critical substrate fertiliser Zn concentrations obtained in the experiments reported here (Table 1) directly to farmers' fields. It is also noteworthy that in soils where the phytoavailability of Zn applied as fertiliser decreases rapidly, the Zn biofortification of crops is generally achieved by foliar application of Zn fertilisers [1,5,9,11,13,15,20,37].

A variety of relationships between substrate Zn concentration and shoot Zn concentration were observed (Figures 1 and 3). A linear relationship between substrate Zn concentration and shoot Zn concentration was often observed but an asymptotic relationship toward a maximum shoot Zn concentration was observed for cabbage genotypes Cape Horn and Red Drumhead in 2014 and an exponential relationship was observed for broccoli genotypes Chevalier, Marathon and Waltham 29 in 2015. Previous studies of cabbage suggest a linear relationship between soil Zn concentration and shoot Zn concentration [39,40]. It is possible that the asymptotic relationship towards a maximum shoot Zn concentration was a consequence of measuring shoot Zn concentrations of only plants that survived the presence of a given substrate Zn concentration, which might have restricted Zn accumulation.

Differences in the gradients of the relationship between substrate Zn concentration and shoot Zn concentration were observed among both cabbage and broccoli genotypes (Table 1). Among the cabbage genotypes, shoot Zn concentrations were greatest in Bison and least in Tundra (Table S1, Figure 1), which is consistent with the observations of Broadley et al. [28] when ample phosphorus was supplied. Among the broccoli genotypes, Waltham 29 generally had the greatest shoot Zn concentration and Belstar the smallest shoot Zn concentration when Zn fertiliser was applied, although this order was reversed when no Zn fertiliser was applied (Table S1, Figure 3). These data reinforce previous studies reporting significant differences in shoot Zn concentrations among both cabbage [26,29] and broccoli genotypes [30,31]. Shoot Zn concentrations determined at the same substrate Zn concentration differed between years for all broccoli genotypes (Table S1, Figure 3). A similar observation was made by [27], who reported that there were significant differences in the shoot Zn concentrations of kale and collard genotypes grown in different years in the field. The differences reported here might be related to the season in which the plants were grown and, therefore, the glasshouse temperature, incident photosynthetically active radiation, day length or another uncontrolled environmental factor. However, no obvious relationships between accumulated temperature ($^{\circ}$C day) or solar radiation were observed for the relationships between substrate Zn concentration, shoot Zn concentration and shoot DW.

Critical shoot Zn concentrations ranged from 0.074 to 1.201 mg Zn g^{-1} DW among the cabbage genotypes studied in 2014 and between 0.117 and 1.666 mg Zn g^{-1} DW among the broccoli genotypes studied in 2015–2017 (Table 1). The smallest values are within the range (0.1–0.3 mg Zn g^{-1} DW) that is commonly quoted for critical shoot Zn concentrations of plants [35,41] and the largest values exceed current estimates of the Zn biofortification potential of leafy vegetables (0.7 mg Zn g^{-1} DW; [13]). The cabbage genotypes studied appeared to have very different critical shoot Zn concentrations, ranging from 0.074 mg Zn g^{-1} DW for Red Drumhead to 1.201 mg Zn g^{-1} DW for Tundra (Table 1; Figure 1). These data are consistent with previous studies indicating that the critical shoot Zn concentration of cabbage approximates 0.05–0.40 mg Zn g^{-1} DW [35,40,42,43] and that cabbage genotypes differ in their critical shoot Zn concentration [44]. Differences in critical shoot Zn concentration appeared to be less pronounced among the broccoli genotypes studied, although Waltham 29 appeared more sensitive and Belstar less sensitive than the other genotypes to increasing shoot Zn concentration (Table 1; Figure 4). Estimates of critical shoot Zn concentrations for broccoli genotypes differed between years (Table 1, Figure 4). This might be attributed to differences in the growth environment in different years. Variation in critical shoot Zn concentration is observed among studies of other brassicaceous species, such as *Brassica napus* L. (compare [45–49]) and *Brassica juncea* (L.) Czern [50–52] that might be related to either the genotypes studied or the experimental conditions.

5. Conclusions

Recent efforts to biofortify edible crops with Zn have focused primarily on staple crops, such as cereals, pulses, cassava and potatoes, and maximum Zn concentrations of 0.02–0.10 mg g^{-1} DW, depending upon the crop, have been achieved without loss of yield [1,6,9,12–21]. The critical shoot Zn concentrations in cabbage and broccoli reported here generally exceed these values (Table 1). This supports the general hypothesis that greater Zn concentrations can be achieved in leafy vegetables than in seeds, roots or tubers [13]. Furthermore, since leafy brassicas, unlike the seeds of legumes and cereals, do not contain large concentrations of phytic acid, the Zn in cabbage and broccoli should be readily bioavailable to humans [8]. This observation could have implications for increasing dietary Zn intake and the alleviation of Zn deficiencies in human populations. If it is assumed that brassicas constitute 5% of the Zn in current diets [3,4], increasing their shoot Zn concentrations from 0.057 mg Zn g^{-1} DW—the average value obtained without the addition of Zn fertilisers in the six experiments reported here (Table S1)—to 0.10–0.30 mg Zn g^{-1} DW through the application of Zn fertilisers to appropriate genotypes could increase dietary Zn intake by 3.8–21.3% without loss of crop yield. This has the potential to raise the Zn status and general health of human populations without any necessity for people to change their diets. However, there might be socioeconomic constraints to developing a strategy to alleviate Zn deficiencies in human populations through Zn biofortification of leafy vegetables because (1) the application of Zn fertilisers for Zn biofortification is an additional production cost and (2) appropriate infrastructure is required to distribute Zn fertilisers to produce Zn biofortified crops and to distribute Zn biofortified crops to populations that lack sufficient Zn in their diets [1].

Supplementary Materials: The following are available online at http://www.mdpi.com/2077-0472/8/3/32/s1, Table S1: The number of replicate plants surviving and the mean fresh weight (FW), dry weight (DW) and zinc (Zn) concentrations of the shoots of five cabbage (*Brassica oleracea* L. var. *capitata*) genotypes (Bison, Cape Horn, Elisa, Red Drumhead and Tundra) grown in 2012, 2013 and 2014 in a standard substrate into which no Zn, 0.075 mg Zn L^{-1} substrate, 0.15 mg Zn L^{-1} substrate, 1.5 mg Zn L^{-1} substrate, 15 mg Zn L^{-1} substrate, 150 mg Zn L^{-1} substrate, 300 mg Zn L^{-1} substrate or 450 mg Zn L^{-1} substrate, 500 mg Zn L^{-1} substrate, 1000 mg Zn L^{-1} substrate, 1500 mg Zn L^{-1} substrate or 3000 mg Zn L^{-1} substrate was incorporated depending on the year of the study, and four broccoli (*Brassica oleracea* L. var. *italica*) genotypes (Belstar, Chevalier, Marathon and Waltham 29) grown in 2015, 2016 and 2017 in a standard substrate into which no Zn, 1.5 mg Zn L^{-1} substrate, 15 mg Zn L^{-1} substrate, 150 mg Zn L^{-1} substrate, 300 mg Zn L^{-1} substrate or 450 mg Zn L^{-1} substrate was incorporated. Data for FW, DW and shoot Zn concentrations are expressed as the mean ± standard error of the mean (SE) for *n* surviving replicate plants.

Acknowledgments: This work was supported by the Rural and Environment Science and Analytical Services (RESAS) Division of the Scottish Government, by an EU Marie Curie Intra-European Fellowship (Grant #623305) to Paula Pongrac, and by a Rank Prize Funds Vacation Studentship to Claire Sneddon.

Author Contributions: All authors conceived and designed the experiments; C.C.S., J.A.T. and G.W. performed the experiments; C.C.S. and P.J.W. analysed the data; P.J.W. drafted the manuscript; all authors commented on the work.

Conflicts of Interest: The authors declare no conflicts of interest.

References

1. White, P.J.; Broadley, M.R. Biofortification of crops with seven mineral elements often lacking in human diets—Iron, zinc, copper, calcium, magnesium, selenium and iodine. *New Phytol.* **2009**, *182*, 49–84. [CrossRef] [PubMed]

2. Wessells, K.R.; Singh, G.M.; Brown, K.H. Estimating the global prevalence of inadequate zinc intake from national food balance sheets: Effects of methodological assumptions. *PLoS ONE* **2012**, *7*, e50565. [CrossRef] [PubMed]

3. Kumssa, D.B.; Joy, E.J.M.; Ander, E.L.; Watts, M.J.; Young, S.D.; Walker, S.; Broadley, M.R. Dietary calcium and zinc deficiency risks are decreasing but remain prevalent. *Sci. Rep.* **2015**, *5*, 10974. [CrossRef] [PubMed]

4. White, P.J. Biofortification of edible crops. In *eLS*; John Wiley & Sons: Chichester, UK, 2016. [CrossRef]

5. Cakmak, I.; McLaughlin, M.J.; White, P. Zinc for better crop production and human health. *Plant Soil* **2017**, *411*, 1–4. [CrossRef]

6. Cakmak, I. Enrichment of cereal grains with zinc: Agronomic or genetic biofortification? *Plant Soil* **2008**, *302*, 1–17. [CrossRef]

7. Khoshgoftarmanesh, A.H.; Schulin, R.; Chaney, R.L.; Daneshbakhsh, B.; Afyuni, M. Micronutrient-efficient genotypes for crop yield and nutritional quality in sustainable agriculture. A review. *Agron. Sustain. Dev.* **2010**, *30*, 83–107. [CrossRef]

8. Joy, E.J.M.; Ander, E.L.; Young, S.D.; Black, C.R.; Watts, M.J.; Chilimba, A.D.C.; Chilimba, B.; Siyame, E.W.P.; Kalimbira, A.A.; Hurst, R.; et al. Dietary mineral supplies in Africa. *Physiol. Plant.* **2014**, *151*, 208–229. [CrossRef] [PubMed]

9. Velu, G.; Ortiz-Monasterio, I.; Cakmak, I.; Hao, Y.; Singh, R.P. Biofortification strategies to increase grain zinc and iron concentrations in wheat. *J. Cereal Sci.* **2014**, *59*, 365–372. [CrossRef]

10. Gartler, J.; Robinson, B.; Burton, K.; Clucas, L. Carbonaceous soil amendments to biofortify crop plants with zinc. *Sci. Total Environ.* **2013**, *465*, 308–313. [CrossRef] [PubMed]

11. Šlosár, M.; Mezeyová, I.; Hegedüsová, A.; Andrejiová, A.; Kováčik, P.; Lošák, T.; Kopta, T.; Keutgen, A.J. Effect of zinc fertilisation on yield and selected qualitative parameters of broccoli. *Plant Soil Environ.* **2017**, *63*, 282–287. [CrossRef]

12. Pfeiffer, W.H.; McClafferty, B. HarvestPlus: Breeding crops for better nutrition. *Crop Sci.* **2007**, *47*, S88–S105. [CrossRef]

13. White, P.J.; Broadley, M.R. Physiological limits to zinc biofortification of edible crops. *Front. Plant Sci.* **2011**, *2*, 80. [CrossRef] [PubMed]

14. Sayre, R.; Beeching, J.R.; Cahoon, E.B.; Egesi, C.; Fauquet, C.; Fellman, J.; Fregene, M.; Gruissem, W.; Mallowa, S.; Manary, M.; et al. The BioCassava Plus Program: Biofortification of cassava for Sub-Saharan Africa. *Ann. Rev. Plant Biol.* **2011**, *62*, 251–272. [CrossRef] [PubMed]

15. Phattarakul, N.; Rerkasem, B.; Li, L.J.; Wu, L.H.; Zou, C.Q.; Ram, H.; Sohu, V.S.; Kang, B.S.; Surek, H.; Kalayci, M.; et al. Biofortification of rice grain with zinc through zinc fertilization in different countries. *Plant Soil* **2012**, *361*, 131–141. [CrossRef]

16. Zou, C.Q.; Zhang, Y.Q.; Rashid, A.; Ram, H.; Savasli, E.; Arisoy, R.Z.; Ortiz-Monasterio, I.; Simunji, S.; Wang, Z.H.; Sohu, V.; et al. Biofortification of wheat with zinc through zinc fertilization in seven countries. *Plant Soil* **2012**, *361*, 119–130. [CrossRef]

17. Saltzman, A.; Birol, E.; Bouis, H.E.; Boy, E.; De Moura, F.F.; Islam, Y.; Pfeiffer, W.H. Biofortification: Progress toward a more nourishing future. *Glob. Food Secur.* **2013**, *2*, 9–17. [CrossRef]

18. Pucher, A.; Høgh-Jensen, H.; Gondah, J.; Hash, C.T.; Haussmann, B.I.G. Micronutrient density and stability in West African pearl millet—Potential for biofortification. *Crop Sci.* **2014**, *54*, 1709–1720. [CrossRef]

19. Mallikarjuna, M.G.; Thirunavukkarasu, N.; Hossain, F.; Bhat, J.S.; Jha, S.K.; Rathore, A.; Agrawal, P.K.; Pattanayak, A.; Reddy, S.S.; Gularia, S.K.; et al. Stability performance of inductively coupled plasma mass spectrometry-phenotyped kernel minerals concentration and grain yield in maize in different agro-climatic zones. *PLoS ONE* **2015**, *10*, e0139067. [CrossRef]

20. Chen, X.-P.; Zhang, Y.-Q.; Tong, Y.-P.; Xue, Y.-F.; Liu, D.Y.; Zhang, W.; Deng, Y.; Meng, Q.-F.; Yue, S.-C.; Yan, P.; et al. Harvesting more grain zinc of wheat for human health. *Sci. Rep.* **2017**, *7*, 7016. [CrossRef] [PubMed]

21. White, P.J.; Thompson, J.A.; Wright, G.; Rasmussen, S.K. Biofortifying Scottish potatoes with zinc. *Plant Soil* **2017**, *411*, 151–165. [CrossRef]

22. Thomson, C.A.; Newton, T.R.; Graver, E.J.; Jackson, K.A.; Reid, P.M.; Hartz, V.L.; Cussler, E.C.; Hakim, I.A. Cruciferous vegetable intake questionnaire improves cruciferous vegetable intake estimates. *J. Am. Diet. Assoc.* **2007**, *107*, 631–641. [CrossRef] [PubMed]

23. Food and Agriculture Organization of the United Nations FAOSTAT. Available online: http://www.fao.org/faostat/en/#home (accessed on 25 October 2017).

24. Moreno, D.A.; Carvajal, M.; López-Berenguer, C.; Garcia-Viguera, C. Chemical and biological characterization of nutraceutical compounds of broccoli. *J. Pharm. Biomed. Anal.* **2006**, *41*, 1508–1522. [CrossRef] [PubMed]

25. Verkerk, R.; Schreiner, M.; Krumbein, A.; Ciska, E.; Holst, B.; Rowland, I.; De Schrijver, R.; Hansen, M.; Gerhauser, C.; Mithen, R.; et al. Glucosinolates in Brassica vegetables: The influence of the food supply chain on intake, bioavailability and human health. *Mol. Nutr. Food Res.* **2009**, *53*, S219. [CrossRef] [PubMed]

26. Harrison, H.C.; Bergman, E.L. Calcium, magnesium, and potassium interrelationships affecting cabbage production. *J. Am. Soc. Hortic. Sci.* **1981**, *106*, 500–503.

27. Kopsell, D.E.; Kopsell, D.A.; Lefsrud, M.G.; Curran-Celentano, J. Variability in elemental accumulations among leafy *Brassica oleracea* cultivars and selections. *J. Plant Nutr.* **2004**, *27*, 1813–1826. [CrossRef]

28. Broadley, M.R.; Lochlainn, S.; Hammond, J.P.; Bowen, H.C.; Cakmak, I.; Eker, S.; Erdem, H.; King, G.J.; White, P.J. Shoot zinc (Zn) concentration varies widely within *Brassica oleracea* L. and is affected by soil Zn and phosphorus (P) levels. *J. Hortic. Sci. Biotechnol.* **2010**, *85*, 375–380. [CrossRef]

29. Singh, B.K.; Sharma, S.R.; Singh, B. Variation in mineral concentrations among cultivars and germplasms of cabbage. *J. Plant Nutr.* **2010**, *33*, 95–104. [CrossRef]

30. Kałużewicz, A.; Bosiacki, M.; Frąszczak, B. Mineral composition and the content of phenolic compounds of ten broccoli cultivars. *J. Elementol.* **2016**, *21*, 53–65. [CrossRef]

31. Šindelářová, K.; Száková, J.; Tremlová, J.; Mestek, O.; Praus, L.; Kaňa, A.; Najmanová, J.; Tlustoš, P. The response of broccoli (*Brassica oleracea* convar. italica) varieties on foliar application of selenium: Uptake, translocation, and speciation. *Food Addit. Contam. A* **2015**, *32*, 2027–2038. [CrossRef]

32. Xiao, Z.L.; Codling, E.E.; Luo, Y.G.; Nou, X.W.; Lester, G.E.; Wang, Q. Microgreens of Brassicaceae: Mineral composition and content of 30 varieties. *J. Food Comp. Anal.* **2016**, *49*, 87–93. [CrossRef]

33. Lewu, F.B.; Lewu, M.N.; Mavengahama, S. Influence of genotype and harvesting age on the mineral dynamics of *Brassica oleracea* var. *acephala* cultivated in South Africa. *J. Food Agric. Environ.* **2012**, *10*, 563–566.

34. Singh, B.K.; Sharma, S.R.; Singh, B. Genetic combining ability for concentration of mineral elements in cabbage head (*Brassica oleracea* var. *capitata* L.). *Euphytica* **2012**, *184*, 265–273. [CrossRef]

35. MacNicol, R.D.; Beckett, P.H.T. Critical tissue concentrations of potentially toxic elements. *Plant Soil* **1985**, *85*, 107–109. [CrossRef]

36. White, P.J.; Broadley, M.R.; Thompson, J.A.; McNicol, J.W.; Crawley, M.J.; Poulton, P.R.; Johnston, A.E. Testing the distinctness of shoot ionomes of angiosperm families using the Rothamsted Park Grass Continuous Hay Experiment. *New Phytol.* **2012**, *196*, 101–109. [CrossRef] [PubMed]

37. White, P.J.; Broadley, M.R.; Gregory, P.J. Managing the nutrition of plants and people. *Appl. Environ. Soil Sci.* **2012**, *2012*, 104826. [CrossRef]

38. Broadley, M.R.; White, P.J.; Hammond, J.P.; Zelko, I.; Lux, A. Zinc in plants. *New Phytol.* **2007**, *173*, 677–702. [CrossRef] [PubMed]

39. Xian, X. Effect of chemical forms of cadmium, zinc, and lead in polluted soils on their uptake by cabbage plants. *Plant Soil* **1989**, *113*, 257–264. [CrossRef]

40. Handreck, K.A. Total and extractable copper and zinc as assessors of phytotoxicity in soilless potting media. *Commun. Soil Sci. Plant Anal.* **1994**, *25*, 2313–2340. [CrossRef]

41. White, P.J.; Brown, P.H. Plant nutrition for sustainable development and global health. *Ann. Bot.* **2010**, *105*, 1073–1080. [CrossRef] [PubMed]

42. Barrameda-Medina, Y.; Montesinos-Pereira, D.; Romero, L.; Ruiz, J.M.; Blasco, B. Comparative study of the toxic effect of Zn in *Lactuca sativa* and *Brassica oleracea* plants: I. Growth, distribution, and accumulation of Zn, and metabolism of carboxylates. *Environ. Exp. Bot.* **2014**, *107*, 98–104. [CrossRef]

43. Paradisone, V.; Barrameda-Medina, Y.; Montesinos-Pereira, D.; Romero, L.; Esposito, S.; Ruiz, J.M. Roles of some nitrogenous compounds protectors in the resistance to zinc toxicity in *Lactuca sativa* cv. Phillipus and *Brassica oleracea* cv. Bronco. *Acta Physiol. Plant.* **2015**, *37*, 137. [CrossRef]

44. Kusznierewicz, B.; Baczek-Kwinta, R.; Bartoszek, A.; Piekarska, A.; Huk, A.; Manikowska, A.; Antonkiewicz, J.; Namiesnik, J.; Konieczka, P. The dose-dependent influence of zinc and cadmium contamination of soil on their uptake and glucosinolate content in white cabbage (*Brassica oleracea* var. capitata f. alba). *Environ. Toxicol. Chem.* **2012**, *31*, 2482–2489. [CrossRef] [PubMed]

45. Lu, Z.G.; Grewal, H.S.; Graham, R.D. Dry matter production and uptake of zinc and phosphorus in two oilseed rape genotypes under differential rates of zinc and phosphorus supply. *J. Plant Nutr.* **1998**, *21*, 25–38. [CrossRef]

46. Bernhard, R.; Verkleij, J.A.C.; Nelissen, H.J.M.; Vink, J.P.M. Plant-specific responses to zinc contamination in a semi-field lysimeter and on hydroponics. *Environ. Pollut.* **2005**, *138*, 100–108. [CrossRef] [PubMed]

47. Hernández-Allica, J.; Becerril, J.M.; Garbisu, C. Assessment of the phytoextraction potential of high biomass crop plants. *Environ. Pollut.* **2008**, *152*, 32–40. [CrossRef] [PubMed]

48. Wang, C.; Zhang, S.H.; Wang, P.F.; Hou, J.; Zhang, W.J.; Li, W.; Lin, Z.P. The effect of excess Zn on mineral nutrition and antioxidative response in rapeseed seedlings. *Chemosphere* **2009**, *75*, 1468–1476. [CrossRef] [PubMed]

49. Peško, M.; Molnárová, M.; Fargašová, A. Effect of lead and zinc treatments on *Brassica napus* L. (cv. Verona) plants: Accumulation and physio-biochemical changes. *Fresenius Environ. Bull.* **2015**, *24*, 3213–3219.

50. Ebbs, S.D.; Kochian, L.V. Toxicity of zinc and copper to *Brassica* species: Implications for phytoremediation. *J. Environ. Qual.* **1997**, *26*, 776–781. [CrossRef]

51. Hamlin, R.L.; Schatz, C.; Barker, A.V. Zinc accumulation in Indian mustard as influenced by nitrogen and phosphorus nutrition. *J. Plant Nutr.* **2003**, *26*, 177–190. [CrossRef]

52. Hamlin, R.L.; Barker, A.V. Influence of ammonium and nitrate nutrition on plant growth and zinc accumulation by Indian mustard. *J. Plant Nutr.* **2006**, *29*, 1523–1541. [CrossRef]

agriculture

MDPI

Review

Macro and Micronutrient Storage in Plants and Their Remobilization When Facing Scarcity: The Case of Drought

Philippe Etienne [1,2], Sylvain Diquelou [1,2,3], Marion Prudent [4], Christophe Salon [4], Anne Maillard [5] and Alain Ourry [1,2,3,*]

[1] Université de Caen Normandie, UMR 950 Ecophysiologie Végétale, Agronomie et Nutritions N, C, S, Esplanade de la Paix, CS 14032, 14032 Caen CEDEX 5, France; philippe.etienne@unicaen.fr (P.E.); sylvain.diquelou@unicaen.fr (S.D.)

[2] INRA, UMR 950 Ecophysiologie Végétale, Agronomie et Nutritions N, C, S, Esplanade de la Paix, CS 14032, 14032 Caen CEDEX 5, France

[3] PLATIN', Plateau d'Isotopie de Normandie, Université de Caen Normandie, Esplanade de la Paix, CS 14032, 14032 Caen CEDEX 5, France

[4] Agroécologie, AgroSup Dijon, INRA, Université de Bourgogne Franche-Comté, 17 Rue Sully, BP 86510, 21065 Dijon CEDEX, France; Marion.Prudent@inra.fr (M.P.); christophe.salon@inra.fr (C.S.)

[5] Centre Mondial d'Innovation, CMI, Groupe Roullier, 18 Avenue Franklin Roosevelt, 35400 Saint-Malo, France; anne.maillard@roullier.com

* Correspondence: alain.ourry@unicaen.fr; Tel.: +33-2-31-56-56-53

Received: 27 November 2017; Accepted: 11 January 2018; Published: 16 January 2018

Abstract: Human mineral malnutrition or hidden hunger is considered a global challenge, affecting a large proportion of the world's population. The reduction in the mineral content of edible plant products is frequently found in cultivars bred for higher yields, and is probably increased by intensive agricultural practices. The filling of grain with macro and micronutrients is partly the result of a direct allocation from root uptake and remobilization from vegetative tissues. The aim of this bibliographic review is to focus on recent knowledge obtained from ionomic analysis of plant tissues in order to build a global appraisal of the potential remobilization of all macro and micronutrients, and especially those from leaves. Nitrogen is always remobilized from leaves of all plant species, although with different efficiencies, while nutrients such as K, S, P, Mg, Cu, Mo, Fe and Zn can be mobilized to a certain extent when plants are facing deficiencies. On the opposite, there is few evidence for leaf mobilization of Ca, Mn, Ni and B. Mechanisms related to the remobilization process (remobilization of mineral forms from vacuolar and organic compounds associated with senescence, respectively) are also discussed in the context of drought, an abiotic stress that is thought to increase and known to modulate the ionic composition of grain in crops.

Keywords: grain filling; nutrient remobilization; senescence; abiotic stress

---

1. Introduction

Mineral malnutrition is considered to be the most serious global challenge to humankind with over 60, 30, 30 and 15% of the world population being Fe, Zn, I or Se deficient, respectively. In addition, Ca, Mg and Cu deficiencies are common in many developed and developing countries (see the extensive review [1]). Because plants are the basis of nearly all food chains, the production of biofortified seeds, fruits or edible vegetative organs with increased concentrations of micronutrients could reduce what is considered "hidden hunger".

Plants are sessile organisms that have to cope with a permanently fluctuating environment, both in space and time, and this includes changes in abiotic factors such as light, temperature, water and nutrient availabilities that are tightly linked to biotic interactions with other plants or rhizosphere microorganisms such as bacteria and fungi. To cope with these variable environments, plants through their evolution have developed multiple mechanisms allowing their survival, growth and reproductive success. For example, the uptake of soil minerals is achieved via many processes to minimize transient deficiency [2–5] including stimulation of root transporter activities, oriented root growth and/or root exudation. Plants may also increase soil mineral availability and improve their nutrient uptake through interactions with rhizospheric microorganisms [6].

Hence, a better understanding of the mechanisms involved in plant nutrient acquisition and distribution in edible products with increased micronutrient concentrations could pave the way to the development of improved plant varieties, and participate in an amelioration of human malnutrition. Seed micronutrient enrichment depends upon the uptake, storage, mobilization and translocation of the micronutrients. All of these complex processes require the coordinated regulation of many genes. For example, a recent study has shown that two chromosome regions were associated with Quantitative Trait Loci for improvement of zinc and iron content in grains [7]. Therefore, breeding for bio-fortification combined with improvement of a well-synchronized supply of various amounts and forms of fertilizers in crop rotations are considered appropriate to alleviate mineral malnutrition [8]. The study by Fan et al. [9] shows that breeding wheat for a better yield could be achieved via enhanced photosynthesis combined with an ample supply of macro-nutrients (N, P, K), but this led to a negative selection for micronutrient content in seeds. These authors analysed mineral concentrations of archived wheat grains and soil samples between 1843 and 2008 in relation to cultivar, yield and harvest index, and reported that the seed micronutrient contents remained stable between 1843 and the mid 1960s but decreased significantly after that time. This coincides with the introduction of semi-dwarf and high-yielding cultivars, while at the same time the soil concentrations of Zn, Fe, Cu and Mg either increased or remained stable. Overall, this indicates that the reduced mineral content of seeds partly resulted from this negative genetic selection and thus manifested as hidden hunger [9].

Additionally, in the context of global changes resulting from increased accumulation of greenhouse gases [10] associated with an increase in temperature and the occurrence of greater risk of aridity [11,12], the mineral contents of seeds may also be adversely affected. For example, a meta-analysis of the literature that considered the effect of increased atmospheric CO_2 on the plant ionome [10] indicated in the majority of cases that leaf and seed N content and most of their mineral nutrients (P, K, Ca, S, Mg, Fe, Zn, Cu, Mn) were negatively impacted as the seed carbon concentration increased. This empirically robust relationship was systemic i.e., independent of plant species, cropping areas and experimental designs, including Free-air CO_2 enrichment (FACE) experiments. Another meta-analysis [11] revealed that drought stress in plants had a short-term (<90 days) negative effect on N and P contents in plants, alleviated either in the long term or by drying-rewetting cycles.

Soil mineral nutrient deficiency may be the consequence of different constraints including intrinsic low soil mineral contents, sub-optimal abiotic conditions like extreme temperatures, pH, low soil water content or anaerobic conditions that will alter mineralization and hence phytoavailability. When nutrient availability is too low, several strategies have been described that enable plants to cope with most macronutrient and some micronutrient deficiencies, and they rely on (i) an over expression of genes encoding for increased or decreased nutrient specific root transporters coupled in the longer term with (ii) an increased exploration of soil due to enhanced root branching and growth [13,14]. Mineral nutrient deficiencies usually elicit mechanisms that rely mostly on an up-regulation of specific root membrane high- or low-affinity transport systems. These mechanisms have been extensively described for N, P, and S [2,3] and the complexity of their regulation is being progressively unravelled. The involvement of multiple transporters in several regulation pathways suggests that higher plants have a sophisticated uptake system finely tuned to external nutrient availability and internal plant needs. However, although up-regulation of nutrient uptake by roots coupled with increased root exploration and/or microbial stimulation via root exudation are among the common coping strategies, they can be unsuccessful under certain circumstances. This may be the case when soil nutrient availability remains too low or when low soil water content restricts nutrient mobility.

Besides its implication in photosynthesis and loss of water through the stomatal aperture [15], plant transpiration conveys many functions such as cooling leaves and driving both the mass flow of nutrients from the soil to the rhizosphere and the xylem flow from roots to the shoot. An increase in transpiratory flux is usually correlated with [15] an increase in photosynthesis activity, leading to loss of water when stomata are open for CO_2 uptake. Mass flow, driven by transpiration, leads to substantial nutrient delivery to the root surface. For example, plants of *Zea mays* grown in a fertile loam had the potential to accumulate 95% of the S, 88% of the Mg, 79% of the N, 73% of the Ca, 18% the K, and the 5% of P by mass flow, the remainder being taken up through diffusion and uptake by the roots [16]. Similar situations where mass flow and hence nutrient mobility in the soil may be reduced can be found when plant transpiration is low. This may occur during the spring bud growth of deciduous trees, regrowth of forage species subjected to mechanical or animal defoliation, or the spring resumption of growth of hemicryptophyte and cryptophyte plant species. Moreover, this could also be the case during the development of annual species. For example, in *Brassica napus* it has been shown that nutrient uptake (especially N) is considerably reduced during the reproductive stage. Consequently, development of reproductive tissues and seed filling relies on monocarpic senescence, which allows degradation of organic compounds associated with internal recycling of nutrients [17].

Whatever the reasons for decreased nutrient uptake (environmental or developmental), short-term use of macro and micronutrients previously stored within tissues may be an effective alternative for sustaining plant growth. Moreover, in the long term, i.e., when nutrient deficiency becomes severe, degradation of organic compounds (which may also contain micronutrients) is associated with leaf senescence [18,19] in order to provide nutrients useful to sustain plant growth. In such situations, plant growth mostly relies on internal stores than can be mobilized from the short term to the long term. However, the importance of nutrient storage and recycling are not always recognized. Thus, some authors [3] consider that recycling of pools of essential macronutrients (N, P, S) previously stored in plant tissues is not sufficient to support normal growth for more than a few days in most higher plants. In contrast, other reports have [19,20] clearly stated that manipulating nutrient remobilization is a major challenge for modern agriculture. For example, some authors suggest that an improvement in nutrient remobilization from senescing tissues constitutes an appropriate lever to increase global nutrient use efficiency (especially N) [17,21] or to biofortify edible products [22–24]. In any case, efficient nutrient mobilization in plants needs to satisfy several conditions, such as the possibility to accumulate large amounts of nutrients in plant parts without reaching a toxicity level, an ability to be moved from storage compartments (e.g., the vacuole) and their capability to be transported [23] between tissues through xylem or phloem vessels (Figure 1). A fourth condition may be a requirement when nutrients have been stored within fairly complex compounds that are either not easily transported by the phloem or the xylem or that require dismantling into mobile and nontoxic forms.

The overall objectives of this review are to focus on the body of knowledge related to plant nutrient mobilization from leaves (the main source organs) to the seeds (the main sink organs) that to our knowledge has never been reviewed at the level of ionome. Moreover, because one of the main effect of drought is to reduce mass flow and hence mineral nutrient uptake, it can be hypothesized that under this abiotic stress, grain filling will mostly rely on leaf nutrient remobilization. In the most part, this review references studies of plant ionome analysis to (i) identify general patterns of leaf accumulation and remobilization according to the nutrient considered and the plant species, (ii) relate these patterns to potential mechanisms (including mineral vacuolar sequestration/mobilization capacity and mobilization of organic compounds associated with senescence), (iii) evaluate from the literature how these mechanisms and the resulting nutrient remobilizations are affected by drought, and finally (iv) estimate from the literature how seed nutrient contents are affected by drought and whether they can be related to leaf remobilization processes.

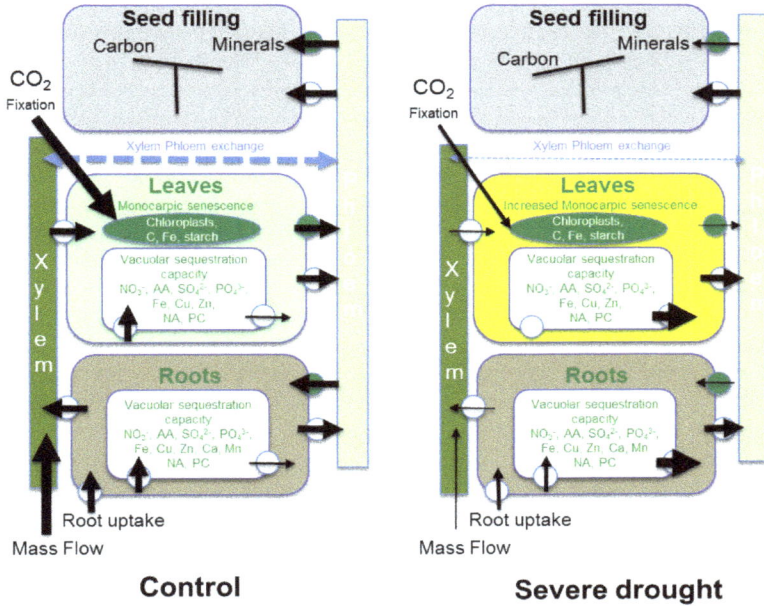

Figure 1. Proposed model for leaf remobilization of macro and micronutrients towards seed filling under normal conditions or during severe drought illustrating the involvement of vacuolar sequestration capacity, coupled under severe drought with increased monocarpic senescence and reduced photosynthesis. NA: nicotianamide, PC: phytochelatin.

2. General Patterns of Nutrient Mobilization from Leaves

Because of the concomitant root uptake, the quantification of nutrient mobilization from vegetative to reproductive tissues usually requires the use of either stable or radioactive isotopes [25]. Numerous studies have used pulse chase labelling of ^{15}N or ^{13}C, and more recently ^{34}S or ^{65}Zn [26,27], and very occasionally ^{54}Mn [28], but only a few other elements have been considered under very specific circumstances (effects of genetic modifications or environmental constraints such as deficiencies or water supply). Alternatively, changes in the ionomic composition of leaves have been used to evaluate (by mass balance) the net remobilization of most leaf nutrients during irreversible processes such as monocarpic senescence [29–33].

Data from a meta-analysis designed to evaluate specific and genetic variations under a wide range of plant culture conditions, tissues, species and genotypes [1], showed that the ratio between the maximum and minimum contents of nutrients such as Fe, Zn, Mg and Cu have a much wider variation in leaves than in seeds, suggesting a remobilization potential in vegetative tissues. The general pattern of nutrient accumulation and remobilization during leaf development is probably nutrient and species dependent (Table 1), being affected by either nutrient availability or plant phenology (reproductive versus vegetative growth). However, according to the previously cited literature, three types of potential recycling can be hypothesized (Figure 2), at least for nutrients for which enough information is available (Table 1). Irrespective of the plant species, N is always remobilized from leaves (Table 1) and requires a proteolytic system to degrade proteins during leaf senescence. However, depending on the plant species, the efficiency of N mobilization is variable, reaching about 90% in *Triticum aestivum*, *Hordeum vulgare* and *Arabidopsis thaliana* against only 39% in *Zea mays*, while *Brassica napus*, *Pisum sativum* and *Glycine max* show intermediate values (Table 1). Other macronutrients such as K,

S, and P are also remobilized from leaves at a rate similar or lower to N, except in *Z. mays* leaves, which only remobilize K with a fairly high efficiency. Magnesium (Mg) is also remobilized from leaves, except in *Z. mays*, *A. thaliana* and *P. sativum*. Calcium (Ca), is a macronutrient that is considered to have a limited mobility in the phloem. Consequently, Ca stored in leaves is not remobilized, at least not in non-poaceae species. Overall, it should be pointed out that for a given species the rates of mobilization of macronutrients (except for Ca) is relative to that of N withdrawal from leaves (Table 1), suggesting that they depend on leaf senescence. The situation for micronutrients (Table 1) is far more complex, showing more interspecies variations. Boron (except in *Z. mays*) and nickel (except in *B. napus*) seem to be fairly immobile in the leaves of all species, while in contrast copper (Cu) is exported from leaves of all species (Table 1). Manganese (Mn), like Ca as previously reported, is only remobilized in *T. aestivum*, *O. vulgare* and *O. sativa*, which may be the result of a rapid release of water during plant dehydration provoking Mn or Ca movement in the xylem [34,35]. Molybdenum (Mo), iron (Fe) and zinc (Zn) show the greatest variation between species, with *B. napus* and *Z. mays* being unable to export Zn and Mo from the leaves, while Fe is not remobilized in poaceae plants (except at a low level in rice) or in *P. sativum*. Similarly, studies in soybean [36] and wheat [37] revealed that levels of Cu, Fe and Zn declined in leaves undergoing senescence. This apparent heterogeneity in the potential remobilization of micronutrients might be linked to the fact that the requirements for these elements in seeds are fairly low, and might be fulfilled by the translocation of nutrient taken up by the roots post anthesis. Accordingly, Sankaran and Grusak [33] suggested that the remobilization of micronutrients would be required only when soil deficiencies limit root uptake.

Based on the previous information, the patterns of leaf nutrient remobilization could be separated into three nutrient groups (Figure 2). Group I corresponds mostly to N, a highly mobile nutrient with a predominantly organic form of storage (especially proteins), and it is accumulated in leaves up to the beginning of leaf senescence (Figure 2). Subsequently N is exported to young developing leaves and/or reproductive tissues in nearly all plant species. When facing N deficiencies during early or late leaf development, the mobilization rates can usually be increased concomitantly with senescence induction [38–41]. Group II (Figure 2) corresponds to nutrients like K, S, P, Mg, Cu, Fe, Zn and Mo, which are more variable in their mobility and mostly found in either inorganic forms or associated with cellular organic molecules such as proteins that are constantly accumulated during leaf development until full expansion. Their remobilization may occur during monocarpic senescence, but can be increased (probably from mineral storage forms) when plants are facing a deficiency, without necessarily inducing senescence. For example, it has been well characterized in *B. napus* that a moderate S deficiency increases leaf S mobilization (mostly from SO_4^{2-}), whereas leaf senescence is delayed [42]. Likewise, in the case of Cu and Fe in wheat [42] and Fe [31] or P [43] in rice, increased remobilization from leaves is triggered by reduced root uptake, which in turn is a consequence of soil mineral deficiencies. Group III (Figure 2) corresponds to fairly immobile nutrients such as Ca, Mn (at least in non poaceae species), B and Ni, which accumulate during the entire leaf life span or reach a plateau after the onset of senescence, without any effect of deficiency on mobilization and/or senescence [32]. Such a lack of remobilization could result from (i) a very low phloem mobility (Ca and Mn), (ii) nutrient sequestration into a fairly complex molecule that is not dismantled during senescence (for example, B and rhamnogalacturonan located in cell walls [44]), or (iii) by potential nutrient toxicity at the cellular level (such as Ni, [45]) precluding its remobilization. In the meantime, whichever nutrient group is considered, the accumulation of all nutrients in leaves can be reduced in a similar way if deficiencies occur in early leaf development (Figure 2) and lead to reduced storage in leaves before seed development. However, while these general patterns of nutrient remobilization from leaves offer an easy way to compare nutrients and plants species, they do not take account of the mechanisms involved. Indeed, processes associated with the mobilization of short- or long-term nutrient storage pools may be affected by the severity of environmental stresses such as drought.

Table 1. Nutrient remobilization from senescent leaves in 8 plant species from studies using multi-element analysis. References are identified by superscript numbers. Percentages indicate the relative decrease in mineral content in senescent leaves during grain filling.

Species	Leaf Remobilization of Macronutrients						Leaf Remobilization of Micronutrients						
	N	K	S	P	Mg	Ca	Mn	Cu	Mo	B	Ni	Fe	Zn
Triticum aestivum	93%[1]	95%[1]	94%[1]	89%[1]	89%[1]	89%[1]	43%[1]	71%[1]	79%[1]	0[1]	0[1]	0[1]	92%[1]
Hordeum vulgare	88%[1]	96%[1]	93%[1]	87%[1]	96%[1]	89%[1]	70%[1]	84%[1]	0[1]	0[1]	0[1]	0[1]	70%[1]
Oryza sativa	-	47%[4]	20%[4]	-	45%[4]	33%[4]	65%[4]	40%[4]	46%[4]	-	0[4]	20%[4]	29%[4]
Zea mays	39%[1]	72%[1]	0[1]	0[1]	0[1]	0[1]	0[1]	0[1]	0[1]	59%[1]	0[1]	0[1]	0[1]
Brassica napus	51%[1]	37%[1]	54%[1]	55%[1]	36%[1]	0[1]	0[1]	50%[1]	78%[2]	0[1]	91%[1]	68%[1]	54%[2]
Arabidopsis thaliana	88%[2]	86%[2]	65%[2]	78%[2]	0[2]	0[2]	0[2]	54%[2]	78%[2]	-	0[2]	55%[2]	0[2]
Glycine max	60%[5]	63%[5]	-	67%[5]	49%[5]	59%[5]	42%[5]	60%[5]	96%[5]	45%[5]	-	38%[5]	48%[5]
Pisum sativum	-	50%[3]	65%[3]	82%[3]	50%[3]	-	21%[3]	73%[3]	-	-	-	70%[3]	32%[3]

References: [1] [32]; [2] [30]; [3] [33]; [4] [31]; [5] [36].

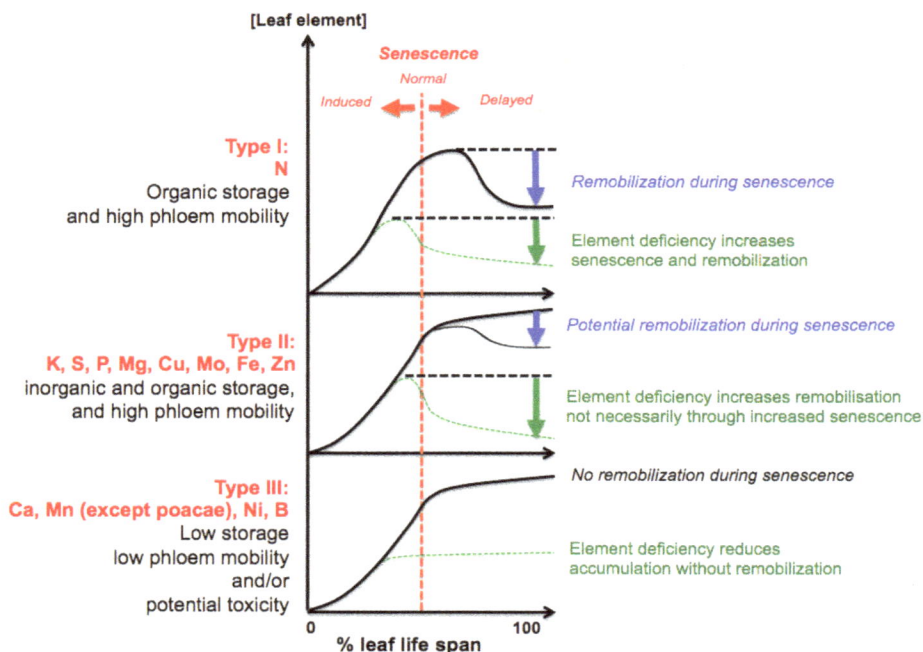

Figure 2. General patterns of macro and micronutrient contents of leaves resulting from their allocation from root uptake before senescence followed by their remobilisation (net loss) as a function of relative leaf lifespan. According to the macro or micronutrient considered, senescence and/or nutrient deficiency induce high (Type I), variable (Type II) or low remobilization (Type III) from the leaves to the seeds.

3. Short-Term Mobilization of Stored Compounds When Facing Nutrient Deficiencies or High Nutrient Needs

For most nutrients, short-term storage compounds are mostly found in the vacuoles (Figure 1), under inorganic (NO_3^-, SO_4^{2-}, K^+, Mg^{2+}, PP, Pi) or complexed forms with organic compounds such as organic acids (Ca, Mo), amino acids (Ni, Se), nicotinamide (NA), specific chelates such as ferritin (Fe, although mostly in chloroplasts), and even proteins (Cu and Zn). Mineral micronutrients are thought to be in chelated or ligand-bound forms when not incorporated into proteins, including during their phloem transport (Figure 1). As reported in the review of Peng and Gong [46], studies on long distance transport of nutrients have focused on transporters expressed in xylem parenchyma cells or in phloem companion cells, as well as the role of phytochelators such as NA and phytochelatins. Considering the vacuole as a pivotal organelle for storage of metabolites and minerals, Peng and Gong [46] proposed the concept of vacuolar sequestration capacity (VSC) acting as a buffering pool not only controlling accumulation in cells but also dynamically mediating transport of nutrients over long distances. Complex regulation of the VSC of metals would involve the modulation of (i) the expression of tonoplast specific transporters (MTP1/MTP3/ZIF1 for Zn, NRAMP3/NRAMP4 for Fe, COPT5 for Cu) or other transporters with lower specificity (VIT1/VIT2 for Fe and Zn, FPN2 for Fe, Co and Ni) and (ii) the synthesis of chelators such as NA that modulate the VSC of Fe and Zn [47] and phytochelatin modulating the VSC of Cu. NA, which is also a precursor of phytosiderophores in grasses, has the capacity to bind Cu, Co, Fe(II) and Fe(III), Mn, Ni, and Zn. This body of evidence strongly suggests that NA is a vital chelator of micronutrients for homeostasis during growth, for translocation within vegetative parts of the plant and also in phloem transport of micronutrients to the seeds. Proteins that transport these NA complexes are the yellow-stripe-like (YSL) proteins, which are members of the oligopeptide transport (OPT) family [48].

While some nutrients such as Ca or Mn are recognized as showing little mobility between tissues, and for which, like Cu and Ni, the transport is poorly understood, others like N, S, P, Fe, Zn are considered to be potentially highly mobile in either xylem or phloem vessels (Figure 1). For example, S is usually recognized as showing variable mobility [49]. In fact, the S content itself in tissues is highly variable and can be correlated with SO_4^{2-} content, which may account for more than 80% of total S in leaves of field grown B. napus [50]. As a consequence, when facing S deficiency, plant growth can be sustained for several weeks [51], provided that enough SO_4^{2-} was accumulated, and thanks to its remobilization and high mobility. In contrast, plants containing only organic S pools (i.e., low mobile compounds) are unable to grow without a significant S supply. A much wider variation was observed for S and SO_4^{2-} concentrations than for N or NO_3^- concentrations, which suggests a smaller contribution of vacuolar NO_3^- to N remobilization, and most of it derived from organic N. However, recent results [52] obtained in B. napus suggest that a decrease in the VSC of NO_3^- in roots enhances N transport to shoots and essentially contributes to higher N use efficiency. This occurs by promoting NO_3^- allocation to aerial parts, likely through coordinated regulation of the NO_3^- tonoplastic transporters, NRT1.5 and NRT1.8. Based on this evidence, there is obviously a relationship between the variable capacity to accumulate a nutrient in vegetative tissue and the potential to remobilize a portion of it when facing a restriction.

4. Monocarpic Senescence and Remobilization of Nutrients

Long-term storage may also occur for some nutrients, mostly in a complex form requiring specific catabolic transformations, first to enabling their remobilization, then their transport. This is the case for nutrients that are components of various polymer structures (e.g., N and S in proteins, S and P in lipids, B in rhamnogalacturonan II, Zn in numerous proteins) or involved in their physiological function (Fe and ferritin, Mo as a co-factor, Mg in chlorophyll, Mn as a catalytically active metal in proteins). It is usually assumed that their mobilization is an intrinsic feature of plant senescence [18]. The complex mobilization process of these nutrients, eventually leading to the death of vegetative organs, involves a well-orchestrated activation of genes encoding catabolic enzymes that gradually dismantle cellular components that are mostly located in chloroplasts. At the same time, basic metabolic activities are kept intact until cell death to ensure the processing of high molecular weight components and the subsequent export of the degradation products and minerals to the phloem [53]. Recent research [17,54] suggests that the link between growth and the ageing process is nutritional in nature, by which resources (most of the time, only N is considered) are recycled from obsolete body parts to newly developing structures. For example, feeding the plant with N fertilizer increases the amount of Rubisco but does not change its turnover, and this N responsive behaviour is characteristic of a storage protein. When the N demand cannot be met by uptake from the rhizosphere alone, N, and potentially its associated nutrients, is withdrawn from older tissues. Similarly, it was suggested that sequential senescence (i.e., acropetal senescence occurring during the vegetative stage) is a useful adaptation because it buffers growth against fluctuation in the supply of N and other elements. Results from studies on legumes demonstrate that the N being remobilized from vegetative parts to fill seeds arises from a common N pool translocated throughout the plant [55]. Although various proteomic and metabolomic studies are being conducted, the detailed catabolic pathways involved are not yet fully characterized. The mechanism of autophagy, consisting of the allocation of unnecessary or damaged cytosolic components (such as organelles and macromolecules) for degradation and recycling by the vacuole (see [56] for review), has been recently characterized during senescence [57]. Such a mechanism might explain how nutrients other than N may also be mobilized, as numerous proteins contain elements such as Zn, Cu, Mn, Fe or N.

The results of Distefeld et al. [58], who compared a wild tetraploid wheat (*Triticum turgidum* ssp. *dicoccoides*) to a cultivated wheat (*Triticum durum*), strengthen the hypothesis that senescence process may increase the mobilization of micronutrients other than N. These authors found a higher mobilization of Zn, Mn and Fe in wild wheat and attributed it in part to allelic variation at a chromosome locus that promotes early senescence and remobilization of proteins from senescing leaves to seeds. Waters and Sankaran [23] considered that because Cu, Zn and Fe transport frequently involves chelation with amino acids, remobilization of these elements might be tightly linked with N catabolism occurring during senescence. This could be further supported by the fact that wheat lines with delayed leaf senescence have lower amounts of Fe and Zn in their seeds [59]. Ricachenevsky et al. [48] suggest a major role of some NAC (acronym derived from three genes that were initially discovered to contain a particular domain (the NAC domain): NAM for no apical meristem, ATAF for *Arabidopsis thaliana* activating factor, and CUC2 for cup-shaped cotyledon) transcription factors in wheat and rice that could regulate the onset of leaf senescence, together with an increased Fe and Zn remobilization, through an increased synthesis of nicotianamine associated with a higher expression of a metal-NA transporter (YSL2). Similarly, Pearce et al. [60] showed that a wheat mutant under-expressing *GPC1*, a transcription factor that reduces leaf senescence, had lower Zn and Fe in the seeds, probably as a result of a down regulation of *Zip* and *YSL* genes coding for Fe and Zn transporters. In legumes, leaf and nodule senescence is a programmed process concomitant with the transition to the reproductive stage in the host plant life cycle [55,61]. Van de Verlde et al. [62] highlighted that the transcriptome of nodule senescence shared a high functional overlap with the transcriptome of leaf senescence, indicating a shift from a carbon sink to a nutrient source tissue. In particular, in soybean, it has been estimated that around 50% of the nodule Fe content was recycled to the seeds [63] and in *Lotus japonicus*, a nodule-specific NA synthase has been identified [64]. Taken together, these results suggest that nodules could also contribute to the mobilization of micronutrients to the seeds [65].

5. How Drought Affects Leaf Accumulation of Nutrients, Leaf Senescence, Mobilization of Nutrients and Seed Filling

Drought reduces the diffusion rate of nutrients in the soil towards the roots, the nutrient uptake by the roots, and their transport to the shoots due to concomitant decreases in transpiration flux, active transport and membrane permeability (see the review from Hu and Schmidhalter [66]. Drought during the vegetative stage in *Arabidopsis thaliana* induces a decrease in the concentration of almost all minerals including Zn, Fe, Mn, Ca and Mg in leaves, except for K [67]. The authors explain these reductions by a preferential allocation of resources to the roots, at the expense of leaf growth. Otherwise, the increase in leaf K content under drought could be related to its role in maintaining plant water potential and in alleviating drought-induced oxidative damage through inhibition of the production of reactive oxygen species (ROS) [68]. Yet from a mechanistic point of view, how vacuolar storage and long distance transport of nutrients is affected by drought during the reproductive stage is fairly unknown. However, it could be postulated that a reduction in mass flow resulting from drought, and hence a reduction in the phytoavailability of soil nutrients, could be similar to a situation of nutrient deficiency (previously described in paragraph 3). Under this assumption, long distance transport would be regulated through modulation of VSC by the changes in both transporter expression and chelate synthesis (Figure 1). For example, in *Medicago truncatula*, genes coding for sulfate transporters involved in sulfate uptake (*SULTR* of Group 1) are down-regulated in roots, while genes coding for sulfate transporters involved in the efflux of sulfate from the vacuole lumen to the cytosol (*SULTR* of Group 4) are up-regulated in shoots [69]. Consequently, short-term vacuolar storage of nutrients would be of prime importance to buffer transient and moderate drought, which could be, at least partly, independent of senescence.

Among environmental stresses, drought is known to induce leaf senescence, especially via changes in hormonal balance related to the concomitant decrease and increase in cytokinin (CKs) and abscisic acid (ABA) contents, respectively [70,71]. Drought-induced senescence is associated with numerous morphological, physiological and molecular modifications in several species including nutrient remobilization from senescing organs (mainly leaves, and to a much smaller extent nodules in legumes) to young tissues (leaves or seeds), thus compensating for the nutrient uptake deficit that results from low soil water content [72,73]. While much work has been conducted to evaluate the consequences of drought on seed nutrient contents, to our knowledge studies focusing on the effect of drought specifically on nutrient remobilization from leaves (estimated from a net balance during senescence) are very scarce.

According to Wang and Frei [74], accelerated senescence under many types of stress, including drought, affects nutrient translocation processes, inducing the remobilization of N from vegetative to reproductive plant parts, and shortening the maturation time, which tends to favour proteins over starch accumulation in cereal grains. Decreased mineral content in seeds can therefore result from reduced root uptake and translocation and/or insufficient remobilization from leaves. This contrasts with reports of increases in some mineral contents in seeds following drought, which are probably a result of reduced photosynthesis leading to reduced C content in the seeds. However, most conclusions appear to be heterogeneous if not opposite (Table 2), even within the same plant species. New experimental design, in which both the severity of drought and the availability of nutrients are finely monitored will be required to obtain more relevant data with different plant species. As pointed out in the reviews of Lawlor [75] and He and Dijkstra [11], many different experimental procedures have been used to subject plants to drought, ranging from field to controlled conditions, using PEG (Polyethylen Glycol) or mannitol in nutrient solution, manipulating soil water content from progressive reductions to complete cessation of watering and finally using cycles of soil drying/rewatering.

Table 2. Effect of drought on the nutrient contents of seeds in different plant species submitted to severe (% decrease in seed yield is given when available) or moderate drought from studies using multi-element analysis. Drought is considered to be severe as opposed to moderate when it significantly reduced grain yield.

Species	Drought	Increased Seed Nutrient Content	Decreased Seed Nutrient Content	Refs.
Hordeum vulgare	Severe	N	P, K, Mg	[76]
	Moderate	No effect	No effect	[76]
	Severe	N, Zn, Mn	No effect	[77]
Triticum turgidum	Severe	N, Fe, Zn	No effect	[78]
	Moderate	No effect	No effect	[78]
Triticum aestivum	Severe	N	Not quantified	[79]
	Severe	(N not quantified), K, Ca	No effect	[80]
	Moderate			[80]
	Moderate	P, Mg, Zn	K	[81]
	Severe	P, Ca, Mg, Zn	K	[81]
Zea mays	Severe	N, Ca, Mg, Cu, N	P, K	[82]
	Severe	N	Fe, Zn, Cu	[83]
	Moderate		Fe, Zn, Cu	[83]
Glycine max	Moderate	No effect	N, K, P, Ca, Fe	[84]
	Severe	P, Ca, Mn, Zn, Mo	No effect	[85]
	Severe	Ca, Fe	K, P, Mn, Cu, Zn	[86]
	Moderate	Ca, K	P, Mn, Cu, Zn	[86]
Oryza sativa	Severe	N	Cu, Fe	[87]
Phaseolus vulgaris	Moderate		Fe, Zn	[88]

For example, it has been shown that high molecular weight PEG can be taken up by roots and accumulated in extracellular spaces inducing cellular but not tissue dehydration [89]. Moreover, Verslues et al. [90] showed that the use of PEG in the nutrient solution also decreased dioxygen movement by increasing solution viscosity, which in turn led to root hypoxia and inhibition of root elongation. It may also be assumed that the use of mannitol or PEG in nutrient solutions have little relevance for natural drought conditions in the fields where drought conditions, and hence, plant perception appear more progressively allowing the plants to acclimate during several days. In order to discriminate between experiments that used moderate (coupled with maintained yield) or severe drought (associated with a yield reduction) and associated mechanisms (Figure 1, i.e., balance between C and mineral nutrients transported to the seeds), Table 2 presents the effect of drought (severe or moderate) on changes to seed nutrient contents in different species. In most species (except for soybean), a severe drought leads to an increased N concentration in grains (Table 2), which may be the result of a balance in favour of leaf N export relative to C, as opposed to a moderate drought that will less strongly affect photosynthesis.

The seed contents of other macronutrients like P, K, Mg, and sometimes Ca, are reduced most of the time (or remain unaffected) by severe drought (Table 2). The situation is more variable for Zn, Fe, Mn or Cu for which seed contents can be decreased or increased by severe or moderate drought, not only as a function of species but also for a given species subjected to different experiments (Table 2). In a hypothesis that parallels the one for N content, seed micronutrient contents are a result of a balance between C export from leaves and starch deposition in grains, and micronutrient leaf remobilization. For such micronutrients, the level of remobilization in leaves could be a function of only the vacuolar sequestration capacity and therefore dependent on root uptake (and storage) before anthesis. However, an extra supply of some micronutrients to the seeds could be provided by catabolism of polymers that may contain such micronutrients, and as a consequence of senescence induced by severe drought conditions. This could be the case for Zn, for example, which is associated with a large diversity of proteins (more than 1200 in *Arabidospis*) and for Fe, which is mostly stored in chloroplasts as ferritin. Mn and Ca are more difficult to understand because they are recognized as having little phloem mobility, and consequently a reduced leaf remobilization. However, it has been shown recently [32] that even though Mn and Ca can be efficiently mobilized from roots (probably through xylem vessels), they would require a significant mass flow, which is not compatible with the main effect of drought.

Other reasons may be invoked to explain changes in nutrient contents in seeds subjected to high or low drought severity, even though, to our knowledge, no literature is yet available. The ionomic content of vegetative tissue can be affected in a very unspecific way by mineral deficiencies or even by drought. The study by Maillard et al. [91] on the effect of individual nutrient deficiencies on the uptake of other nutrients in *B. napus*, found about 18 different situations where the uptake of other nutrients was increased. Such cross talk between nutrients was found, for example, during S and K deficiencies, which increased Mo and Na uptake, respectively. In both cases, this could be explained by the lack of specificity of some transporters that are up-regulated during deficiency. For example, sulfate transporters are mostly responsible for higher molybdate uptake under S deficiency, while K transporters, which are up-regulated by K deficiency, increased Na uptake. Similarly, Acosta-Gamboa et al. [92] showed that drought increases the leaf content of Mn and Cu, whereas it decreased Fe content and it was hypothesized that this was partly due to a higher requirement for Mn superoxide dismutase, which is involved in the ROS detoxification induced by drought stress. However, alternative explanations can be provided from [91] who showed that increases in Cu and Mn uptake in response to Fe deficiency can also be explained by the non-specific transport of these nutrients being shared by similar up-regulated transporters [93]. Ghandilyan et al. [67] highlighted that depending on the water availability, correlations between leaf nutrient contents were not necessarily maintained. In particular, correlations between K and Ca concentrations were positive under optimal water conditions, and negative under drought. Taken together, these results suggest that drought, while affecting root nutrient uptake may also modify cross talk between micronutrients (uptake and storage) and hence modify the composition of the leaf with potential consequences for seed composition.

A better knowledge of how ionomic composition of grain, and hence their quality, are modified by drought requires future experiments in which both drought conditions (repeated and moderated or severe) and mineral nutrient supplies are finely controlled under soil culture. A precise estimation of the modification of grain ionomic signature by drought should be done in relation to a kinetic analysis of leaf remobilization of each nutrient. Such experimental design was not previously used to our knowledge and should be considered. In such a way, it should be possible to distinguish (i) nutrients for which grain content is decreased as a result of limited leaf remobilization, from (ii) nutrient whose grain content is increased such as N (Table 2 and Figure 1), as a result of an efficient leaf remobilization coupled with a yield reduction and finally, (iii) nutrients that are indirectly increased in grain by drought as a result of nutrient crosstalk (see previous paragraph for examples). This would also require a better knowledge of all nutrient crosstalks induced by individual mineral deficiency as previously described [91] and for a larger range of plant species. This approach could also include the analysis of ionomic signatures combined with molecular analysis that would give access to metabolic pathways modulated by deficiencies and/or drought.

Moreover, the plant nutrient composition as well as the plant physiological response to drought are both genotype-dependent [67,83,92,94,95]. This emphasizes the need to take advantage of genetic variability, first to understand processes underlying micronutrient remobilization under drought and secondly to breed crops with higher seed micronutrient contents under water fluctuating environments. For the moment, strategies for enhancing drought stress tolerance by delaying senescence either via increasing chloroplast stability [96] or via induced production of CKs and the alteration in C and N homeostasis [97] have been tested and resulted in an increased grain yield, but could also lead to a limited N remobilization efficiency, implying a decreased protein content [54,98].

6. Conclusions and Perspectives

Under normal growth conditions, the storage and remobilization of macronutrients in particular have been extensively studied, but the handling of micronutrients remains less well known. Further, the mechanisms underlying nutrient remobilization to the seeds under drought conditions have been poorly addressed. In the context of human micronutrient malnutrition and climate change involving more frequent drought episodes, the need to breed for biofortified crops [99] adapted to water deprivation should become a major focus. Efforts are necessary concerning the understanding of micronutrient fluxes under limited mass flow due to decreased plant transpiration. These efforts may be coupled on the one hand with the screening of large collections of genotypes, and on the other hand with the exploitation of plant microbe interactions [100], which could enhance micronutrient availability, plant uptake, storage and remobilization to the seeds.

Acknowledgments: This work has been partly conducted through the EAUPTIC project, through funding from the Regional Council of Normandy and The "Fond Unique Interministériel". The authors thank Laurence Cantrill for improving the English of this manuscript.

Author Contributions: Philippe Etienne, Sylvain Diquelou, Marion Prudent, Christophe Salon, Anne Maillard and Alain Ourry performed the literature analysis and wrote the paper.

Conflicts of Interest: The authors declare no conflict of interest.

References

1. White, P.J.; Broadley, M.R. Biofortification of crops with seven mineral elements often lacking in human diets-iron, zinc, copper, calcium, magnesium, selenium and iodine. *New Phytol.* **2009**, *182*, 49–84. [CrossRef] [PubMed]

2. Amtmann, A.; Armengaud, P. Effects of N, P, K and S on metabolism: New knowledge gained from multi-level analysis. *Curr. Opin. Plant Biol.* **2009**, *12*, 275–283. [CrossRef] [PubMed]

3. Gojon, A.; Nacry, P.; Davidian, J.C. Root uptake regulation: A central process for NPS homeostasis in plants. *Curr. Opin. Plant Biol.* **2009**, *12*, 328–338. [CrossRef] [PubMed]

4. Hänsch, R.; Mendel, R.R. Physiological functions of mineral micronutrients (Cu, Zn, Mn, Fe, Ni, Mo, B, Cl). *Curr. Opin. Plant Biol.* **2009**, *12*, 259–266. [CrossRef] [PubMed]

5. Tejada-Jiménez, M.; Galván, A.; Fernández, E.; Llamas, A. Homeostasis of the micronutrients Ni, Mo and Cl with specific biochemical functions. *Curr. Opin. Plant Biol.* **2009**, *12*, 358–363. [CrossRef] [PubMed]

6. Philippot, L.; Raaijmakers, J.M.; Lemanceau, P.; van der Putten, W.H. Going back to the roots: The microbial ecology of the rhizosphere. *Nat. Rev. Microbiol.* **2013**, *11*, 789–799. [CrossRef] [PubMed]

7. Jin, T.; Zhou, J.; Chen, J.; Zhu, L.; Zhao, Y.; Huang, Y. The genetic architecture of zinc and iron content in maize grains as revealed by QTL mapping and meta-analysis. *Breed. Sci.* **2013**, *63*, 317–324. [CrossRef] [PubMed]

8. Rengel, Z.; Batten, G.D.; Crowler, D.E. Agronomic approaches for improving the micronutrient density in edible portions of field crops. *Field Crops Res.* **1999**, *60*, 27–40. [CrossRef]

9. Fan, M.S.; Zhao, F.J.; Fairweather-Tait, S.J.; Poulton, P.R.; Dunham, S.J.; McGrath, S.P. Evidence of decreasing mineral density in wheat grain over the last 160 years. *J. Trace Elem. Med. Biol.* **2008**, *22*, 315–324. [CrossRef] [PubMed]

10. Loladze, I. Hidden shift of the ionome of plants exposed to elevated CO_2 depletes minerals at the base of human nutrition. *eLife* **2014**, *3*, e02245. [CrossRef] [PubMed]

11. He, M.; Dijkstra, F.A. Drought effect on plant nitrogen and phosphorus: A meta-analysis. *New Phytol.* **2014**, *204*, 924–931. [CrossRef] [PubMed]

12. Seneviratne, S.I.; Luthi, D.; Litschi, M.; Schar, C. Land-atmosphere coupling and climate change in Europe. *Nature* **2006**, *443*, 205–209. [CrossRef] [PubMed]

13. López-Bucio, J.; Cruz-Ramírez, A.; Herrera-Estrella, L. The role of nutrient availability in regulating root architecture. *Curr. Opin. Plant Biol.* **2003**, *6*, 280–287. [CrossRef]

14. Giehl, R.F.H.; Gruber, B.D.; von Wirén, N. It is time to make changes: Modulation of root system architecture by nutrient signals. *J. Exp. Bot.* **2013**, *65*, 769–778. [CrossRef] [PubMed]

15. Cramer, M.D.; Hawkins, H.J.; Verboom, G.A. The importance of nutritional regulation of plant water flux. *Oecologia* **2009**, *161*, 15–24. [CrossRef] [PubMed]

16. Barber, S.A. *Soil Nutrient Bioavailability: A Mechanistic Approach*, 2nd ed.; Wiley: New York, NY, USA, 1995.

17. Avice, J.-C.; Etienne, P. Leaf senescence and nitrogen remobilization efficiency in oilseed rape (*Brassica Napus* L.). *J. Exp. Bot.* **2014**, *65*, 3813–3824. [CrossRef] [PubMed]

18. Thomas, H. Senescence, ageing and death of the whole plant. *New Phytol.* **2013**, *197*, 696–711. [CrossRef] [PubMed]

19. Pottier, M.; Masclaux Daubresse, C.; Yoshimoto, K.; Thomine, S. Autophagy as a possible mechanism for micronutrient remobilization from leaves to seeds. *Front. Plant Sci.* **2014**, *5*, 11. [CrossRef] [PubMed]

20. Schippers, J.H.M.; Schmidt, R.; Wagstaff, C.; Jing, H.C. Living to die and dying to live: The survival strategy behind leaf senescence. *Plant Physiol.* **2015**, *169*, 914–930. [CrossRef] [PubMed]

21. Masclaux-Daubresse, C.; Daniel-Vedele, F.; Dechorgnat, J.; Chardon, F.; Gaufichon, L.; Suzuki, A. Nitrogen uptake, assimilation and remobilization in plants: Challenges for sustainable and productive agriculture. *Ann. Bot.* **2010**, *105*, 1141–1157. [CrossRef] [PubMed]

22. Grusak, M.A.; Pearson, J.N.; Marentes, E. The physiology of micronutrient homeostasis in field crops. *Field Crops Res.* **1999**, *60*, 41–56. [CrossRef]

23. Waters, B.; Sankaran, R.P. Moving micronutrients from the soil to the seeds: Genes and physiological processes from a biofortification perspective. *Plant Sci.* **2011**, *180*, 562–574. [CrossRef] [PubMed]

24. Sperotto, R.A.; Ricachenevsky, F.K.; Williams, L.E.; Vasconcelos, M.W.; Menguer, P.K. From soil to seed: Micronutrient movement into and within the plant. *Front. Plant Sci.* **2014**, *5*, 438. [CrossRef] [PubMed]

25. Salon, C.; Avice, J.C.; Colombi, S.; Dieuaide-Noubhani, M.; Gallardo, K.; Jeudy, C.; Ourry, A.; Prudent, M.; Voisin, A.S.; Rolin, D. Fluxomics links cellular functional analyses to whole-plant phenotyping. *J. Exp. Bot.* **2017**, *68*, 2083–2098. [CrossRef] [PubMed]

26. Haslett, B. Zinc mobility in wheat: Uptake and distribution of zinc applied to leaves or roots. *Ann. Bot.* **2001**, *87*, 379–386. [CrossRef]

27. Impa, S.M.; Johnson-Beebout, S.E. Mitigating zinc deficiency and achieving high grain Zn in rice through integration of soil chemistry and plant physiology research. *Plant Soil* **2012**, *361*, 3–41. [CrossRef]

28. Riesen, O.; Feller, U. Redistribution of nickel, cobalt, manganese, zinc, and cadmium via the phloem in young and maturing wheat. *J. Plant Nutr.* **2005**, *28*, 421–430. [CrossRef]

29. Drossopoulos, J.B.; Bouranis, D.L.; Bairaktari, B.D. Patterns of mineral nutrient fluctuations in soybean leaves in relation to their position. *J. Plant Nutr.* **1995**, *117*, 1017–1035. [CrossRef]

30. Himelblau, E.; Amasino, R.M. Nutrients mobilized from leaves of *Arabidopsis thaliana* during leaf senescence. *J. Plant Physiol.* **2001**, *158*, 1317–1323. [CrossRef]

31. Sperotto, R.A.; Vasconcelos, M.W.; Grusak, M.A.; Fett, J.P. Effects of different Fe supplies on mineral partitioning and remobilization during the reproductive development of rice (*Oryza Sativa* L.). *Rice* **2012**, *5*, 27. [CrossRef] [PubMed]

32. Maillard, A.; Diquélou, S.; Billard, V.; Laîné, P.; Garnica, M.; Prudent, M.; Garcia-Mina, J.M.; Yvin, J.C.; Ourry, A. Leaf mineral nutrient remobilization during Leaf senescence and modulation by nutrient deficiency. *Front. Plant Sci.* **2015**, *6*, 317. [CrossRef] [PubMed]

33. Sankaran, R.P.; Grusak, M.A. Whole shoot mineral partitioning and accumulation in pea (*Pisum sativum*). *Front. Plant Sci.* **2014**, *5*, 149. [CrossRef] [PubMed]

34. Nable, R.; Loneragan, J. Translocation of manganese in subterranean clover (*Trifolium subterraneum* L. Cv. Seaton Park) I. redistribution during vegetative growth. *Funct. Plant Biol.* **1984**, *11*, 101–111.

35. Dayod, M.; Tyerman, S.D.; Leigh, R.A.; Gilliham, M. Calcium storage in plants and the implications for calcium biofortification. *Protoplasma* **2010**, *247*, 215–231. [CrossRef] [PubMed]

36. Mauk, C.S.; Noodén, L.D. Regulation of mineral redistribution in pod-bearing soybean explants. *J. Exp. Bot.* **1992**, *43*, 1429–1440. [CrossRef]

37. Hocking, P.J. Dry-matter production, mineral nutrient concentrations, and nutrient distribution and redistribution in irrigated spring wheat. *J. Plant Nutr.* **1994**, *17*, 1289–1308. [CrossRef]

38. Gombert, J.; Etienne, P.; Ourry, A.; Le Dily, F. The expression patterns of SAG12/Cab genes reveal the spatial and temporal progression of leaf senescence in *Brassica napus* L. with sensitivity to the environment. *J. Exp. Bot.* **2006**, *57*, 1949–1956. [CrossRef] [PubMed]

39. Etienne, P.; Desclos, M.; Le Gou, L.; Gombert, J.; Bonnefoy, J.; Maurel, K.; Le Dily, F.; Ourry, A.; Avice, J.-C. N-protein mobilisation associated with the leaf senescence process in oilseed rape is concomitant with the disappearance of trypsin inhibitor activity. *Funct. Plant Biol.* **2007**, *34*, 895–906. [CrossRef]

40. Desclos, M.; Etienne, P.; Coquet, L.; Jouenne, T.; Bonnefoy, J.; Segura, R.; Reze, S.; Ourry, A.; Avice, J.C. A combined ^{15}N tracing/proteomics study in *Brassica napus* reveals the chronology of proteomics events associated with N remobilisation during leaf senescence induced by nitrate limitation or starvation. *Proteom* **2009**, *9*, 3580–3608. [CrossRef] [PubMed]

41. Dubousset, L.; Abdallah, M.; Desfeux, A.; Etienne, P.; Meuriot, F.; Hawkesford, M.J.; Gombert, J.; Segura, R.; Bataille, M.-P.; Reze, S.; et al. Remobilization of leaf S compounds and senescence in response to restricted sulphate supply during the vegetative stage of oilseed rape are affected by mineral N availability. *J. Exp. Bot.* **2009**, *60*, 3239–3253. [CrossRef] [PubMed]

42. Garnett, T.P.; Graham, R.D. Distribution and remobilization of iron and copper in wheat. *Ann. Bot.* **2005**, *95*, 817–826. [CrossRef] [PubMed]

43. Jeong, K.; Baten, A.; Waters, D.L.E.; Pantoja, O.; Julia, C.C.; Wissuwa, M.; Heuer, S.; Kretzschmar, T.; Rose, T.J. Phosphorus remobilization from rice flag leaves during grain filling: An RNA-seq study. *Plant Biotechnol. J.* **2016**, *15*, 15–26. [CrossRef] [PubMed]

44. Blevins, D.G.; Lukaszewski, K.M. Boron in plant structure and function. *Annu. Rev. Plant Physiol. Plant Mol. Biol.* **1998**, *49*, 481–500. [CrossRef] [PubMed]

45. Yusuf, M.; Fariduddin, Q.; Hayat, S.; Ahmad, A. Nickel: An overview of uptake, essentiality and toxicity in plants. *Bull. Environ. Contam. Toxicol.* **2011**, *86*, 1–17. [CrossRef] [PubMed]

46. Peng, J.S.; Gong, J.M. Vacuolar sequestration capacity and long-distance metal transport in plants. *Front. Plant Sci.* **2014**. [CrossRef] [PubMed]

47. Curie, C.; Cassin, G.; Couch, D.; Divol, F.; Higuchi, K.; Jean, M.; Misson, J.; Schikora, A.; Czernic, P.; Mari, S. Metal movement within the plant: Contribution of nicotianamine and yellow stripe 1-like transporters. *Ann. Bot.* **2009**, *103*, 1–11. [CrossRef] [PubMed]

48. Ricachenevsky, F.K.; Koprovski Menguer, P.; Sperotto, R.A. kNACking on heaven's door: How important are NAC transcription factors for leaf senescence and Fe/Zn remobilization to seeds? *Front. Plant Sci.* **2013**, *4*, 226. [CrossRef] [PubMed]

49. Hill, J. The remobilization of nutrients from leaves. *J. Plant Nutr.* **1980**, *2*, 407–444. [CrossRef]

50. Sarda, X.; Diquelou, S.; Abdallah, M.; Nesi, N.; Cantat, O.; Le Gouee, P.; Avice, J.C.; Ourry, A. Assessment of sulphur deficiency in commercial oilseed rape crops from plant analysis. *J. Agri. Sci.* **2013**, 1–18. [CrossRef]

51. Abdallah, M.; Dubousset, L.; Meuriot, F.; Etienne, P.; Avice, J.C.; Ourry, A. Effect of mineral sulphur availability on nitrogen and sulphur uptake and remobilization during the vegetative growth of *Brassica napus* L. *J. Exp. Bot.* **2010**, *61*, 2635–2646. [CrossRef] [PubMed]

52. Han, Y.L.; Song, H.X.; Liao, Q.; Yu, Y.; Jian, S.F.; Lepo, J.E.; Liu, Q.; Rong, X.M.; Tian, C.; Zeng, J.; et al. Nitrogen use efficiency is mediated by vacuolar nitrate sequestration capacity in roots of *Brassica napus*. *Plant Physiol.* **2016**, *170*, 1684–1698. [CrossRef] [PubMed]

53. Gregersen, P.L.; Holm, P.B.; Krupinska, K. Leaf senescence and nutrient remobilisation in barley and wheat. *Plant Biol.* **2008**, *10*, 37–49. [CrossRef] [PubMed]

54. Masclaux-Daubresse, C.; Reisdorf-Cren, M.; Orsel, M. Leaf nitrogen remobilisation for plant development and grain filling. *Plant Biol.* **2008**, *10*, 23–36. [CrossRef] [PubMed]

55. Schiltz, S.; Munier-Jolain, N.; Jeudy, C.; Burstin, J.; Salon, C. Dynamics of exogenous nitrogen partitioning and nitrogen remobilisation from vegetative organs in pea (*Pisum sativum* L.) revealed by ^{15}N in vivo labelling throughout the seed filling. *Plant Physiol.* **2005**, *137*, 1463–1473. [CrossRef] [PubMed]

56. Michaeli, S.; Galili, G.; Genschik, P.; Fernie, A.R.; Avin-Wittenberg, T. Autophagy in plants—What's new on the menu? *Trends Plant Sci.* **2016**, *21*, 134–144. [CrossRef] [PubMed]

57. Ono, Y.; Wada, S.; Izumi, M.; Makino, A.; Ishida, H. Evidence for contribution of autophagy to Rubisco degradation during leaf senescence in *Arabidopsis thaliana*. *Plant Cell Environ.* **2013**, *36*, 1147–1159. [CrossRef] [PubMed]

58. Distelfeld, A.; Cakmak, I.; Peleg, Z.; Ozturk, L.; Yazici, A.M.; Budak, H.; Saranga, Y.; Fahima, T. Multiple QTL-effects of wheat Gpc-B1 locus on grain protein and micronutrient concentrations. *Physiol. Plant.* **2007**, *129*, 635–643. [CrossRef]

59. Uauay, C.; Brevis, J.C.; Dubcovsky, J. The high grain protein content gene Gpc-B1 accelerates senescence and has pleiotropic effects on protein content in wheat. *J. Exp. Bot.* **2006**, *57*, 2785–2794. [CrossRef] [PubMed]

60. Pearce, S.; Tabbita, F.; Cantu, D.; Buffalo, V.; Avni, R.; Vazquez-Gross, H.; Zhao, R.; Conley, C.J.; Distelfeld, A.; Dubcovksy, J. Regulation of Zn and Fe transporters by the GPC1 gene during early wheat monocarpic senescence. *BMC Plant Biol.* **2014**, *14*, 368. [CrossRef] [PubMed]

61. Puppo, A.; Groten, K.; Bastian, F.; Carzaniga, R.; Soussi, M.; Lucas, M.M.; de Felipe, M.R.; Harrison, J.; Vanacker, H.; Foyer, C.H. Legume nodule senescence: Roles for redox and hormone signalling in the orchestration of the natural aging process. *New Phytol.* **2005**, *165*, 683–701. [CrossRef] [PubMed]

62. Van de Velde, W.; Guerra, J.C.P.; Keyser, A.D.; De Rycke, R.; Rombauts, S.; Maunoury, N.; Mergaert, P.; Kondorosi, E.; Holsters, M.; Goormachtig, S. Aging in legume symbiosis. A molecular view on nodule senescence in *Medicago truncatula*. *Plant Physiol.* **2006**, *141*, 711–720. [CrossRef] [PubMed]

63. Burton, J.W.; Harlow, C.; Theil, E.C. Evidence for reutilization of nodule iron in soybean seed development. *J. Plant Nutr.* **1998**, *21*, 913–927. [CrossRef]

64. Hakoyama, T.; Watanabe, H.; Tomita, J.; Yamamoto, A.; Sato, S.; Mori, Y.; Kouchi, H.; Suganuma, N. Nicotianamine synthase specifically expressed in root nodules of Lotus japonicus. *Planta* **2009**, *230*, 309–317. [CrossRef] [PubMed]

65. González-Guerrero, M.; Matthiadis, A.; Sáez, Á.; Long, T.A. Fixating on metals: New insights into the role of metals in nodulation and symbiotic nitrogen fixation. *Front. Plant Sci.* **2014**, *5*, 45. [CrossRef] [PubMed]

66. Hu, Y.; Schmidhalter, U. Drought and salinity: A comparison of their effects on mineral nutrition of plants. *J. Plant Nutr. Soil Sci.* **2005**, *168*, 541–549. [CrossRef]

67. Ghandilyan, A.; Barboza, L.; Tisné, S.; Granier, C.; Reymond, M.; Koornneef, M.; Schat, H.; Aarts, M.G.M. Genetic analysis identifies quantitative trait loci controlling rosette mineral concentrations in Arabidopsis thaliana under drought. *New Phytol.* **2009**, *184*, 180–192. [CrossRef] [PubMed]

68. Cakmak, I. The role of potassium in alleviating detrimental effects of abiotic stresses in plants. *J. Plant Nutr. Soil Sci.* **2005**, *168*, 521–530. [CrossRef]

69. Gallardo, K.; Courty, P.-E.; Le Signor, C.; Wipf, D.; Vernoud, V. Sulfate transporters in the plant's response to drought and salinity: Regulation and possible functions. *Front. Plant Sci.* **2014**, *5*, 580. [CrossRef] [PubMed]

70. Ali, M.; Jensen, C.R.; Mogensen, V.O.; Andersen, M.N.; Henson, I.E. Root signalling and osmotic adjustment during the intermittent soil drying sustain grain yields of field grown wheat. *Field Crops Res.* **1999**, *62*, 35–52. [CrossRef]

71. Yang, J.C.; Zhang, J.H.; Wang, Z.Q.; Zhu, Q.S.; Liu, L.J. Involvement of abscisic acid and cytokinins in the senescence and remobilization of carbon reserves in wheat subjected to water stress during grain filling. *Plant Cell Environ.* **2003**, *26*, 1621–1631. [CrossRef]

72. Farooq, M.; Wahid, A.; Kobayashi, N.; Fujita, D.; Basra, S.M.A. Plant drought stress: Effects, mechanisms and management. *Agron. Sustain. Dev.* **2009**, *29*, 185–212. [CrossRef]

73. Simova-Stoilova, L.; Vassileva, V.; Feller, U. Selection and breeding of suitable crop genotypes for drought and heat periods in a changing climate: Which morphological and physiological properties should be considered? *Agriculture* **2016**, *6*, 26. [CrossRef]

74. Wang, Y.; Frei, M. The impact of abiotic environmental stresses on crop quality. *Agric. Ecosyst. Environ.* **2011**, *141*, 271–286. [CrossRef]

75. Lawlor, D.W. Genetic engineering to improve plant performance under drought: Physiological evaluation of achievements, limitations, and possibilities. *J. Exp. Bot.* **2013**, *64*, 83–108. [CrossRef] [PubMed]

76. Lawlor, D.W.; Day, W.; Johnston, A.E.; Legg, B.J.; Parkinson, K.J. Growth of spring barley under drought: Crop development, photosynthesis, dry-matter accumulation and nutrient content. *J. Agric. Sci.* **1981**, *96*, 167–186. [CrossRef]

77. Maleki Farahani, S.M.; Chaichi, M.R.; Mazaheri1, D.; Tavakkol Afshari, R.; Savaghebi, G.H. Barley grain mineral analysis as affected by different fertilizing systems and by drought stress. *J. Agric. Sci. Technol.* **2011**, *13*, 315–326.

78. Guzmán, C.; Autrique, J.E.; Mondal, S.; Singh, R.P.; Govindan, V.; Morales-Dorantes, A.; Posadas-Romano, G.; Crossa, J.; Ammar, K.; Peña, R.J. Response to drought and heat stress on wheat quality, with special emphasis on bread-making quality, in durum wheat. *Field Crops Res.* **2016**, *186*, 157–165. [CrossRef]

79. Haberle, J.; Svoboda, P.; Raimanová, I. The effect of post-anthesis water supply on grain nitrogen concentration and grain nitrogen yield of winter wheat. *Plant Soil Environ.* **2008**, *7*, 304–312.

80. Rose, T.J.; Raymond, C.A.; Bloomfield, C.; King, G.J. Perturbation of nutrient source-sink relationships by post-anthesis stresses results in differential accumulation of nutrients in wheat grain. *J. Plant Nutr. Soil Sci.* **2015**, *178*, 89–98. [CrossRef]

81. Zhao, C.-X.; He, M.-R.; Wang, Z.-L.; Wang, Y.-F.; Lin, Q. Effects of different water availability at post-anthesis stage on grain nutrition and quality in strong-gluten winter wheat. *C. R. Biol.* **2009**, *332*, 759–764. [CrossRef] [PubMed]

82. Ti, D.G.; Fang, G.S.; San'an, N.; Ning, B.S.; He'ai, X.; Cheng, L.T. Differential responses of yield and selected nutritional compositions to drought stress in summer maize grains. *J. Plant Nutr.* **2010**, *33*, 1811–1818.

83. Oktem, A. Effect of water shortage on yield, and protein and mineral compositions of drip-irrigated sweet corn in sustainable agricultural systems. *Agric. Water Manag.* **2008**, *95*, 1003–1010. [CrossRef]

84. Bellaloui, N.; Gillen, A.M.; Mengistu, A.; Kebede, H.; Fisher, D.K.; Smith, J.R.; Reddy, K.N. Responses of nitrogen metabolism and seed nutrition to drought stress in soybean genotypes differing in slow-wilting phenotype. *Front. Plant Sci.* **2013**, *4*, 498. [CrossRef] [PubMed]

85. Samarah, N.; Mullen, R.; Cianzio, S. Size distribution and mineral nutrients of soybean seeds in response to drought stress. *J. Plant Nutr.* **2004**, *27*, 815–835. [CrossRef]

86. Kresović, B.; Gajic, B.A.; Tapanarova, A.; Dugalić, G. Yield and chemical composition of soybean seed under different irrigation regimes in the Vojvodina region. *Plant Soil Environ.* **2017**, *63*, 34–39.

87. Nam, K.H.; Kim, D.Y.; Shin, H.J.; Nam, K.J.; An, J.H.; Pack, I.S.; Park, J.H.; Jeong, S.C.; Kim, H.B.; Kim, C.G. Drought stress-induced compositional changes in tolerant transgenic rice and its wild type. *Food Chem.* **2014**, *153*, 145–150. [CrossRef] [PubMed]

88. Pereira, H.S.; Del Peloso, M.J.; Bassinello, P.Z.; Guimarães, C.M.; Melo, L.C.; Faria, L.C. Genetic variability for iron and zinc content in common bean lines in interaction with water availability. *Gen. Mol. Res.* **2014**, *13*, 6773–6785. [CrossRef] [PubMed]

89. Jacomini, E.; Bertani, A.; Mapelli, S. Accumulation of polyethylene glycol 6000 and its effects on water content and carbohydrate level in water-stressed tomato plants. *Can. J. Bot.* **1988**, *66*, 970–973. [CrossRef]

90. Verslues, P.E.; Ober, E.S.; Sharp, R.E. Root growth and oxygen relations at low water potentials. Impact of oxygen availability in polyethylene glycol solutions. *Plant Physiol.* **1998**, *116*, 1403–1412. [CrossRef] [PubMed]

91. Maillard, A.; Etienne, P.; Diquélou, S.; Trouverie, J.; Billard, V.; Yvin, J.C.; Ourry, A. Nutrient deficiencies modify the ionomic composition of plant tissues: A focus on cross-talk between molybdenum and other nutrients in *Brassica napus*. *J. Exp. Bot.* **2016**, *67*, 5631–5641. [CrossRef] [PubMed]

92. Acosta-Gamboa, L.M.; Liu, S.; Langley, E.; Campbell, Z.; Castro-Guerrero, N.; Mendoza-Cozatl, D.; Lorence, A. Moderate to severe water limitation differentially affects the phenome and ionome of *Arabidopsis*. *Funct. Plant Biol.* **2017**, *44*, 94–106. [CrossRef]

93. Baxter, I.R.; Vitek, O.; Lahner, B.; Muthukumar, B.; Borghi, M.; Morrissey, J.; Guerinot, M.L.; Salt, D.E. The leaf ionome as a multivariable system to detect a plant's physiological status. *Proc. Nat. Acad. Sci. USA* **2008**, *105*, 12081–12086. [CrossRef] [PubMed]

94. Sankaran, R.P.; Huguet, T.; Grusak, M. A. Identification of QTL affecting seed mineral concentrations and content in the model legume *Medicago truncatula*. *Theor. Appl. Gen.* **2009**, *119*, 241–253. [CrossRef] [PubMed]

95. Sade, N.; Rubio-Wilhelmi, M.M.; Umnajkitikorn, K.; Blumwald, E. Stress-induced senescence and plant tolerance to abiotic stress. *J. Exp. Bot.* **2017**. [CrossRef] [PubMed]

96. Sade, N.; Umnajkitikorn, K.; del Mar Rubio Wilhelmi, M.; Wright, M.; Wang, S.; Blumwald, E. Delaying chloroplast turnover increases water-deficit stress tolerance through the enhancement of nitrogen assimilation in rice. *J. Exp. Bot.* **2017**. [CrossRef] [PubMed]

97. Peleg, Z.; Blumwald, E. Hormone balance and abiotic stress tolerance in crop plants. *Curr. Opin. Plant Biol.* **2011**, *14*, 290–295. [CrossRef] [PubMed]

98. Havé, M.; Marmagne, A.; Chardon, F.; Masclaux-Daubresse, C. Nitrogen remobilization during leaf senescence: Lessons from *Arabidopsis* to crops. *J. Exp. Bot.* **2017**, *68*, 2513–2529. [PubMed]

99. Mayer, J.E.; Pfeiffer, W.H.; Beyer, P. Biofortified crops to alleviate micronutrient malnutrition. *Curr. Opin. Plant Biol.* **2008**, *11*, 166–170. [CrossRef] [PubMed]

100. Bagayoko, M.; George, E.; Romheld, V.; Buerkert, A.B. Effects of mycorrhizae and phosphorus on growth and nutrient uptake of millet, cowpea and sorghum on a West African soil. *J. Agric. Sci.* **2000**, *135*, 399–407. [CrossRef]

agriculture

MDPI

Article

Interactive Effects of N-, P- and K-Nutrition and Drought Stress on the Development of Maize Seedlings

Christoph Studer [1,2], Yuncai Hu [1] and Urs Schmidhalter [1,*]

[1]   Department of Plant Science, Technical University of Munich, Freising D-85354, Germany;
      christoph.studer@bfh.ch (C.S.); hu@wzw.tum.de (Y.H.)
[2]   Swiss College of Agriculture, Zollikofen CH-3052, Switzerland
*   Correspondence: schmidhalter@wzw.tum.de; Tel.: +49-8161-713390

Received: 8 August 2017; Accepted: 20 October 2017; Published: 28 October 2017

Abstract: Global climate change is likely to increase the risk of frequent drought. Maize, as the principal global cereal, is particularly impacted by drought. Nutrient supply may improve plant drought tolerance for better plant establishment during seedling growth stages. Thus, this study investigated the interactive effects of drought and the application of the nutrients N, P and K either individually or in combination. The maize seedlings were harvested between 12 and 20 days after sowing, and the leaf area, shoot fresh and dry weight and root dry weight were determined, and shoot water content and root/shoot dry weight ratio were calculated. Among the N, P and K fertilization treatments applied individually or in combination, the results showed that there was generally a positive effect of combined NPK and/or NP nutrient supply on shoot growth such as leaf area, shoot fresh and dry weight at day 20 after sowing under both well-watered and drought conditions compared with no nutrient supply. Compared with the effect of N and P nutrient supply, it seems that K was not limiting to plant growth due to the mineralogical characteristics of the illitic-chloritic silt loam used, which provided sufficient K, even though soil tests showed a low K nutrient status. Interestingly, the root/shoot ratio remained high and constant under drought regardless of NPK application, while it decreased with NPK applications in the well-watered treatment. This suggests that the higher root/shoot ratios with N, NP, PK and NPK under drought could be exploited as a strategy for stress tolerance in crop plants.

Keywords: drought stress; maize; nitrogen; phosphorus; potassium; root growth; shoot growth

1. Introduction

Drought is a primary constraint to global crop production, and global climate change is likely to increase the risk of frequent drought, especially in rain-fed agriculture [1,2]. Maize, as the principal global cereal, is particularly impacted by frequent drought spells. Under drought stress, reduced nutrient availability is one of the most important factors limiting plant growth [3]. Drought reduces nutrient uptake by the roots, in part, because the decline in soil moisture results in a decreased rate of diffusion of nutrients from the soil matrix to the absorbing root surface [3–5]. Moreover, nutrient transport from the roots to the shoots is also restricted by reduced transpiration rates and impaired active transport and membrane permeability, together resulting in a reduced root absorbing power by crop plants [6,7]. Thus, nutrient supply strongly affects crop productivity under drought conditions, but this is also very complex [8–10]. The positive effects of N and P on plant growth under drought conditions are attributed to an increase in water-use efficiency, stomatal conductance [11–14], photosynthesis and ATPase activity [5,15], as well as higher cell membrane stability and improved osmotic adjustment [16,17]. However, K increases a plant's drought tolerance through its functions in stomatal regulation, energy status, charge balance, protein synthesis and homeostasis [18–20]. Many physiological mechanisms are involved in the improvement of plant growth through nutrient supply. However, the different studies on the benefit of N, P and K fertilization on plant growth under drought stress are controversial [14,21,22]. Thus, it is necessary to compare the effect of a single nutrient supply, such as N, P and K, and/or combined nutrients on plant growth within the same experiment.

Maize growth is most sensitive to drought during the early growth stages [23–30]. At seedling growth stages, maize growth is characterized by leaf initiation and elongation and by changes in relative root growth maintenance and root architecture [31,32]. Such growth processes react sensitively to drought [31,33–36]. In addition, the seedling growth stage of maize is highly relevant for P deficiency, since most P-deficiencies in maize are often observed in western Europe at an early growth stage, which has been shown in long-term experimentation [37]. Therefore, it is important to understand how nutrient supply can maintain shoot and root growth, especially during the early growth of plants under drought stress.

The objective of this study was to investigate the interactive effects of different N-, P- and K-nutrition supplementations singly and in combination, as well as the effect of drought stress on shoot and root development of young maize plants, and to understand the different sensitivities of root and shoot growth to drought stress during the early growth stages under different nutrient supplies.

2. Materials and Methods

2.1. Plant Materials and Growth Conditions

Maize seeds (*Zea mays* L. cv Issa), pre-germinated for one day in distilled water, were sown in a pot (10.5 cm in diameter and 20 cm tall), containing 1.5 L of silty soil. The wetted soil was mixed thoroughly and allowed to equilibrate for more than two days. Thereafter, the soil was sieved and filled into the pots. To minimize water loss through evaporation, a 2-cm layer of quartz sand (ϕ = 2 mm) covered the soil surface. Four days after sowing, the seedlings were thinned from four to three plants per pot. There were 3 pots per treatment, i.e., all treatments were replicated three times. The properties of the soil are shown in Table 1 [38]. The nitrate, P, K and Mg contents of the soil were determined in NH4-acetate-EDTA extracts before the experiment started (Table 2). For the nutrient treatments, modified Hoagland's nutrient solution was added to the dry soil to obtain the following: (a) a gravimetric soil water content of 27% and (b) eight nutrient treatments, i.e., no nutrients (control), single nutrient treatments consisting of N, P and K and combined nutrient treatments consisting of NP, NK, PK and NPK. The soil nutrient status in the different treatments is presented in Table 2. The experiment was conducted in a growth chamber with a 12 h photoperiod. The photosynthetic photon flux density (PPFD) was approximately 450 μmol m^{-2} s^{-1}. The air temperature was 20/18 °C day/night and the relative humidity was maintained at 50–65%.

The soil moisture content in all treatments was maintained at the initial content by adding tap water until day 10 after sowing. Then, the soil water content for the well-watered treatments (half of the pots) was continually maintained at the initial water content by adding tap water daily. The drought treatments for the other half of pots were obtained by drying the soil out without further watering.

During the experiment, pots were weighed daily before watering. The bulk soil water content was determined from gravimetric measurements of the pots (plant weight was estimated and considered in the calculations). At each harvest, the soil water content was determined gravimetrically using soil samples (mixed samples from the whole soil volume). Soil matric potential was calculated using a soil retention curve that was previously established (data not shown) and is presented in Table 3.

Table 1. Soil characteristics (CEC: cation exchange capacity).

Clay (%)	Silt (%)	Sand (%)	Organic Matter (%)	pH	CEC (mmol kg^{-1})	Ca (g kg^{-1})
9.1	59.5	31.4	0.85	8.2	48	3.07

Table 2. Composition of the different levels of the NPK nutrients based on an EDTA extraction for the well-watered and drought treatments.

Nutrient Treatments	Composition of Nutrients (mg kg^{-1} soil)			
	NO$_3$$^-$	P	K	Mg
NPK	295	98	70	1045
NP	288	106	30	965
NK	312	66	82	937
PK	41	110	82	924
N	255	69	31	948
P	48	101	30	687
K	52	74	88	677
0	51	69	30	1022

Table 3. Bulk soil matric potentials in pots of the different nutrient treatments on different harvest days.

Nutrient Treatments	Soil Matric Potential (MPa)					
	Well-Watered	Drought				
		Days after Sowing				
		12	14	16	18	20
NPK		−0.055	−0.072	−0.134	−0.207	−0.637
NP		−0.058	−0.08	−0.126	−0.138	−0.503
NK		−0.056	−0.078	−0.095	−0.116	−0.162
PK	−0.04–0.07	−0.052	−0.071	−0.084	−0.097	−0.135
N		−0.062	−0.077	−0.112	−0.135	−0.278
PK		−0.051	−0.079	−0.099	−0.129	−0.189
K		−0.054	−0.066	−0.088	−0.094	−0.127
0		−0.055	−0.074	−0.091	−0.113	−0.166

2.2. Determination of the Leaf Area, the Shoot and Root Biomass and the Shoot Water Content

The maize plants were harvested on days 12, 14, 16, 18 and 20 after sowing. At each harvest date, the leaf area was measured with a leaf area meter (accuracy: ±2%) (LI-300A, Bioscienses, Lincoln, Nebraska, USA). The shoot fresh weight (FW) was weighed using a balance (accuracy: 0.01 g) (Sartorius AG, Goettingen, Germany), and then the shoots were dried in an oven for 24 h for the determination of the dry weight (DW). The roots were washed on a sieve and were subsequently dried in an oven for 24 h for the determination of the DW. The root/shoot ratios, based on the shoot and root DW, were calculated. The shoot water content was calculated from the shoot FW and DW using the equation: WC (%) = (FW − DW)/FW.

2.3. Statistical Analysis of the Data

A randomized complete design was used. The data were analysed by an analysis of variance (ANOVA) using SAS (SAS, Institute Inc., Cary, NC, USA) to test the significance of the main effects. Duncan's test was applied for the post hoc multiple comparisons within the well-watered or drought treatment. The terms were considered to be significant at nominal $p < 0.05$.

3. Results

3.1. Analysis of Variance (ANOVA)

The results from the analysis of variance (Tables 4–6) show that the single and combined nutrient treatments were significant as main effect in the ANOVAs, while the water supply was significant as a main effect for leaf area, whole plant DW, shoot water content, root DW and root/shoot ratio at day

20 after sowing and for shoot FW at days 18 and 20 after sowing. An interaction between nutrient treatments and soil water conditions was found for the shoot water content, root/shoot ratio ($p < 0.001$) and shoot FW ($p < 0.01$) at day 20 after sowing (Table 4).

Table 4. Analysis of variance (ANOVA) results for leaf area, whole plant dry weight (DW), shoot water content (WC), root dry weight (DW) and root/shoot DW ratios at day 20 after sowing.

Source of Variation	df	Significance of F Ratio				
		Leaf Area	Whole Plant DW	Shoot WC	Root DW	Root/Shoot Ratio
Nutrient treatments (N)	7	***	***	***	***	***
Water supply (W)	1	***	ns	***	*	***
N × W	7	ns	ns	***	ns	***

* Significant at the 0.05 level; *** Significant at the 0.001 level; ns—not significant. *df*: the degrees of freedom in the source

Table 5. Analysis of variance (ANOVA) results for shoot fresh weight (FW) at days 12, 14, 16, 18 and 20 after sowing.

Source of Variation	df	Significance of F Ratio				
		Days after Sowing				
		12	14	16	18	20
Nutrient treatments (N)	7	***	*	***	***	***
Water supply (W)	1	ns	ns	ns	**	***
N × W	7	ns	ns	ns	ns	**

* Significant at the 0.05 level; ** Significant at the 0.01 level; *** Significant at the 0.001 level; ns—not significant.

Table 6. Analysis of variance (ANOVA) results for shoot dry weight (DW) at days of 12, 14, 16, 18 and 20.

Source of Variation	df	Significance of F Ratio				
		Days after Sowing				
		12	14	16	18	20
Nutrient treatments (N)	7	***	ns	***	***	***
Water supply (W)	1	ns	ns	ns	ns	ns
N × W	7	ns	ns	ns	ns	ns

*** Significant at the 0.001 level; ns—not significant.

3.2. Interactive Effects of NPK Nutrients and Drought on the Total Plant Dry Weight and Leaf Area of Maize Seedlings

Compared to the control treatment (without nutrient supply), the leaf area of the maize seedlings under both drought and well-watered conditions was significantly increased with the supplementation of NP and NPK (Figure 1). However, nitrogen supply only significantly enhanced leaf area under well-watered conditions. The results in Figure 1 also show a significant difference between the drought and well-watered conditions with single N and combined NP and NPK fertilization at day 20 after sowing.

Figure 2 presents the interactive effects of drought and nutrients on the total plant dry weight (DW) of the maize seedlings (shoot DW + root DW) at day 20 after sowing. The NPK supply significantly enhanced the total dry weight under well-watered conditions, and the NP and NPK significantly increased the total dry weight under drought compared with the no nutrient supply condition. In contrast to leaf area, there was no significant difference in the total plant DW between well-watered and drought conditions regardless of nutrient treatment.

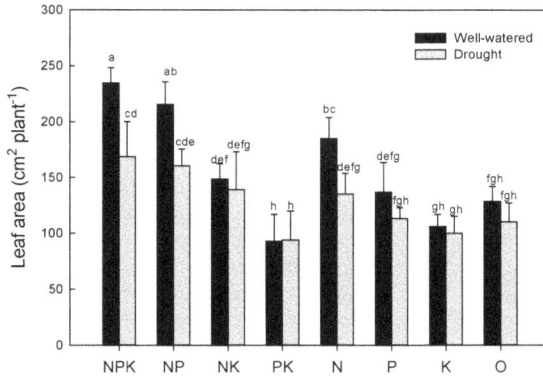

Figure 1. Interactive effects of drought and the N, P and K nutrients applied individually or in combination, or with no nutrient application (0) on the leaf area of the maize seedlings at day 20 after sowing. The error bars represent the standard errors of the mean (n = 3) and fit within the plot symbol if not visible. Means accompanied by different letters are significantly different at $p < 0.05$.

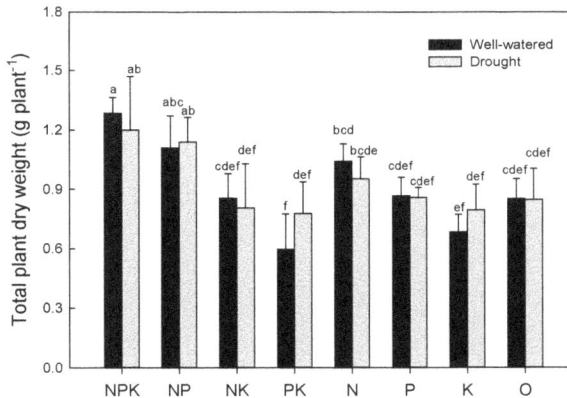

Figure 2. Interactive effects of drought and the N, P and K nutrients applied individually or in combination, or with no nutrient application (0) on the total dry weight (shoots + roots) of the maize seedlings at day 20 after sowing. The error bars represent the standard errors of the mean (n = 3) and fit within the plot symbol if not visible. Means accompanied by different letters are significantly different at $p < 0.05$.

3.3. Interactive Effects of NPK Nutrients and Drought on the Fresh and Dry Weights of the Shoots and the Water Content of the Maize Seedlings

Figures 3 and 4 reveal different effects of the nutrient supply on the development of the shoot FW and DW of the maize seedlings under both the drought and well-watered conditions. At day 12 after sowing, the results in Figure 3 show that NPK fertilization already significantly increased shoot FW compared with no nutrient supply. At day 20 after sowing, the shoot FW was significantly enhanced by N, NP and NPK in the well-watered treatments and NP and NPK in the drought treatment compared with no nutrient supply. A significant difference between well-watered and drought conditions was found for the nutrient supply with NP and NPK at day 18 after sowing and N, NP and NPK at day 20 after sowing.

Although N, NP and NPK fertilization significantly increased shoot DW at day 12 after sowing for the well-watered treatment and N and NP nutrient supply for the drought treatment compared with no nutrient supply (Figure 4), the effect of nutrient supply individually and in combination on the shoot DW under both well-watered and drought conditions was not consistent between day 12 and day 20 after sowing. At day 20, the NP and NPK fertilization significantly increased shoot DW for both well-watered and drought conditions compared with no nutrient supply. In contrast to shoot FW, there was no significant difference in shoot DW between well-watered and drought treatments regardless of harvest time (Figure 4).

Nutrient supply increased shoot water content compared with no nutrient supply treatment, especially under well-watered conditions (Figure 5). For example, except for P and PK, N, K, NP, NK, NPK fertilization significantly increased the shoot water content compared with no nutrient supply under well-watered conditions, whereas, under drought stress, a significant increase in shoot water content was found for K, NK and NP treatments. The shoot water content was significantly reduced by drought stress for all nutrient treatments (Figure 5).

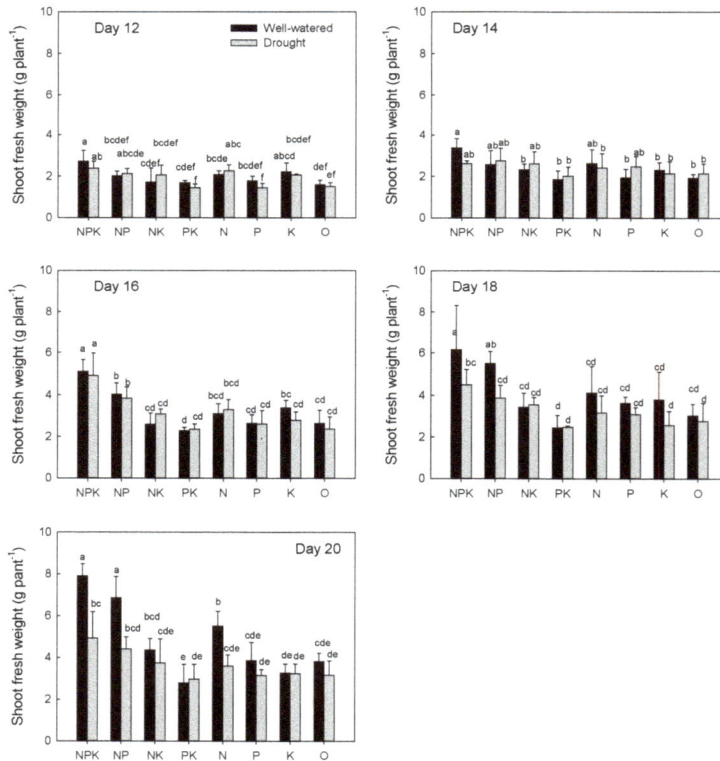

Figure 3. Interactive effects of drought and the N, P and K nutrients applied individually or in combination, or with no nutrient application (0) on the shoot fresh weight of the maize seedlings at days 12, 14, 16, 18 and 20 after sowing. The error bars represent the standard errors of the mean ($n = 3$) and fit within the plot symbol if not visible. Means accompanied by different letters are significantly different at $p < 0.05$.

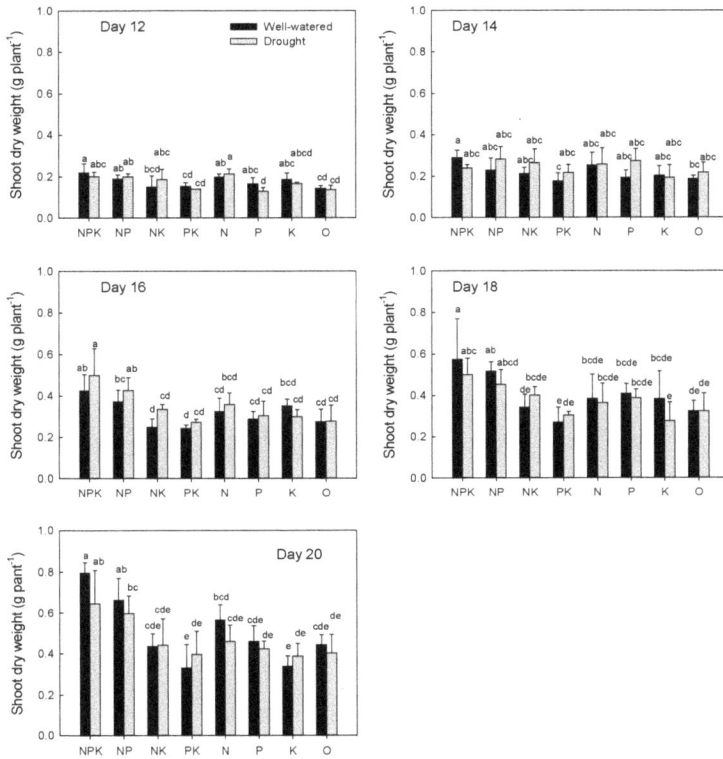

Figure 4. Interactive effects of drought and the N, P and K nutrients applied individually or in combination, or with no nutrient application (0) on the shoot dry weight of the maize seedlings at days 12, 14, 16, 18 and 20 after sowing. The error bars represent the standard errors of the mean ($n = 3$) and fit within the plot symbol if not visible. Means accompanied by different letters are significantly different at $p < 0.05$.

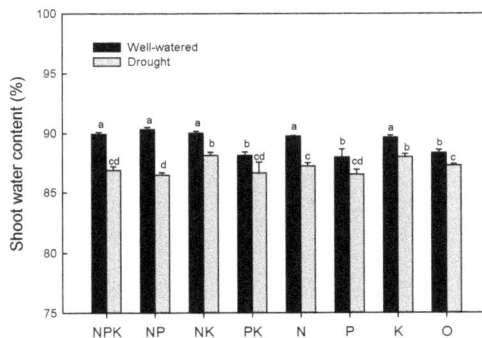

Figure 5. Interactive effects of drought and the N, P and K nutrients applied individually or in combination, or with no nutrient application (0) on the shoot water content of the maize seedlings at day 20 after sowing. The error bars represent the standard errors of the mean ($n = 3$) and fit within the plot symbol if not visible. Means accompanied by different letters are significantly different at $p < 0.05$.

3.4. Interactive Effects of NPK Nutrients and Drought on the Root Dry Weight and Root/Shoot Dry Weight Ratios of the Maize Seedlings

Compared with no nutrient supply treatments, there was no significant change in root DW with N, P individually and NP, NK and NPK in combination (Figure 6). Similar to the effect of drought on the shoot DW, there was no significant effect of drought on the root DW at day 20 after sowing for all other nutrient treatments except NK supply.

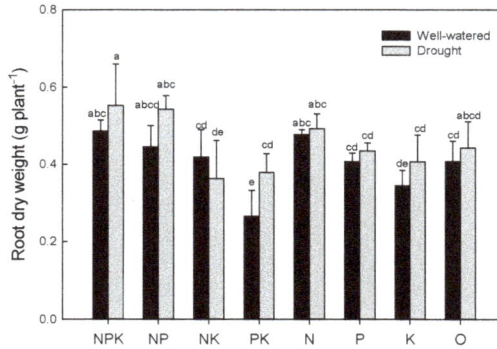

Figure 6. Interactive effects of drought and the N, P and K nutrients applied individually or in combination, or with no nutrient application (0) on the root dry weight of the maize seedlings at day 20 after sowing. The error bars represent the standard errors of the mean (n= 3) and fit within the plot symbol if not visible. Means accompanied by different letters are significantly different at $p < 0.05$.

Compared with no nutrient supply treatments, the root/shoot DW ratio tended to be reduced by nutrient supply under well-watered conditions and to remain unchanged under drought conditions. However, a significant decrease in the root/shoot DW ratio was found in response to NP and NPK under well-watered conditions and NK and NPK under drought stress compared with no nutrient supply (Figure 7). Drought enhanced the root/shoot DW ratio at day 20 after sowing for all other nutrient treatments except NK supply. A significant increase in the root/shoot DW ratio under drought was found with N, NP, PK and NPK fertilization (Figure 7).

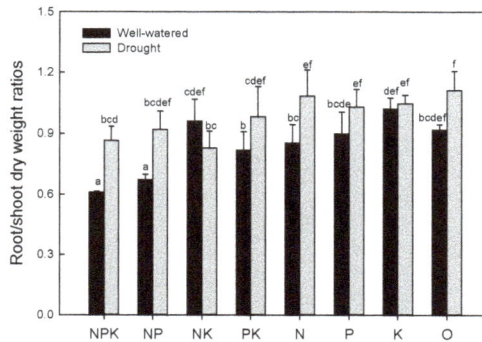

Figure 7. Interactive effects of drought and the N, P and K nutrients applied individually or in combination, or with no nutrient application (0) on the root/shoot dry weight ratios of the maize seedlings at day 20 after sowing. The error bars represent the standard errors of the mean (n = 3) and fit within the plot symbol if not visible. Means accompanied by different letters are significantly different at $p < 0.05$.

4. Discussion

The results in this study showed that there was generally a positive effect of combined NPK and/or NP nutrient supply on shoot growth, such as leaf area, shoot FW and DW at day 20 after sowing, not only under well-watered conditions, but also under drought compared with no nutrient supply treatment (Figures 1, 3 and 4). Nitrogen, P and K are the major nutrients required for plant growth. According to Liebig's law of the minimum [4], any nutrient at a limiting concentration in soil will inhibit plant growth. Compared with N and P nutrient supply, it seems that K was not limiting to plant growth in this study. This might be because the mineralogical characteristics of the illitic-chloritic silt loam (fine mixed mesic Aquic Ustifluvent [32]) used in this investigation makes it plausible that a sufficient supply is still rendered possible without further K supply. This might indicate that current soil test methods used do not adequately indicate the soil nutrient supply for the given soil. Optimal nutrient levels of fertilization are even more important under water deficit conditions, since reduced nutrient availability is one of the major factors limiting plant growth under drought [5,9,14]. To our knowledge, this is the first report of a comparison of maize seedling growth with N, P and K nutrient application individually or in combination under drought and well-watered conditions. Thus, the results here may suggest that supplying nutrients under deficient conditions could increase the drought tolerance of plants by increasing plant biomass for early growth vigour and establishment, which is in agreement with reports from the literature [9,10,31,34,35]. According to the literature [9], however, caution should be taken when considering an increase in the supply of nutrients to alleviate the adverse effects of drought stress only if the nutrient is present in the soil in insufficient amounts and the drought stress is not severe. The positive effects of nutrients on plant growth under drought conditions can be explained by their physiological functions. For example, such positive effects of N and P are attributed to an increase in photosynthesis [15], stomatal conductance and water-use efficiency [11] and higher cell membrane stability and osmotic adjustment [16]. Nitrogen fertilization, particularly using nitrate [12], regulates water flux through plants by N-flux-linked signalling mechanisms [39–41]. In most cases, the reports in the literature show that an improved K nutritional status in plants is of great importance for maintaining the osmotic potential and turgor of the cells [5,22] and for regulating the stomatal function and water use efficiency [4,19].

Drought stress at different growth stages causes various morpho-physiological changes in plants [42]. At the seedling growth stage, drought stress might result in higher root dry weights and longer roots and/or reduced shoot growth [1,24–26,43–45]. Although our study did not show a significant increase in root DW or a significant decrease in shoot DW under drought stress for all nutrient treatments, the results in Figures 4 and 6 demonstrated that there was a tendency of higher root DW for all nutrient treatments except for NK fertilization at day 20 after sowing and of lower shoot DW for single N and combined NP and NPK under drought than under well-watered conditions. Consequently, the root/shoot ratios were significantly increased with N, NP, PK and NPK supply under drought conditions compared with well-watered conditions (Figure 7). Possibly, with the development of less soil moisture, plants respond to drought by increasing root biomass or a root absorptive surface relative to the shoot biomass [25,26,30]. The enhanced root growth may be able to explore and absorb more water and nutrients from the soil [8,24,46]. The lower shoot DW with N, NP and NPK fertilization may be due to an increase in assimilate allocation to roots for maintaining root growth, for osmotic adjustment and for turgor maintenance [10,25,26]. Similar results were reported by other authors [21,47,48]. Therefore, higher root/shoot ratios could be exploited as a strategy for stress tolerance in crop plants.

In conclusion, the results in this study showed that there was generally a positive effect of combined NPK and/or NP nutrient supply on shoot growth, such as leaf area, shoot FW and DW, at day 20 after sowing under both well-watered and drought conditions compared with no nutrient supply treatment. Compared with the effect of N and P nutrient supply, it seems that K was not limiting plant growth due to the mineralogical characteristics of the illitic-chloritic silt loam (fine mixed mesic Aquic Ustifluvent) used, which contained sufficient K. Compared with well-watered conditions, the higher root/shoot ratios with N, NP, PK and NPK under drought could be exploited as a strategy for stress tolerance in crop plants.

Acknowledgments: This work was supported by the German Research Foundation (DFG) and the Technical University of Munich (TUM) in the framework of the Open Access Publishing Program.

Author Contributions: U.S. and C.S. conceived and designed the experiments; C.S. performed the experiments; C.S. and Y.H. analyzed the data; Y.H. and U.S. wrote the paper.

Conflicts of Interest: The authors declare no conflict of interest.

References

1. Intergovernmental Panel on Climate Change (IPCC). *Climate Change 2014: Synthesis Report*; Contribution of Working Groups I, II and III to the 5th Assessment Report of the Intergovernmental Panel on Climate Change; IPCC: Geneva, Switzerland, 2014.
2. Farooq, M.; Wahid, A.; Kobayashi, N.; Fujita, D.; Basra, S.M.A. Plant drought stress: Effects, mechanisms and management. *Agron. Sustain. Dev.* **2009**, *29*, 185–212. [CrossRef]
3. Viets, F.G. Water deficits and nutrient availability. In *Water Deficits and Plant Growth. Bd III: Plant Responses and Control of Water Balance*; Kozlowski, T.T., Ed.; Academic Press: New York, NY, USA; London, UK, 1972.
4. Marschner, H. *Mineral Nutrition of Higher Plants*; Academic Press: London, UK, 1995; pp. 3–70.
5. Waraich, E.A.; Ahmad, R.S.; Ashraf, M.Y. Role of mineral nutrition in alleviation of drought stress in plants. *Aust. J. Crop Sci.* **2011**, *5*, 764–777.
6. Hsiao, T.C. Plant responses to water stress. *Ann. Rev. Plant Physiol. Plant Mol. Biol.* **1973**, *24*, 519–570. [CrossRef]
7. Kramer, P.; Boyer, J. *Water Relations of Plants and Soils*; Academic Press: New York, NY, USA; London, UK, 1995.
8. Turner, B.L.; Haygarth, P.M. Biogeochemistry—Phosphorus solubilisation in rewetted soils. *Nature* **2011**, *411*, 258. [CrossRef] [PubMed]
9. Hu, Y.C.; Schmidhalter, U. Drought and salinity: A comparison of their effects on mineral nutrition of plants. *J. Plant Nutr. Soil Sci.* **2005**, *168*, 541–549. [CrossRef]
10. Studer, C.; Hu, Y.C.; Schmidhalter, U. Evaluation of the differential osmotic adjustments between roots and leaves of maize seedlings with single or combined NPK-nutrient supply. *Funct. Plant Biol.* **2007**, *34*, 228–236. [CrossRef]
11. Brück, H.; Payne, W.A.; Sattelmacher, B. Effects of phosphorus and water supply on yield, transpirational water-use efficiency and carbon isotope discrimination of pearl millet. *Crop Sci.* **2000**, *40*, 120–125. [CrossRef]
12. Cramer, M.D.; Hawkins, H.J.; Verboom, G.A. The importance of nutritional regulation of plant water flux. *Oecologia* **2009**, *161*, 15–24. [CrossRef] [PubMed]
13. He, M.Z.; Dijkstra, F.A. Drought effect on plant nitrogen and phosphorus: A meta-analysis. *New Phytol.* **2014**, *204*, 924–931. [CrossRef] [PubMed]
14. Saud, S.; Fahad, S.; Chen, Y.J.; Ihsan, M.Z.; Hammad, H.M.; Nasim, W.; Amanullah, Jr.; Arif, M.; Alharby, H. Effects of nitrogen supply on water stress and recovery mechanisms in Kentucky bluegrass plants. *Front. Plant Sci.* **2017**, *8*. [CrossRef] [PubMed]
15. Ackerson, R.C. Osmoregulation in cotton in response to water-stress. 3. Effects of phosphorus fertility. *Plant Physiol.* **1985**, *77*, 309–312. [CrossRef] [PubMed]
16. Sawwan, J.; Shibli, R.A.; Swaidat, I.; Tahat, M. Phosphorus regulates osmotic potential and growth of African violet under in vitro-induced water deficit. *J. Plant Nutr.* **2000**, *23*, 759–771. [CrossRef]
17. Garg, B.K.; Burman, U.; Kathju, S. The influence of phosphorus nutrition on the physiological response of moth bean genotypes to drought. *J. Plant Nutr. Soil Sci.* **2004**, *167*, 503–508. [CrossRef]
18. Beringer, H.; Trolldenier, G. Influence of K nutrition on the response to environmental stress. In *Potassium Research—Review and Trents*; International Potash Institute (IPI): Bern, Switzerland, 1978; pp. 189–222.
19. Grzebisz, W.; Gransee, A.; Szczepaniak, W.; Diatta, J. The effects of potassium fertilization on water-use efficiency in crop plants. *J. Plant Nutr. Soil Sci.* **2013**, *176*, 355–374. [CrossRef]

20. Martineau, E.; Domec, J.C.; Bosc, A.; Denoroy, P.; Fandino, V.A.; Lavres, J.; Jordan-Meille, L. The effects of potassium nutrition on water use in field-grown maize (*Zea mays* L.). *Environ. Exp. Bot.* **2017**, *134*, 62–71. [CrossRef]

21. Bennett, J.M.; Mutti, L.S.M.; Rao, P.S.C.; Jones, J.W. Interactive effects of nitrogen and water stresses on biomass accumulation, nitrogen uptake and seed yield of maize. *Field Crops Res.* **1989**, *19*, 297–311. [CrossRef]

22. Raza, M.A.S.; Saleem, M.F.; Shah, G.M.; Jamil, M.; Khan, I.H. Potassium applied under drought improves physiological and nutrient uptake performances of wheat (*Triticum aestivum* L.). *J. Soil Sci. Plant Nutr.* **2013**, *13*, 175–185.

23. Sharp, R.E.; Davies, W.J. Solute regulation and growth by roots and shoots of water-stressed maize plants. *Planta* **1979**, *147*, 43–49. [CrossRef] [PubMed]

24. Schmidhalter, U.; Oertli, J.J. Germination and seedling growth of carrots under salinity and moisture stress. *Plant Soil* **1991**, *132*, 243–251. [CrossRef]

25. Schmidhalter, U.; Burucs, Z.; Camp, K.H. Sensitivity of root and leaf water status in maize (*Zea mays*) subjected to mild soil dryness. *Aust. J. Plant Physiol.* **1998**, *25*, 307–316. [CrossRef]

26. Schmidhalter, U.; Evequoz, M.; Camp, K.H.; Studer, C. Sequence of drought response of maize seedlings in drying soil. *Physiol. Plant.* **1998**, *104*, 159–168. [CrossRef]

27. Lynch, J.P.; Brown, K.M. Topsoil foraging—An architectural adaptation of plants to low phosphorus availability. *Plant Soil* **2001**, *237*, 225–237. [CrossRef]

28. Jaramillo, R.E.; Nord, E.A.; Chimungu, J.G.; Brown, K.M.; Lynch, J.P. Root cortical burden influences drought tolerance in maize. *Ann. Bot.* **2013**, *112*, 429–437. [CrossRef] [PubMed]

29. Chimungu, J.G.; Brown, K.M.; Lynch, J.P. Reduced root cortical cell file number improves drought tolerance in maize. *Plant Physiol.* **2014**, *166*, 1943–1955. [CrossRef] [PubMed]

30. Lambers, H.; Chapin, F.S.; Pons, T.L. *Plant Physiological Ecology*; Springer: New York, NY, USA, 2008.

31. Hu, Y.C.; Burucs, Z.; von Tucher, S.; Schmidhalter, U. Short-term effects of drought and salinity on mineral nutrient distribution along growing leaves of maize seedlings. *Environ. Exp. Bot.* **2007**, *60*, 268–275. [CrossRef]

32. Saab, I.N.; Sharp, R.E. Non-hydraulic signals from maize roots in drying soil—Inhibition of leaf elongation but not stomatal conductance. *Planta* **1989**, *179*, 466–474. [CrossRef] [PubMed]

33. Tardieu, F.; Reymond, M.; Hamard, P.; Granier, C.; Muller, B. Spatial distributions of expansion rate, cell division rate and cell size in maize leaves: A synthesis of the effects of soil water status, evaporative demand and temperature. *J. Exp. Bot.* **2000**, *51*, 1505–1514. [CrossRef] [PubMed]

34. Hu, Y.C.; Burucs, Z.; Schmidhalter, U. Short-term effect of drought and salinity on growth and mineral elements in wheat seedlings. *J. Plant Nutr.* **2006**, *29*, 2227–2243. [CrossRef]

35. Hu, Y.C.; Burucs, Z.; Schmidhalter, U. Effect of foliar fertilization application on the growth and mineral nutrient content of maize seedlings under drought and salinity. *Soil Sci. Plant Nutr.* **2008**, *54*, 133–141. [CrossRef]

36. Clauw, P.; Coppens, F.; De Beuf, K.; Dhondt, S.; Van Daele, T.; Maleux, K.; Storme, V.; Clement, L.; Gonzalez, N.; Inze, D. Leaf responses to mild drought stress in natural variants of arabidopsis. *Plant Physiol.* **2015**, *167*, 800–816. [CrossRef] [PubMed]

37. Von Tucher, S.; Hörndl, D.; Schmidhalter, U. Interaction of soil pH and phosphorus fertilizer efficacy: Long-term effects of varying phosphorus fertilizer and lime applications on yield and phosphorus uptake of winter wheat, winter barley and sugar beet. *Ambio* **2018**, in press.

38. Schmidhalter, U.; Selim, H.M.; Oertli, J.J. Measuring and modeling root water uptake based on 36chloride discrimination in a silty soil affected by groundwater. *Soil Sci.* **1994**, *158*, 97–105. [CrossRef]

39. Wilkinson, S.; Bacon, M.A.; Davies, W.J. Nitrate signalling to stomata and growing leaves: Interactions with soil drying, ABA and xylem sap pH in maize. *J. Exp. Bot.* **2007**, *58*, 1705–1716. [CrossRef] [PubMed]

40. Gloser, V.; Zwieniecki, M.A.; Orians, C.M.; Holbrook, N.M. Dynamic changes in root hydraulic properties in response to nitrate availability. *J. Exp. Bot.* **2007**, *58*, 2409–2415. [CrossRef] [PubMed]

41. Kant, S.; Kafkafi, U. Potassium and abiotic stresses in plants. In *Role of Potassium in Nutrient Management for Sustainable Crop Production in India*; Pasricha, N.S., Bansal, S.K., Eds.; Potash Research Institute of India: Gurgaon, Haryana, 2002.

42. Ali, Z.I.; Golombek, S.D. Effect of drought and nitrogen availability on osmotic adjustment of five pearl millet cultivars in the vegetative growth stage. *J. Agron. Crop Sci.* **2016**, *202*, 433–444. [CrossRef]

43. Sharp, R.E.; Davies, W.J. Root-growth and water-uptake by maize plants in drying soil. *J. Exp. Bot.* **1985**, *36*, 1441–1456. [CrossRef]

44. Westgate, M.E.; Boyer, J.S. Osmotic adjustment and the inhibition of leaf, root, stem and silk growth at low water potentials in maize. *Planta* **1985**, *164*, 540–549. [CrossRef] [PubMed]

45. Dhanda, S.S.; Sethi, G.S.; Behl, R.K. Indices of drought tolerance in wheat genotypes at early stages of plant growth. *J. Agron. Crop Sci.* **2004**, *190*, 6–12. [CrossRef]

46. Subbarao, G.V.; Johansen, C.; Slinkard, A.E.; Rao, R.C.N.; Saxena, N.P.; Chauhan, Y.S. Strategies for improving drought resistance in grain legumes. *Crit. Rev. Plant Sci.* **1995**, *14*, 469–523. [CrossRef]

47. Morgan, J.A. The effects of n-nutrition on the water relations and gas-exchange characteristics of wheat (*Triticum-aestivum* L). *Plant Physiol.* **1986**, *80*, 52–58. [CrossRef] [PubMed]

48. Nnoham, O.I.; Odurukwe, S.O. Effects of n-fertilizer rate, soil-water tension and soil texture on growth and n-uptake by maize (*Zea mays* L). *Fertil. Res.* **1987**, *13*, 241–254. [CrossRef]

![agriculture logo] *agriculture*

MDPI

Article

Nitrogen Use Efficiency and the Genetic Variation of Maize Expired Plant Variety Protection Germplasm

Adriano T. Mastrodomenico [†], C. Cole Hendrix [‡] and Frederick E. Below *

Department of Crop Sciences, University of Illinois, Urbana, IL 61801, USA;
adriano.mastrodomenico@limagrain.com (A.T.M.); c.cole.hendrix@gmail.com (C.C.H.)
* Correspondence: fbelow@illinois.edu; Tel.: +1-217-333-9745
† Current address: PR-445 Road, km 56.5, Limagrain, Londrina, PR 86115-000, Brazil.
‡ Current address: 718 Forest Park Blvd. Apt. D108, Oxnard, CA 93036, USA.

Received: 3 November 2017; Accepted: 21 December 2017; Published: 1 January 2018

Abstract: Nitrogen use efficiency (NUE) in maize (*Zea mays* L.) is an important trait to optimize yield with minimal input of nitrogen (N) fertilizer. Expired Plant Variety Protection (ex-PVP) Act-certified germplasm may be an important genetic resource for public breeding sectors. The objectives of this research were to evaluate the genetic variation of N-use traits and to characterize maize ex-PVP inbreds that are adapted to the U.S. Corn Belt for NUE performance. Eighty-nine ex-PVP inbreds (36 stiff stalk synthetic (SSS), and 53 non-stiff stalk synthetic (NSSS)) were genotyped using 26,769 single-nucleotide polymorphisms, then 263 single-cross maize hybrids derived from these inbreds were grown in eight environments from 2011 to 2015 at two N fertilizer rates (0 and 252 kg N ha^{-1}) and three replications. Genetic utilization of inherent soil nitrogen and the yield response to N fertilizer were stable across environments and were highly correlated with yield under low and high N conditions, respectively. Cluster analysis identified inbreds with desirable NUE performance. However, only one inbred (PHK56) was ranked in the top 10% for yield under both N-stress and high N conditions. Broad-sense heritability across 12 different N-use traits varied from 0.11 to 0.77, but was not associated with breeding value accuracy. Nitrogen-stress tolerance was negatively correlated with the yield increase from N fertilizer.

Keywords: expired Plant Variety Protection (ex-PVP); maize; nitrogen stress; nitrogen use efficiency (NUE); U.S. Corn Belt germplasm

1. Introduction

World-wide, producers used approximately 109 million tons of nitrogen (N) fertilizer in 2014 [1]. Of that amount, more than 5 million tons are used for maize (*Zea mays* L.) production in the United States (U.S.) [2]. Nitrogen is the mineral macronutrient required in the greatest amount by the maize crop, with uptake values being measured at 280 kg N ha^{-1} for a crop producing 14.4 Mg ha^{-1} of grain [3]. Although supplemental N fertilizer is often necessary to increase maize grain yield, N fertilizer consumption has remained constant in the U.S. for the last 20 years [1]. The maize yield increases observed, despite the constant N fertilizer consumption in the United States during the last two decades, were a result of both genetic improvement and better agronomic practices [4]. In contrast, in some regions, such as sub-Saharan Africa, limited N fertilizer use and soil availability prevent achieving yields that are similar to the United States [5]. The world population growth will require increased grain production, and therefore more N fertilizer efficiency will be necessary to meet the world's demand [6]. Innovative agricultural technologies, such as new N fertilizer sources, precision agriculture, and crop genetic improvement will be important to increase nitrogen use efficiency in maize production [7].

Nitrogen use efficiency (NUE) is defined as the ratio of grain yield to N fertilizer that is supplied [8], and is the product of nitrogen uptake efficiency (NUpE, the ratio of the additional plant N content due to fertilizer N to the amount of fertilizer-applied N) and nitrogen utilization efficiency (NUtE, the ratio of yield increase to the difference in plant N content compared to those of an unfertilized crop). In addition, NUE is a complex phenotypic trait influenced by several plant physiological mechanisms [9]. Nitrogen uptake has been correlated to increased yield via many physiological factors, including: increased plant biomass, root architecture, photosynthesis, leaf area index, nitrate content, glutamine synthetase, Rubisco, PEP carboxylase, and asparagine [10–13]. Additionally, NUtE has been associated with factors, such as increased photosynthesis, remobilization, transport, and the balance of carbon and nitrogen assimilates, as well as kernel set, [14–17]. Since most maize breeding programs developed their germplasm under high soil N conditions, genetic selection for improved NUE is often ignored [18]. The genetic improvement of NUE in maize up to now was mainly achieved through indirect selection for increased hybrid yield performance. Nonetheless, large genotypic differences in maize NUE have been reported [18–20].

Over the past few decades, maize hybrids in North America have increased yield performance under both low and high N availability conditions [21], but the genetic gain of maize performance when grown under low N was almost twice the genetic gain found when hybrids were grown with high N fertility [20]. Genetic variation of NUE in maize has been attributed to hybrids expressing NUpE and NUtE at different levels [20,22]. These N-responsive traits contribute differently to NUE depending on the germplasm [23], the soil N status [7,9], and the progeny seed quality composition [19]. Using the Illinois Protein Strain collection, strain-hybrids with high seed protein concentration exhibited greater NUpE and lesser NUtE than the strain-hybrids with low seed protein concentration [20]. Phenotypic evaluation of NUpE and NUtE in a breeding population may be an important method to characterize and identify maize genotypes with desirable NUE performance [20,24]. Genetic improvement of NUE in U.S. germplasm using conventional or molecular breeding will require the simultaneous enhancement of both NUpE and NUtE. As a result, more research is needed to evaluate the genetic characteristics underlying NUE in the U.S. Corn Belt germplasm.

Since the U.S. Plant Variety Protection (PVP) Act was passed in 1970, which protects seed-bearing varieties for 20 years, plant breeders have been generating new genetic combinations using only the most elite material available, thereby decreasing the genetic diversity of commercial breeding programs in the U.S. [25]. Expired PVP Act-certified germplasm, named ex-PVP, are publically available and may represent an important genetic resource for both public and private breeding programs. Current U.S. maize germplasm has reduced allelic diversity; most of the current germplasm originated from only seven progenitor lines: B73, Mo17, PH207, PHG39, LH123Ht, LH82, and PH595 [26]. However, elite ex-PVP inbreds may be genetically diverse and an important genetic resource for maize breeding programs [27]. Although ex-PVP germplasm may not be integrated directly into a commercial breeding program, these genotypes can be used to originate new genetic combinations with desirable traits [28]. Up to now, little agronomic and quantitative breeding research has been done using a representative number of maize ex-PVP parental lines and hybrid combinations.

The objectives of this research were to characterize ex-PVP maize hybrids for N-use traits, evaluate the genetic variation and the phenotypic correlation of different N-responsive traits across different maize heterotic groups, and identify the parental lines and hybrid combinations with desirable NUE performance.

2. Materials and Methods

2.1. Germplasm and Genomic Data

A collection of 89 ex-PVP and two public maize inbreds, B73 and Mo17, were selected for this study (Table S1). All of the germplasm seed was obtained from the North Central Regional Plant Introduction Station (http://www.ars-grin.gov/npgs, verified 24 August 2016). Twelve ex-PVP inbreds were selected that contain the majority of allelic diversity encountered in current U.S. maize germplasm [27]. In addition, a random set of inbreds adapted to the U.S. Corn Belt with more recently expired PVP certificates from a selection of seed companies were included. Findings from these most recently-released ex-PVP lines may reveal the genetic diversity shifts observed during the past 20 years in germplasm usage by different breeding programs [29]. Overall, the ex-PVP collection used for this study contains genotypes that were released from 1972 to 2011, as developed by six different seed companies.

Leaf samples from all the inbreds (14-day old seedlings) were collected for DNA extraction using the CTAB protocol [30]. Inbreds were genotyped using the genotype-by-sequencing method [31] and two enzyme combinations were used to reduce genomic complexity: PstI-HF-Bfal and PstI-HF-HinP1I. The enzyme PstI-HF is considered a rare cutter, while HinP1I and BfaI are common cutters. These enzyme combinations were used to obtain adequate genome coverage. Sequenced data were obtained using an Illumina HiSeq2000 (W.M. Keck Center for Comparative and Functional Genomics, Urbana, IL, USA) and single-nucleotide-polymorphism (SNP) data were called using the GBS pipeline in TASSEL 3.0 [32]. Minor allele frequency cutoff was set to 10%, and SNPs with more than 50% missing data were removed. A total of 26,769 SNPs were used for the analyses.

Discriminant analysis of principal components (DAPC) was performed for all of the inbred lines using the Adegenet package [33] in R Studio [34]. Since pedigrees from ex-PVP's are often vague [26], DAPC is well suited to define genetic clusters in these situations [33]. Genotyping revealed that the ex-PVP germplasm used in this study was composed of 36 stiff-stalk synthetic (SSS) lines and 53 non-SSS (NSSS) lines; the latter of which included 19 lines from the Iodent sub-heterotic group, and 34 lines from the Lancaster sub-heterotic group (Figure 1). Knowledge of genetic relatedness between parental inbreds is fundamental for hybrid heterosis, due to dominance and epistatic effects [35]. Therefore, all of the single cross maize hybrids evaluated in this study were generated between SSS and NSSS parental lines.

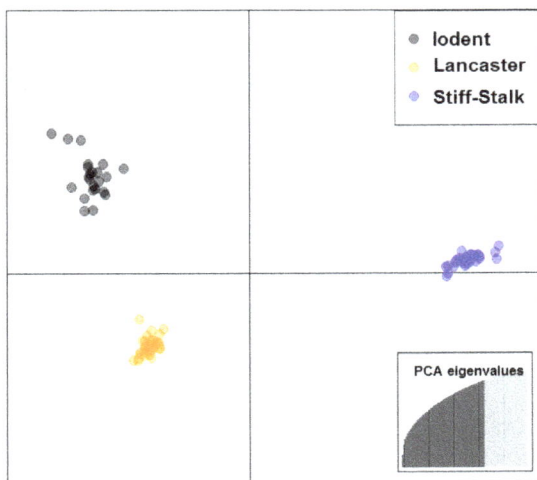

Figure 1. Scatterplots of the discriminant analysis of principal components of 89 ex-Plant Variety Protection (ex-PVP) maize inbred lines. Scatterplot displays the first two components using 26,768 single nucleotide polymorphism markers. Heterotic groups are represented by different colors: Iodent (black), Lancaster (yellow), and Stiff-stalk synthetic (blue), and each dot represents an individual inbred line.

Hybrid seed were created in an incomplete factorial design between SSS and NSSS inbred lines from 2011 to 2014 at Champaign, IL, USA. A total of 263 single cross maize hybrids that were derived from a random combination between SSS and NSSS parental lines were evaluated. On average, each SSS line was combined in 20 (range 3–57), and each NSSS line was combined in 13 (range 3–38) different hybrid combinations. A heatmap view of the incomplete factorial hybrid combination evaluated can be found online as Supplemental Figure S1.

2.2. Research Sites and Crop Management

Maize hybrids were grown in eight field environments from 2011 through 2015. Data from 2012 of the original experiment was excluded from the analysis due to severe drought stress. Research sites were planted in one environment at DeKalb, IL, USA (41°47′ N, 88°50′ W; 19 May 2014), five environments at Champaign, IL, USA (40°3′ N, 88°14′ W; 17 May 2011, 20 May 2013, 22 April 2014, 24 April 2015, and 19 May 2015), and two environments at Harrisburg, IL, USA (37°43′ N, 88°27′ W; 29 May 2013, and 23 May 2014). Soil types at the research sites were primarily Flanagan silt loam at DeKalb, IL, USA, Drummer silty clay loam at Champaign, IL, USA, and Patton silty clay loam at Harrisburg, IL, USA. The previous crop planted in each environment was soybean (*Glycine max* (L.) Merr.). The experiment was planted using a precision plot planter (SeedPro 360, ALMACO, Nevada, IA, USA) and plots were 5.6 m in length with 0.76 m row spacing and two rows in width. The target plant density was 79,000 plants ha^{-1}. All seeds were treated with Maxim® XL fungicide (Fludioxonil and Mefenoxam at 0.07 mg active ingredient kernel^{-1}; Syngenta Crop Protection, Greensboro, NC, USA), and Cruiser® 5FS insecticide (Thiamethoxam at 0.80 mg active ingredient kernel^{-1}; Syngenta Crop Protection, Greensboro, NC, USA) to prevent early season disease and insect damage, respectively. In addition, Force 3G® insecticide (Tefluthrin 2,3,5,6-tetrafluoro-4-methylphenyl)methyl-(1α,3α)-(Z)-(±)-3-(2-chloro-3,3,3-trifluoro-1-propenyl)-2, 2-dimethylcyclopropanecarboxylate; Syngenta Crop Protection, Greensboro, NC, USA) was applied at planting in-furrow (0.15 kg active ingredient ha^{-1}) in order to control soil pests. Pre-emergence herbicide Lumax® EZ (mixture of S-Metolachlor, Atrazine, and Mesotrione; Syngenta Crop Protection, Greensboro, NC, USA) was applied at a rate of 7 L ha^{-1} to control early season weeds.

At maturity, plots were harvested with a two-row plot combine (SPC40, ALMACO, Nevada, IA, USA). Grain yield is reported as Mg ha^{-1} at 15.5% grain moisture. Grain protein concentrations were estimated from a representative grain subsample from each plot that was collected during harvest using near infrared transmittance (NIT) spectroscopy (Infratec 1241, FOSS, Eden Prairie, MN, USA).

2.3. Experimental Treatments and Design

The 263 single-cross maize hybrids were grown as part of a randomized complete block design, with three replications and two N fertilizer rates (0 and 252 kg N ha^{-1}; designated low and high N, or −N and +N, respectively) in a split-plot arrangement. The main-plot was hybrid and the split-plot was N fertilizer rate. On average, 83 hybrids were tested in each environment (Supplemental Figure S1). Nitrogen stress tolerance was measured by yield of the check plot (0 kg N ha^{-1}), while 252 kg N ha^{-1} was used to maximize the yield response to N from all of the hybrids, regardless of their yield potential. Nitrogen fertilizer was hand applied in a diffuse band as urea (46-0-0) during the V2 to V3 growth stages [36]. Nitrogen application dates were 17 June 2014 at DeKalb, IL, USA, 2 June 2011, 4 June 2013, 4 June 2014, 18 May 2015, and 10 June 2015 at Champaign, IL, USA, and 25 June 2013, and 13 June 2014 at Harrisburg, IL, USA.

2.4. Phenotype Measurements

Aboveground plant biomass from each plot was sampled at the R6 growth stage (physiological maturity), when the maximum biomass accumulation for maize is achieved [36]. Six representative plants (visual assessment) from each plot were sampled and separated into stover (leaf, stem, and husks) and ear (grain and cob). The sampling criteria established consisted of selecting two adjacent plants near one end of the plot (1.2 m along the length of the first row), two adjacent plants at the center of the plot (approximately 2.7 m from the origin), and two adjacent plants at the other end of the plot (approximately 4.1 m along the length of the second row). Whole stover fresh weight was determined before shredding in a brush chipper (Vermeer BC600XL; Vermeer Midwest, Goodfield, IL, USA). A representative subsample of the fresh shredded material was weighed and dried in a forced-draft oven (75 °C) for approximately five days. Total stover dry weight was calculated using the fresh stover weight and the moisture level of the shredded material. Individual plant dry total biomass (g plant^{-1}) was the sum of the dry stover, cob, and grain weights (adjusted to 0% moisture). Dried stover samples were ground in a Wiley mill (Thomas Scientific, Swedesboro, NJ, USA) to pass a 20-mesh screen, and N concentration (g kg^{-1}) was analyzed by using a combustion technique (EA1112 N-Protein analyzer; CE Elantech, Inc., Lakewood, NJ, USA). Grain protein concentration was estimated by multiplying N concentration by a factor of 6.25, and was abbreviated as Protein$_{-N}$ or Protein$_{+N}$, from plants grown at 0 or 252 N ha^{-1}, respectively). Stover N content (g N plant^{-1}) was calculated by multiplying stover dry weight (g plant^{-1}) by stover N concentration. Similarly, grain N content (g N plant^{-1}) was calculated by multiplying grain dry weight (g plant^{-1}) by grain N concentration. Individual plant N content (g N plant^{-1}) was calculated as the sum of stover and grain N contents. Shelled grain weights from the ears sampled at R6 were combined with the remaining plot grain weight for yield determination.

In combination with grain yield and plant N content, NUE, N-uptake efficiency (NUpE), N-utilization efficiency (NUtE), harvest index (HI), and N-harvest index (NHI) were calculated according to Equations (1)–(7), with the expressed units shown:

$$\text{NUE} = (\text{Yield}_{+N} - \text{Yield}_{-N})/\text{NR} = (\text{kg yield}) \, (\text{kg N})^{-1}, \tag{1}$$

$$\text{NUpE} = (\text{PN}_{+N} - \text{PN}_{-N})/\text{NR} = (\text{kg plant N}) \, (\text{kg N})^{-1}, \tag{2}$$

$$\text{NUtE} = (\text{Yield}_{+N} - \text{Yield}_{-N})/(\text{PN}_{+N} - \text{PN}_{-N}) = (\text{kg yield}) \, (\text{kg plant N})^{-1}, \tag{3}$$

$$\text{HI}_{+N} = (\text{kg grain}_{+N} \, \text{plant}^{-1})/(\text{kg dry weight}_{+N} \, \text{plant}^{-1}) = \text{kg kg}^{-1}, \tag{4}$$

$$\text{HI}_{-N} = (\text{kg grain}_{-N} \, \text{plant}^{-1})/(\text{kg dry weight}_{-N} \, \text{plant}^{-1}) = \text{kg kg}^{-1}, \tag{5}$$

$$\text{NHI}_{+N} = \text{kg grainN}_{+N}/\text{PN}_{+N} = (\text{kg grain N}) \, (\text{kg plant N})^{-1}, \tag{6}$$

$$\text{NHI}_{-N} = \text{kg grainN}_{-N}/\text{PN}_{-N} = (\text{kg grain N}) \, (\text{kg plant N})^{-1}, \tag{7}$$

in which Yield$_{+N}$ corresponds to grain yield (kg ha^{-1}) at 252 kg N ha^{-1}, Yield$_{-N}$ corresponds to grain yield at 0 kg N ha^{-1}, NR is the N fertilizer rate (kg N, 252 kg N ha^{-1}), PN represents the total plant N content (kg plant N ha^{-1}) at 252 kg N ha^{-1} (PN$_{+N}$) and at 0 kg N ha^{-1} (PN$_{-N}$). In addition, genetic utilization (GU) (kg yield kg^{-1} plant N), which measures the physiological efficiency of plants to produce grain utilizing the plant N accumulated when grown without N fertilizer [18–20] was calculated according to Equation (8), with the expressed units shown:

$$\text{GU} = \text{PG}_{-N}/\text{PN}_{-N} = (\text{kg}) \, (\text{kg plant N})^{-1}, \tag{8}$$

in which PG$_{-N}$ is the individual plant grain mass (kg plant^{-1}) at 0 kg N ha^{-1} and PN$_{-N}$ represents the total per plant N content (kg plant N) at physiological maturity derived from residual or mineralized soil N.

2.5. Statistical Analysis

Since there is a weak correlation between the performances of inbred parents and their hybrid progeny's performance for NUE [37], the effects of general combining ability (GCA) and specific combining ability (SCA) of inbreds were evaluated using a random combination of ex-PVP hybrids. Moreover, the genetic variance and covariances between hybrids were calculated separately for each heterotic group [38]. Best linear unbiased predictions (BLUPs) were calculated for each phenotypic trait using the restricted maximum likelihood method to account for the unbalanced data. In addition, year-location combinations were considered environments. General and specific combining abilities were obtained in PROC MIXED SAS version 9.4 [39]. A linear model for an incomplete factorial design, according to Equation (9), was used:

$$Y_{ijklm} = \mu + E_i + B_{j(i)} + S_k + N_l + SN_{kl} + ES_{ik} + EN_{il} + ESN_{ikl} + \varepsilon_{ijklm},\qquad (9)$$

where Y_{ijklm} is the mth observation of the klth hybrid in the jth block in the ith environment; μ is the grand mean, E_i is the random effect of ith environment (I = 1 to 8); $B_{j(i)}$ is the random effect of jth block nested within the ith environment (j = 1 to 3); S_k is the GCA effect of kth SSS inbred (k = 1 to 36); N_l is the random GCA effect of lth NSSS inbred (l = 1 to 53); SN_{kl} is the SCA effect of klth hybrid (kl = 1 to 522); ES_{ik} is the random environment by SSS interaction; EN_{il} is the random environment by NSSS interaction; ESN_{ikl} is the random environment by hybrid interaction; and, ε_{ijklm} is the random error term. Genotypic variance was calculated by multiplying the sum of the genetic variance components (SSS, NSSS, and hybrid) by two. Phenotypic variance was calculated as the sum of all the variance components, except for the variance component for block effect [40]. Broad-sense heritability was calculated as the ratio of genotypic and phenotypic variance. The estimated breeding value of each hybrid was calculated, according to Equation (10):

$$EBV_{kl} = \mu + GCA_k + GCA_l + SCA_{kl},\qquad (10)$$

where EBV_{kl} is the estimated breeding value of klth hybrid; μ is the grand mean; GCA_k is the GCA effect of kth inbred; GCA_l is the GCA effect of lth inbred; and, SCA_{kl} is the SCA effect of klth hybrid. Estimated breeding value (EBV) measures the average effect of an individual's genotypic value on the mean performance of its progeny [41] and is a widely-used measurement in maize breeding programs for the selection of superior genotypes.

Pearson's correlation coefficients were calculated in SAS version 9.4 [39] between the GCA's of different N-use traits. Hierarchical cluster analysis was conducted on each heterotic group across different N-use traits, using the Euclidean method in R Studio [34]. The estimated breeding value (EBV) accuracy of the phenotypic traits was calculated according to Equation (11), [42]:

$$EBV_{Accuracy} = \sqrt{((1 - SE)/((1 + F) \times \sigma_A^2))},\qquad (11)$$

where SE is the standard error of the inbred GCA, F is the inbreeding coefficient of the individual (assumed to be zero), and σ_A^2 is the additive variance component of the heterotic group (SSS or NSSS).

3. Results and Discussion

3.1. Phenotypic Variation of N-Use Traits

Yield under low N conditions (Yield$_{-N}$) accounted for 54% of the yield that was produced by the hybrids under high N conditions (Yield$_{+N}$) (Table 1). In addition, N fertilizer increased the mean harvest index (HI), the nitrogen harvest index (NHI), and the grain protein concentration. Average NUE, NUpE, NUtE, and GU values of 16.7 kg kg_{Nfert}^{-1}, 0.43 kg_{plantN} kg_{Nfert}^{-1}, 41.8 kg kg_{plantN}^{-1}, and 59.0 kg kg_{plantN}^{-1}, respectively, are similar to other reports using U.S. Corn Belt germplasm [19,20]. Moreover, the additive effect distribution (range in GCA) of the two maize heterotic groups were similar for most N-use traits. In contrast, the NSSS group exhibited a greater additive effect range for NUE than the SSS group. The large additive effect variation that was observed among different N-use traits indicates that an opportunity exists for selecting maize genotypes with an improved NUE.

The relative importance of the genotypic and phenotypic variation to broad-sense heritability was dependent on the N-use trait and the N fertilizer rate (Table 1). Yield at high N exhibited greater genetic variance (within heterotic groups and hybrids) and environmental variance, but lower residual variance than Yield$_{-N}$. Greater genetic variance under high N when compared to low N has also been documented previously [43,44]. Conversely, genetic and environmental variance for harvest index at low N (HI$_{-N}$) were greater than at high N (HI$_{+N}$). Additionally, the genotype by environment interaction was greater under high N for yield and grain protein concentration, but greater at low N for HI and NHI.

Broad-sense heritability (H^2) ranged from 0.11 to 0.77 across phenotypic traits (Table 1), indicating a difference in additive and dominant effects among N-use traits (Table 1). Relatively large residual variances for Yield$_{-N}$, HI$_{-N}$, NHI at low N (NHI$_{-N}$), NUpE, and NUtE resulted in low H^2 of these traits. However, heritability was higher for GU than NUpE or NUtE. The large genotypic variance of GU found is consistent with previous studies [20].

Pearson's pairwise correlations between the GCA effects of different N-use traits are presented in Table 2. Yield at high N is generally positively correlated with Yield$_{-N}$, but the correlation tends to be less under greater N stress [22,45]. Similarly, in this study, the correlation between Yield$_{+N}$ and Yield$_{-N}$ was +0.31. Hybrid correlation coefficients between Yield$_{+N}$ and NUE, NUpE, and NUtE were +0.74, +0.64, and +0.44, respectively, which is in agreement with reports that these traits are frequently positively correlated [20,46]. On the other hand, Yield$_{-N}$ was positively correlated with HI$_{-N}$, HI$_{+N}$, NHI$_{+N}$, and GU.

While significant genetic gains in maize yield have been documented over the past 60 years, the grain protein concentration has consistently decreased during the same period [47]. When averaged over hybrids and environments, grain protein concentration was negatively correlated to yield within each N fertilizer rate (Table 2). In addition, NUpE was positively correlated with grain protein concentration at low N ($r = 0.22$, $p \leq 0.05$) and NUtE was negatively correlated with grain protein concentration at high N (Protein$_{+N}$), ($r = -0.47$, $p \leq 0.001$) (Table 2). This finding reinforces the concept of the inverse relationship of starch and protein in maize grain, with greater N utilization underlying a greater proportion of starch than protein accumulation in the grain. Under high N fertility conditions, NHI was positively correlated to Protein$_{+N}$. Hybrids of the Illinois Protein-Strains germplasm, generating low or high grain protein concentration, exhibited the same overall NUE; while hybrids with high grain protein concentration exhibited high NUpE and NHI, and hybrids with low grain protein concentration exhibited high NUtE [19]. Therefore, maize hybrids with high NUpE may exhibit greater root development and N uptake, while hybrids with high NUtE will show more ability to utilize N for starch production.

Genetic improvements have increased maize yield under low and high N conditions, yet plant N uptake levels have only increased under high N [20]. As such, the genotypic correlations between N-use traits indicate that traits that are related to N fertilizer response (NUE, NUpE, and NUtE) are associated with yield performance under high N conditions, and traits related to the efficiency of nutrient or biomass partitioning to the grain (HI$_{-N}$, HI$_{+N}$, NHI$_{+N}$, and GU) are associated with yield performance under N stress conditions. Although Yield$_{-N}$ and Yield$_{+N}$ are positively correlated, developing maize genotypes with high yield performance under high and low N conditions may be challenging, since the desirable traits for each of these N conditions are negatively correlated (HI, NHI, and GU vs. NUE, NUpE, and NUtE) (Table 2).

Table 1. Mean estimates and range for yield at low N (Yield$_{-N}$, Mg ha^{-1}), yield at high N (Yield$_{+N}$, Mg ha^{-1}), harvest index at high N (HI$_{+N}$, kg kg^{-1}), N harvest index at low N (NHI$_{-N}$, kg$_{grainN}$ kg$_{plantN}$$^{-1}$), N harvest index at high N (NHI$_{+N}$, kg$_{grainN}$ kg$_{plantN}$$^{-1}$), grain protein concentration at low N (Protein$_{-N}$, g kg^{-1}), grain protein concentration at high N (Protein$_{+N}$, g kg^{-1}), N-use efficiency (NUE, kg kg^{-1}), N-uptake efficiency (NUpE, kg$_{plantN}$ kg$_{NR}$$^{-1}$), N-utilization efficiency (NUtE, kg kg$_{plantN}$$^{-1}$), and genetic utilization (GU, kg kg$_{plantN}$$^{-1}$). Variance components for general and specific combining ability effects (GCA and SCA) were calculated using 36 stiff-stalk synthetic (SSS) and 53 non-SSS (NSSS) ex-PVP parental inbred lines across different N conditions (0 and 252 kg N ha^{-1}, respectively).

Trait	Mean ± SE [‡]	GCA$_{SSS}$ [†] Range Min./Max.	σ^2_{SSS}	GCA$_{NSSS}$ Range Min./Max.	σ^2_{NSSS}	SCA Range Min./Max.	σ^2_{SCA}	σ^2_{E}	$\sigma^2_{SCA×E}$	σ^2_{R}	H^2
Yield$_{-N}$	4.9 ± 0.19	−0.7/0.7	0.13	−0.8/0.5	0.12	-	0.00	0.70	0.01	1.31	0.31
Yield$_{+N}$	9.1 ± 0.28	−0.9/+0.9	0.25	−1.2/0.9	0.32	−0.2/+0.3	0.06	2.12	0.36	1.07	0.61
HI$_{-N}$	0.36 ± 0.01	−0.05/+0.09	6 × 10^{-4}	−0.10/+0.06	1 × 10^{-3}	−0.01/+0.01	1 × 10^{-4}	2 × 10^{-3}	3 × 10^{-4}	4 × 10^{-3}	0.63
HI$_{+N}$	0.47 ± 0.01	−0.02/+0.03	2 × 10^{-4}	−0.06/+0.03	4 × 10^{-4}	−0.01/+0.01	5 × 10^{-5}	6 × 10^{-4}	1 × 10^{-5}	1 × 10^{-3}	0.73
NHI$_{-N}$	0.56 ± 0.01	−0.01/+0.01	2 × 10^{-4}	−0.01/+0.01	1 × 10^{-5}	−0.03/+0.02	4 × 10^{-4}	9 × 10^{-3}	2 × 10^{-4}	8 × 10^{-3}	0.11
NHI$_{+N}$	0.68 ± 0.01	−0.04/+0.02	3 × 10^{-4}	−0.05/+0.03	5 × 10^{-4}	−0.02/+0.01	1 × 10^{-4}	3 × 10^{-4}	1 × 10^{-4}	3 × 10^{-3}	0.44
Protein$_{-N}$	62 ± 1.3	−6.1/+5.9	0.8	−5.2/+5.4	0.8	−3.2/+2.3	0.4	1.7	1.3	2.1	0.74
Protein$_{+N}$	85 ± 1.3	−3.8/+4.9	0.7	−6.0/+4.6	1.3	−2.5/+2.5	0.2	5.5	0.5	3.0	0.77
NUE	16.7 ± 1.14	−3.6/+3.9	3.81	−5.7/+5.2	5.50	−0.95/+1.20	0.80	10.13	4.97	18.56	0.60
NUpE	0.43 ± 0.03	−0.03/+0.05	6 × 10^{-4}	−0.08/+0.09	1 × 10^{-3}	−0.01/+0.42	2 × 10^{-4}	2 × 10^{-3}	1 × 10^{-3}	0.01	0.27
NUtE	41.8 ± 1.79	−3.4/+2.9	5.5	−2.7/+4.2	5.2	−0.58/+0.60	1.3	59.3	7.8	201.7	0.11
GU	59.0 ± 2.2	−7.8/+8.9	17.8	−9.9/+7.4	16.0	−3.2/+2.9	5.5	29.1	7.8	88.7	0.58

[†] σ^2_{SSS}, σ^2_{NSSS}, σ^2_{SCA}, σ^2_{E}, $\sigma^2_{SCA×E}$, σ^2_{R} represent variance components for stiff-stalk lines, non-stiff-stalk lines, hybrid, environment, hybrid × environment interaction, and residual effects, respectively (Equation (9)); [‡] SE, standard error of the mean; Min./Max., Minimum and maximum observed values compared to the respective means.

Table 2. Pearsons's pairwise correlations between the GCA effects of the N-use traits of yield at low N (Yield$_{-N}$), yield at high N (Yield$_{+N}$), harvest index at low N (HI$_{-N}$), harvest index at high N (HI$_{+N}$), N harvest index at low N (NHI$_{-N}$), N harvest index at high N (NHI$_{+N}$), grain protein concentration at low N (Protein$_{-N}$), grain protein concentration at high N (Protein$_{+N}$), N-uptake efficiency (NUpE), N-use efficiency (NUE), N-utilization efficiency (NUtE), and genetic utilization (GU) for 263 single-cross maize hybrids grown from 2011 to 2015 under low and high N conditions (0 and 252 kg N ha^{-1}, respectively).

	Yield$_{-N}$	Yield$_{+N}$	HI$_{-N}$	HI$_{+N}$	NHI$_{-N}$	NHI$_{+N}$	Protein$_{-N}$	Protein$_{+N}$	NUE	NUpE	NUtE
Yield$_{+N}$	0.31 **	-									
HI$_{-N}$	0.63 ***	−0.33 **	-								
HI$_{+N}$	0.49 ***	NS	0.77 ***	-							
NHI$_{-N}$	NS	NS	NS	NS	-						
NHI$_{+N}$	0.51 ***	NS	0.65 ***	0.78 ***	NS	-					
Protein$_{-N}$	−0.38 ***	−0.22 *	NS	NS	NS	0.37 ***	-				
Protein$_{+N}$	NS	−0.39 ***	NS	NS	NS	NS	0.73 ***	-			
NUE	−0.33 **	0.74 ***	−0.73 ***	−0.42 ***	NS	−0.35 ***	NS	−0.26 *	-		
NUpE	NS	0.64 ***	−0.59 ***	−0.43 ***	NS	−0.27 *	0.22 *	NS	0.77 ***	-	
NUtE	−0.29 *	0.44 ***	−0.46 ***	NS	NS	−0.21 *	NS	−0.47 ***	0.66 ***	NS	-
GU	0.67 ***	NS	0.82 ***	0.59 ***	NS	0.50 ***	−0.51 ***	NS	−0.59 ***	−0.48 ***	−0.32 **

* Significant at $p \leq 0.05$. ** Significant at $p \leq 0.01$. *** Significant at $p \leq 0.001$.

3.2. Genotype × Environment Interaction of N-Use Traits

In addition to the genotypic correlation between traits, another major challenge for breeding programs is to model the effect of the genotype × environment interaction (G × E) on the desirable phenotypic traits [48]. While the genetic correlation of some N-use traits may be correlated to yield at low or high N conditions, their relationship might differ depending on other environmental conditions influencing yield. A way to compare the effect of an environment on yield is by measuring the average yield of multiple hybrids in each environment receiving similar crop management, termed the 'environmental index'. Several studies have investigated the genetic variability of N-use traits across different N soil conditions [8,20,43,49], but few studies have investigated the effect of G × E on N-use traits. Therefore, regression analysis between an inbreds' EBV at each environment (GCA + GCA × E + E) and the environmental index (E) was performed using the phenotypic traits that correlated to yield at low and high N conditions, respectively (Figure 2). Under low N conditions, GU was stable across environmental indices, and HI_{-N} (0.04 kg kg^{-1}/Mg ha^{-1}), HI_{+N} (0.02 kg kg^{-1}/Mg ha^{-1}), and NHI_{+N} (0.02 kg kg^{-1}/Mg ha^{-1}) increased as the environmental index increased (Figure 2A). Under high N conditions, NUE was stable across environmental indices, while NUtE decreased (-3.60 kg kg$_{plantN}$$^{-1}$/Mg ha^{-1}) and NUpE increased ($+0.03$ kg$_{plantN}$ kg$_{Nfert}$$^{-1}$/Mg ha^{-1}) as the environmental index increased (Figure 2B). The relationship between the G × E effect on N-use traits and the environmental index indicates the degree of trait dominance effects across different environmental yield conditions. A stable additive effect of NUE and GU across environmental indices is desirable for breeding selection in a wide range of environments.

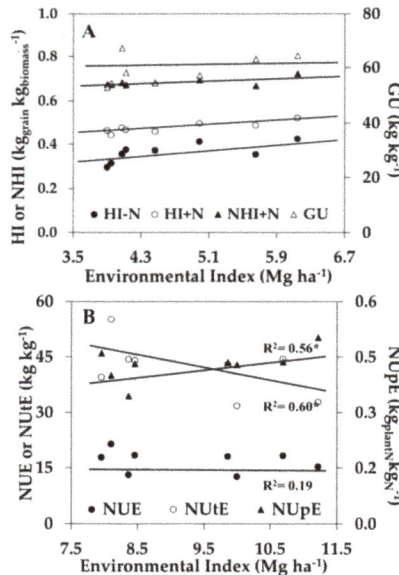

Figure 2. Influence of N supply and environment on selected N-use traits. (**A**) Changes in harvest index at low and high N (HI_{-N} and HI_{+N}), N-harvest index at high N (NHI_{+N}), and genetic utilization (GU) due to the environmental index for maize hybrids grown at low N (0 kg N ha^{-1}); and (**B**) Changes in N-use efficiency (NUE), N-utilization efficiency (NUtE), and N-uptake efficiency (NUpE) due to the environmental index for maize hybrids grown with high N (252 kg N ha^{-1}). Values shown for each phenotypic trait are averaged over all the hybrids grown in each of the eight environments from 2011 to 2015. * Indicates significant slopes at $p \leq 0.001$.

3.3. Identification of Maize Genotypes with Improved NUE

Hybrid NUE performance is determined by the plant's ability to take up nitrogen from the soil (NUpE), the physiological capacity to generate and partition N to the grain (HI and NHI), and the sink strength to set kernels and accumulate starch under high or low N conditions (NUtE and GU, respectively). Consequently, the aim of NUE breeding should be to integrate multiple desirable N-use traits into the same maize genotype. Hierarchical cluster analysis using the GCA effect of different phenotypic traits have categorized SSS (Group 1) and NSSS lines (Group 2) based on their NUE performance (Figure 3). Clusters within heterotic groups consisted of inbreds exhibiting correlated N-use traits (Table 3).

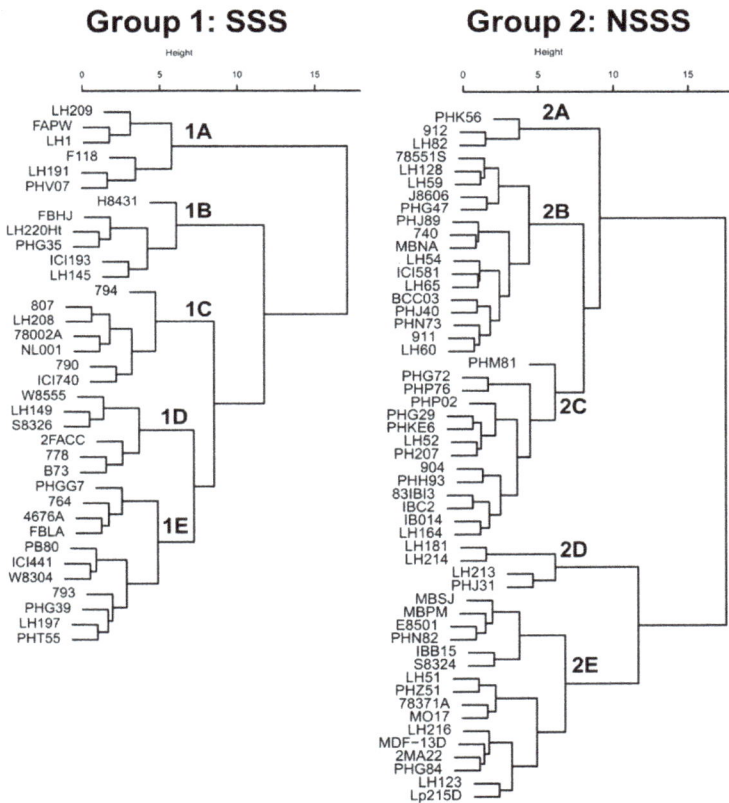

Figure 3. Hierarchical cluster analysis using different N-use traits of 36 stiff-stalk synthetic (SSS, Group 1) and 53 non-SSS (NSSS, Group 2) inbred lines. Clusters A, B, C, D, and E represent groups of inbreds with similar N-use trait performances. Clusters were generated using the inbreds' GCA from 12 N-use traits. Inbred GCAs were calculated from 263 maize hybrids grown from 2011 to 2015 under low and high N conditions (0 and 252 kg N ha^{-1}, respectively).

Table 3. Yield at low and high N (Yield$_{-N}$ and Yield$_{+N}$), harvest index at low and high N (HI$_{-N}$ and HI$_{+N}$), N-harvest index at low and high N (NHI$_{-N}$ and NHI$_{+N}$), grain protein concentration at low and high N (Protein$_{-N}$ and Protein$_{+N}$), N-use efficiency (NUE), N-uptake efficiency (NUpE), N-utilization efficiency (NUtE), and genetic utilization (GU) based on parental inbred cluster groups for the stiff-stalk synthetic lines (SSS, Groups 1A–1E) and non-stiff-stalk synthetic lines (NSSS, Groups 2A–2E). A total of 263 maize hybrids were grown from 2011 to 2015 under low and high N conditions (0 and 252 kg N ha^{-1}, respectively).

Group	N†	Yield		HI		NHI		Grain Protein		NUE	NUpE	NUtE	GU
		Low N	High N	Low N	High N	Low N	High N	Low N	High N				
		Mg ha^{-1}		kg kg^{-1}		kg$_{grainN}$ kg$_{plantN}$$^{-1}$		g kg^{-1}		kg kg$_{Nfert.}$$^{-1}$	kg$_{plantN}$ kg$_{Nfert}$$^{-1}$	kg kg$_{plantN}$$^{-1}$	kg kg$_{plantN}$$^{-1}$
SSS													
1A	6	4.57	9.15	0.33	0.47	0.57	0.67	86.4	68.3	18.29	0.44	42.90	53.26
1B	6	5.05	8.82	0.39	0.48	0.57	0.69	85.0	64.1	15.37	0.42	41.10	64.41
1C	7	5.24	9.13	0.38	0.48	0.57	0.69	85.9	66.1	15.45	0.42	40.51	60.55
1D	6	4.93	9.29	0.37	0.49	0.57	0.69	82.2	64.8	17.90	0.42	43.42	60.44
1E	11	4.88	9.05	0.36	0.48	0.57	0.68	86.1	67.1	16.87	0.43	41.47	57.56
LSD ‡ $p \leq 0.05$		0.29	0.50	0.01	0.01	0.01	0.02	2.6	2.3	1.39	0.02	1.30	1.40
NSSS													
2A	3	5.31	9.32	0.42	0.50	0.57	0.70	83.5	64.0	16.21	0.40	42.70	64.92
2B	16	5.02	9.17	0.37	0.48	0.57	0.69	83.3	65.0	16.71	0.42	42.22	60.18
2C	14	5.08	8.74	0.40	0.49	0.57	0.70	88.1	67.2	14.85	0.42	40.22	61.26
2D	4	4.31	9.52	0.29	0.45	0.57	0.66	86.3	69.1	21.11	0.48	43.75	51.91
2E	16	4.78	9.18	0.34	0.47	0.57	0.67	84.5	66.1	17.47	0.44	42.06	56.65
LSD $p \leq 0.05$		0.21	0.52	0.02	0.02	0.00	0.02	3.2	2.3	1.38	0.03	1.05	1.47

† N, number of ex-PVP inbreds categorized within each cluster group. ‡ LSD, Least significant difference was estimated from different cluster groups within each phenotypic trait.

In the SSS cluster, groups 1A and 1B exhibited unique characteristics with the lowest Yield$_{-N}$ and the lowest Yield$_{+N}$, respectively (Table 3). Group 1A also exhibited high grain protein concentration (under low and high N conditions), NUE, and NUpE, but the lowest GU within the SSS group. In contrast, groups 1B and 1C exhibited high Yield$_{-N}$, but group 1B had the highest GU. Lastly, group 1D exhibited high Yield$_{+N}$, NUE, and NUtE, while group 1E had an average performance for most of the N-use traits. In the NSSS cluster, groups 2A, 2B, and 2C exhibited higher HI and GU than groups 2D and 2E, but groups 2A and 2B had the lowest grain protein concentrations. Group 2A exhibited high Yield$_{-N}$ and Yield$_{+N}$ and the highest GU within the NSSS group. In contrast, group 2D presented high Yield$_{+N}$ and the lowest GU.

Across heterotic groups, only seven inbreds (78551S, B73, LH128, ICI740, PHK56, W8304, and W8555) ranked in the top 25% GCA for both Yield$_{-N}$ and Yield$_{+N}$, and only one inbred (PHK56) ranked in the top 10% for high yield performance under both N conditions (data not shown). Inbred PHK56 was one of the most referenced lines in the U.S. Patent database, and it was derived from PHG35 (from recombination of PHG47 and Oh07-Midland) from the Oh43 background [25]. In addition, inbreds that are genetically related exhibited similar NUE performance (Figure 3). As such, inbreds Mo17 and LH51 (97% identical by descent from Mo17), which are important progenitors of the Lancaster germplasm [25], were categorized in the same cluster (group 2E). Likewise, inbred PH207 is the main founder of the Iodent heterotic group and is an ancestor of several Pioneer Hi-Bred inbreds such as PHG29 and PHG50 [25]. These inbreds exhibited high tolerance to N deficiency and high GU (Group 2C).

One breeding strategy for NUE improvement could be to utilize new inbred or hybrid combinations from the cluster groups with desirable N-use traits. Interestingly, group 2A was the only group exhibiting the combination of high Yield$_{-N}$ and Yield$_{+N}$. Group 2A represents approximately 5% of all NSSS lines tested in this study and could be used as a potential genetic resource for the development of maize genotypes with an improved performance under high N or under N-stress conditions. Inbred combinations between groups 1C × 2A and 1D × 2D, in theory would produce single cross hybrids with high NUE performance under low and high N conditions, respectively.

The identification of maize genotypes with high N-deficiency tolerance and/or high yield performance under sufficient soil N conditions is important for better hybrid placement and agronomic management positioning for maximum and efficient yields. Among the 263 hybrids that were evaluated, only 22 produced yields ranked in the top 25% for both Yield$_{-N}$ and Yield$_{+N}$, and only five hybrids obtained yields ranked in the top 10% for both of the N conditions. Moreover, hybrid ICI740 × PHK56 (combination between groups 1C × 2A) exhibited high yield performance under low and high N conditions (Figure 4). This hybrid exhibited the highest average EBV for Yield$_{-N}$ (6.2 Mg ha^{-1}) and the 9th highest EBV for Yield$_{+N}$ (10.3 Mg ha^{-1}). Hybrid LH145×83IBI3 (groups 1B × 2C) exhibited high tolerance to N deficiency (Yield$_{-N}$ = 5.2 Mg ha^{-1}), but low EBV for Yield$_{+N}$ (8.2 Mg ha^{-1}). This hybrid also combined above average EBV for HI and GU, and below average EBV for NUE and NUtE. In contrast, hybrid F118 × LH214 (groups 1A × 2D) presented the highest average EBVs for Yield$_{+N}$ (11.1 Mg ha^{-1}), NUE, and NUpE, but low EBV for Yield$_{-N}$ (4.4 Mg ha^{-1}) and GU.

Estimated breeding value accuracy is an important method to compare the prediction reliability of desirable traits. Estimated breeding value accuracy ranged from 0.12 to 0.92, and, with the exception of NHI$_{-N}$, EBV accuracies were similar among heterotic groups (Figure 5). While the majority of the inbreds exhibited high EBV accuracy, some of the genotypes did not. Skewness of EBV accuracy may be related to unbalanced data and genotypes with low yield stability across environments.

While precise estimates of H^2 and EBV accuracy are a function of genetic and residual variance, there was no relationship between EBV accuracy averaged across heterotic groups and H^2 (Table 1 and Figure 5). While the H^2 for NUtE and NHI$_{-N}$ was both 0.11, their EBV accuracies were 0.61 and 0.28, respectively. Broad-sense heritability for Yield$_{-N}$ was almost 50% less than H^2 for Yield$_{+N}$. However, these traits presented similar EBV accuracy (approximately 0.82). Discrepancies between H^2 and EBV accuracy can be associated with the genetic architecture of complex traits. Though large residual variance reduced the H^2 of some phenotypic traits (e.g., Yield$_{-N}$, NUpE, and NUtE), large additive variances increased their EBV accuracies.

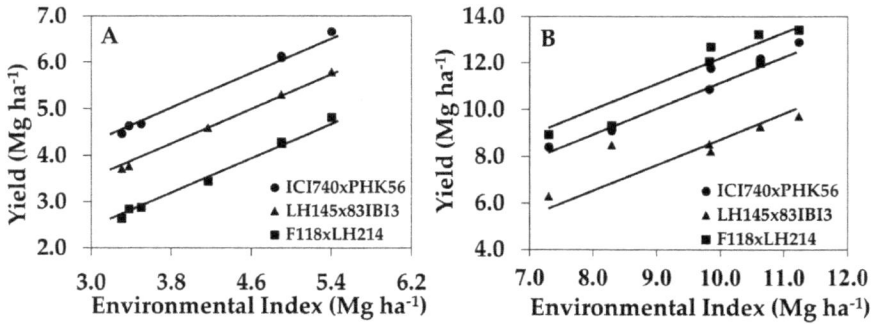

Figure 4. Yield of select hybrids across environmental indices when grown with (**A**) low N (0 kg N ha^{-1}), and (**B**) high N (252 kg N ha^{-1}). Data values are the average yields within an environment for ICI740 × PHK56 (high tolerance to N-deficiency and high positive response to N fertilizer), LH145 × 83IBI3 (high tolerance to N-deficiency and low positive response to N fertilizer), and F118 × LH214 (low tolerance to N-deficiency and high positive response to N fertilizer).

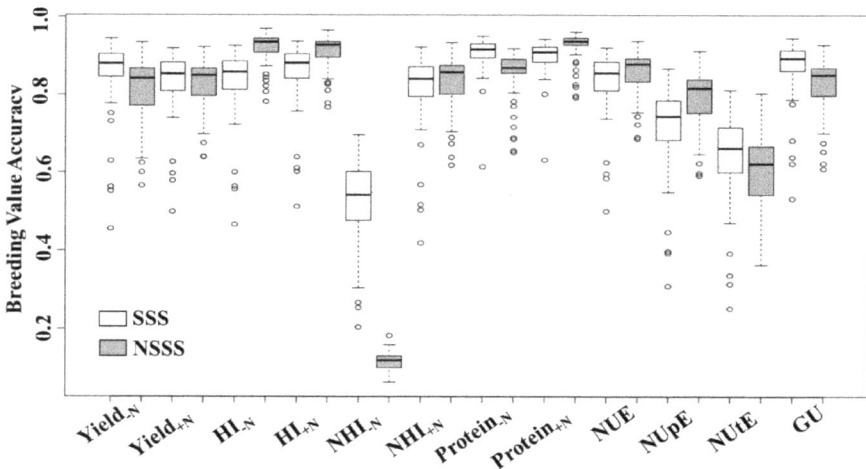

Figure 5. Box-plot of breeding value accuracies for yield at low and high N (Yield$_{-N}$ and Yield$_{+N}$), harvest index at low and high N (HI$_{-N}$ and HI$_{+N}$), N-harvest index at low and high N (NHI$_{-N}$ and NHI$_{+N}$), grain protein concentration at low and high N (Protein$_{-N}$ and Protein$_{+N}$), N-use efficiency (NUE), N-uptake efficiency (NUpE), N-utilization efficiency (NUtE), and genetic utilization (GU) in stiff-stalk synthetic (SSS) and non-stiff-stalk synthetic (NSSS) maize lines. Breeding value accuracy was estimated according to Equation (11). Values are based on the yield performance of 263 hybrids developed from these lines and grown in eight environments from 2011 to 2015 under low and high N conditions (0 and 252 kg N ha^{-1}, respectively).

4. Conclusions

Although 89 inbred lines were evaluated, there were certainly more ex-PVP lines available at the National Plant Germplasm System. Even so, this subset was able to display large genetic variation among ex-PVP lines for most N-use traits. The large range of broad-sense heritabilities that were found for phenotypic traits highlights the importance of accurate phenotypic selection under field conditions. In addition, differences in the stability of N-use traits across environments will have important implications for phenotypic selection. Genetic utilization and NUE were stable across environments and were highly correlated with yield under low and high N conditions, respectively. Hybrids with high N-deficiency tolerance or high yield response to N fertilizer were associated with different phenotypic traits, allowing for breeders to more easily select genotypes that would optimize yields in the respective expected situations. However, because of this dichotomy, less than 2% of the hybrids evaluated exhibited high yield performance under both low and high N conditions. Increasing yields worldwide will require a combination of developing hybrids that have tolerance to low-N situations, as well as hybrids that can take advantage of optimum conditions for a grower's best crop production in their environment. Nitrogen use efficiency is the end result of highly polygenic and complex traits, therefore, deciding which traits are relevant can speed hybrid selection. Future genetic improvement of NUE will require effective integration between accurate field phenotyping and marker-assisted breeding strategies, such as genome-wide prediction and metabolic profiling studies.

Supplementary Materials: The following are available online at http://www.mdpi.com/2077-0472/8/1/3/s1, Table S1: Maize line name, year of release, heterotic group, and proprietary company name of Ex-Plant Variety Protection (PVP) inbreds used as parents in this study, Figure S1: Heatmap showing maize hybrid combinations between 36 stiff-stalk synthetic and 53 non-stiff-stalk synthetic lines developed with the corresponding number of environments tested over three locations in Illinois from 2011 to 2015.

Acknowledgments: This research was made possible with partial funding from the National Institute of Food and Agriculture project NC1200 "Regulation of Photosynthetic Processes" and the Illinois AES project 802-908. The authors wish to thank Juliann Seebauer for her assistance in manuscript preparation, and past and present staff of the crop physiology laboratory for their support.

Author Contributions: All authors contributed in the conception and development of the research. C.C.H. performed initial crossings and experiments. A.T.M. extended the analysis and wrote the paper.

Conflicts of Interest: The authors declare no conflicts of interest.

References

1. FAOSTAT. Production. Food and Agricultural Organization of the United Nations, Statistics Division. 2014. Available online: http://faostat3.fao.org/home/E (accessed on 13 October 2017).
2. USDA-NASS (USDA National Agricultural Statistics Service). *National Statistics for Corn: Corn Grain—Yield, Measured in Bu/acre. Statistics by Subject*; USDA-NASS: Washington, DC, USA, 2014. Available online: http://www.nass.usda.gov/Statistics_by_Subject/index.php (accessed on 14 February 2016).
3. Bender, R.R.; Haegele, J.W.; Ruffo, M.L.; Below, F.E. Nutrient uptake, partitioning, and remobilization in modern, transgenic insect-protected maize hybrids. *Agron. J.* **2013**, *105*, 161–170. [CrossRef]
4. Duvick, D.N. The contribution of breeding to yield advances in maize (*Zea mays* L.). *Adv. Agron.* **2005**, *86*, 83–145.
5. Kihara, J.; Nziguheba, G.; Zingore, S.; Coulibaly, A.; Esilaba, A.; Kabambe, V.; Njoroge, S.; Palm, C.; Huising, J. Understanding variability in crop response to fertilizer and amendments in sub-Saharan Africa. *Agric. Ecosyst. Environ.* **2016**, *229*, 1–12. [CrossRef] [PubMed]
6. Cassman, K.G.; Dobermann, A.; Walters, D.T. Agroecosystems, nitrogen-use efficiency, and nitrogen management. *Ambio* **2002**, *31*, 132–140. [CrossRef] [PubMed]
7. Raun, W.R.; Johnson, G.V. Improving nitrogen use efficiency for cereal production. *Agron. J.* **1999**, *91*, 357–363. [CrossRef]
8. Moll, R.H.; Kamprath, E.J.; Jackson, W.A. Analysis and interpretation of factors which contribute to efficiency of nitrogen utilization. *Agron. J.* **1982**, *74*, 562–564. [CrossRef]

9. Moose, S.P.; Below, F.E. Biotechnology approaches to improving maize nitrogen use efficiency. In *Molecular Genetic Approaches to Maize Improvement. Biotechnology in Agriculture and Forestry Series*; Kriz, A.L., Larkins, B.A., Eds.; Springer: Berlin/Heidelberg, Germany, 2009; Volume 63, pp. 65–77, ISBN 978-3-540-68919-5.

10. Uribelarrea, M.; Crafts-Brandner, S.J.; Below, F.E. Physiological N response of field-grown maize hybrids (*Zea mays* L.) with divergent yield potential and grain protein concentration. *Plant Soil* **2009**, *316*, 151–160. [CrossRef]

11. Seebauer, J.R.; Moose, S.P.; Fabbri, B.J.; Crossland, L.D.; Below, F.E. Amino acid metabolism in maize earshoots. Implications for assimilate preconditioning and nitrogen signaling. *Plant Physiol.* **2004**, *136*, 4326–4334. [CrossRef] [PubMed]

12. Hirel, B.; Bertin, P.; Quillere, I.; Bourdoncle, W.; Attagnant, C.; Dellay, C.; Gouy, A.; Cadiou, S.; Retaillau, C.; Falque, M.; et al. Towards a better understanding of the genetic and physiological basis for nitrogen use efficiency in maize. *Plant Physiol.* **2001**, *125*, 1258–1270. [CrossRef] [PubMed]

13. Hammer, G.L.; Dong, Z.; McLean, G.; Doherty, A.; Messina, C.; Schussler, J.; Zinselmeier, C.; Paskiewicz, S.; Cooper, M. Can changes in canopy and/or root system architecture explain historical maize yield trends in the U.S. Corn Belt? *Crop Sci.* **2009**, *49*, 299–312. [CrossRef]

14. Leach, K.A.; Tran, T.M.; Slewinski, T.L.; Meeley, R.B.; Braun, D.M. Sucrose transporter2 contributes to maize growth, development, and crop yield. *J. Integr. Plant Biol.* **2017**, *59*, 390–408. [CrossRef] [PubMed]

15. Peng, Y.; Li, C.; Fritschi, F.B. Diurnal dynamics of maize leaf photosynthesis and carbohydrate concentrations in response to differential N availability. *Environ. Exp. Bot.* **2014**, *99*, 18–27. [CrossRef]

16. Bihmidine, S.; Hunter, C.T., III; Johns, C.E.; Koch, K.E.; Braun, D.M. Regulation of assimilate import into sink organs: Update on molecular drivers of sink strength. *Front. Plant Sci.* **2013**, *4*, 177. [CrossRef] [PubMed]

17. Weiland, R.T.; Ta, T.C. Allocation and retranslocation of N-15 by maize (*Zea mays* L.) hybrids under field conditions of low and high-N fertility. *Aust. J. Plant Physiol.* **1992**, *19*, 77–88. [CrossRef]

18. Kamprath, E.J.; Moll, R.H.; Rodriguez, N. Effects of nitrogen fertilization and recurrent selection on performance of hybrid populations of corn. *Agron. J.* **1982**, *74*, 955–958. [CrossRef]

19. Uribelarrea, M.; Moose, S.P.; Below, F.E. Divergent selection for grain protein affects nitrogen use in maize hybrids. *Field Crop. Res.* **2007**, *100*, 82–90. [CrossRef]

20. Haegele, J.W.; Cook, K.A.; Nichols, D.M.; Below, F.E. Changes in nitrogen use traits associated with genetic improvement for grain yield of maize hybrids released in different decades. *Crop Sci.* **2013**, *53*, 1256–1268. [CrossRef]

21. Tollenaar, M.; Nissanka, S.P.; Rajcan, I.; Bruulsema, T.W. Yield response of old and new corn hybrids to nitrogen. *Better Crops.* **1997**, *81*, 3–5.

22. Presterl, T.; Groh, S.; Landbeck, M.; Seitz, G.; Schmidt, W.; Geiger, H.H. Nitrogen uptake and utilization efficiency of European maize hybrids developed under conditions of low and high nitrogen input. *Plant Breed.* **2002**, *121*, 480–486. [CrossRef]

23. Gallais, A.; Coque, M. Genetic variation and selection for nitrogen use efficiency in maize: A synthesis. *Maydica* **2005**, *50*, 531–537.

24. Laftte, H.R.; Edmeades, G.O. Stress tolerance in tropical maize is linked to constitutive changes in ear growth characteristics. *Crop Sci.* **1995**, *35*, 820–826. [CrossRef]

25. Mikel, M.A.; Dudley, J.W. Evolution of North American dent corn from public to proprietary germplasm. *Crop Sci.* **2006**, *46*, 1193–1205. [CrossRef]

26. Nelson, P.T.; Coles, N.D.; Holland, J.B.; Bubeck, D.M.; Smith, S.; Goodman, M.M. Molecular characterization of maize inbreds with expired US plant variety protection. *Crop Sci.* **2008**, *48*, 1673–1685. [CrossRef]

27. Hauck, A.L.; Johnson, G.R.; Mikel, M.A.; Mahone, G.S.; Morales, A.J.; Rocheford, T.R.; Bohn, M.O. Generation means analysis of elite ex-plant variety protection commercial inbreds: A new public maize genetics resource. *Crop Sci.* **2014**, *54*, 174–189. [CrossRef]

28. Bari, M.A.A.; Carena, M.J. Can expired proprietary maize (*Zea mays* L.) industry lines be useful for short-season breeding programs? II. Agronomic traits. *Euphytica* **2016**, *207*, 69–77. [CrossRef]

29. Smith, J.S.C.; Duvick, D.N.; Smith, O.S.; Cooper, M.; Feng, L. Changes in pedigree backgrounds of Pioneer brand maize hybrids widely grown from 1930 to 1999. *Crop Sci.* **2004**, *44*, 1935–1946. [CrossRef]

30. Doyle, J.J.; Doyle, J.L. A rapid DNA isolation procedure for small quantities of fresh leaf tissue. *Phytochem. Bull.* **1987**, *19*, 11–15.

31. Elshire, R.J.; Glaubitz, J.C.; Sun, Q.; Poland, J.A.; Kawamoto, K.; Buckler, E.S.; Mitchell, S.E. A robust, simple genotyping-by-sequencing (GBS) approach for high diversity species. *PLoS ONE* **2011**, *6*. [CrossRef] [PubMed]

32. Bradbury, P.J.; Zhang, Z.; Kroon, D.E.; Casstevens, T.M.; Ramdoss, Y.; Buckler, E.S. TASSEL: Software for association mapping of complex traits in diverse samples. *Bioinformatics* **2007**, *23*, 2633–2635. [CrossRef] [PubMed]

33. Jombart, T.; Devillard, S.; Balloux, F. Discriminant analysis of principal components: A new method for the analysis of genetically structured populations. *BMC Genet.* **2010**, *11*, 94. [CrossRef] [PubMed]

34. The R Project for Statistical Computing. Available online: http://www.r-project.org (accessed on 10 December 2016).

35. Bernardo, R. Molecular markers and selection for complex traits in plants: Learning from the last 20 years. *Crop Sci.* **2008**, *48*, 1649–1664. [CrossRef]

36. Ritchie, S.W.; Hanway, J.J.; Benson, G.O. *How a Corn Plant Develops*; Spec. Rep. No. 48; Iowa State University of Science and Technology Cooperative Extension Service: Ames, IA, USA, 1997.

37. Betrán, F.J.; Beck, D.; Bänziger, M.; Edmeades, G.O. Genetic analysis of inbred and hybrid grain yield under stress and nonstress environments in tropical maize. *Crop Sci.* **2003**, *43*, 807–817. [CrossRef]

38. Stuber, C.W.; Cockerham, C.C. Gene effects and variances in hybrid populations. *Genetics* **1966**, *54*, 1279–1286. [PubMed]

39. SAS Institute. *SAS 9.2 User's Guide*; SAS Institute: Cary, NC, USA, 2013.

40. Holland, J.B.; Nyquist, W.E.; Cervantes-Martinez, C.T. Estimating and interpreting heritability for plant breeding: An update. *Plant Breed. Rev.* **2003**, *22*, 9–112.

41. Falconer, D.S.; Mackay, T.F. *Introduction to Quantitative Genetics*, 4th ed.; Longman Technical: Essex, UK, 1996.

42. Gilmour, A.; Cullis, B.; Welham, S.; Gogel, B.; Thompson, R. An efficient computing strategy for prediction in mixed linear models. *Comput. Stat. Data Anal.* **2004**, *44*, 571–586. [CrossRef]

43. Brun, E.L.; Dudley, J.W. Nitrogen responses in the USA and Argentina of corn populations with different proportions of flint and dent germplasm. *Crop Sci.* **1989**, *29*, 565–569. [CrossRef]

44. Trachsel, S.; Leyva, M.; Lopez, M.; Suarez, E.A.; Mendoza, A.; Montiel, N.G.; Macias, M.S.; Burgueno, J.; San Vicente, F. Identification of tropical maize germplasm with tolerance to drought, nitrogen deficiency, and combined heat and drought stresses. *Crop Sci.* **2016**, *56*, 3031–3045. [CrossRef]

45. Bänziger, M.; Betrán, F.J.; Lafitte, H.R. Efficiency of high-nitrogen selection environments for improving maize for low-nitrogen target environments. *Crop Sci.* **1997**, *37*, 1110–1117. [CrossRef]

46. Abe, A.; Menkir, A.; Moose, S.P.; Adetimirin, V.O.; Olaniyan, A.B. Genetic variation for nitrogen-use efficiency among selected tropical maize hybrids differing in grain yield potential. *J. Crop Improv.* **2012**, *27*, 31–52. [CrossRef]

47. Duvick, D.N.; Cassman, K.G. Post-green revolution trends in yield potential of temperate maize in the northcentral United States. *Crop Sci.* **1999**, *39*, 1622–1630. [CrossRef]

48. Van Eeuwijk, F.A.; Bustos-Korts, D.V.; Malosetti, M. What should students in plant breeding know about the statistical aspects of genotype × environment Interactions? *Crop Sci.* **2016**, *56*, 2119–2140. [CrossRef]

49. Smiciklas, K.D.; Below, F.E. Influence of heterotic pattern on nitrogen use and yield of maize. *Maydica* **1990**, *35*, 209–213.

![agriculture logo] *agriculture*

MDPI

Article

Impact of the Genetic–Environment Interaction on the Dynamic of Nitrogen Pools in Arabidopsis

Giorgiana Chietera [1], Sylvain Chaillou [1], Magali Bedu [1,2], Anne Marmagne [1], Céline Masclaux-Daubresse [1] and Fabien Chardon [1,*]

[1]   Institut Jean-Pierre Bourgin, INRA, AgroParisTech, CNRS, Université Paris-Saclay, 78000 Versailles, France; giorgiana.chietera@genmils.com (G.C.); sylvain.chaillou1@gmail.com (S.C.); magali.bedu@agroparistech.fr (M.B.); anne.marmagne@inra.fr (A.M.); celine.masclaux-daubresse@inra.fr (C.M.-D.)

[2]   AgroParisTech, UMR GENIAL, 1 Avenue des Olympiades, 91744 MASSY CEDEX, France

*   Correspondence: fabien.chardon@inra.fr; Tel.: +33-1-30-83-33-76

Received: 15 December 2017; Accepted: 12 February 2018; Published: 14 February 2018

Abstract: Mineral nutrient availability and in particular nitrogen abundance has a huge impact on plant fitness and yield, so that plants have developed sophisticated adaptive mechanisms to cope with environmental fluctuations. The vast natural variation existing among the individuals of a single species constitutes a great potential to decipher complex traits such as nutrient use efficiency. By using natural accessions of *Arabidopsis thaliana* that differ for their pattern of adaptation to nitrogen stress, we investigated the plant response to nitrate supplies ranging from 0.01 mM up to 50 mM nitrate. The biomass allocation and the different nitrogen pools in shoot and in roots were monitored to establish the nutrition status of each plant. Analysis of variation for these traits revealed genetic differences between accessions for their sensibility to nitrate availability and for their capacity to produce shoot biomass with the same nitrogen nutrition index. From the correlation matrix of all traits measured, a statistical model was formulated to predict the shoot projected area from the nitrate supply. The proposed model points out the importance of genetic variation with respect to the correlation between root thickness and amino acids content in roots. The model provides potential new targets in plant breeding for nitrogen use efficiency.

Keywords: nitrogen use efficiency; G × E interaction; natural variation; nitrogen stress; *Arabidopsis thaliana*

1. Introduction

In addition to light, water and temperature, mineral nutrient availability has a huge impact on plant growth and fitness, and in an agricultural context, on crop yield. As sessile organisms, plants cannot escape unfavorable atmospheric or edaphic conditions; therefore, they have developed sophisticated adaptive mechanisms allowing them to cope with the dramatic fluctuations of their environment. This is particularly well illustrated by the strategic acquisition of mineral nutrients from the soil by the roots. For instance, the availability of nitrate (NO_3), the main nitrogen (N) source for nutrition in most high plant species, can vary dramatically in both time and space. Plants are able to react to these variations thanks to specific NO_3 sensing systems, making this ion one of the most potent signal molecule affecting plant physiology and development [1]. The biochemical mechanisms involved in NO_3 uptake, assimilation and remobilization have been widely studied in order to identify the features that determine the nitrogen use efficiency (NUE) of a plant [2–6]. The effect of low N availability on plant biomass, NO_3 uptake, ion contents and root architecture has been widely investigated [7–10]. There is evidence that plants modify their root architecture, changing the lateral to primary roots ratio and decreasing the shoot-to-root ratio at the same time, to forage the soil for nutrients. This foraging response normally involves increased proliferation of lateral roots within nutrient-rich soil patches [11]. Genome-wide microarray analysis in the model plant *Arabidopsis thaliana* under limiting nitrogen conditions shows extensive changes in primary and secondary metabolisms, protein synthesis, and cellular growth processes, with numerous changes to regulatory genes and other cellular pathways [8]. The phenotypic description of Arabidopsis Recombinant Inbred Lines (RIL) built-up to study the plant response to low and high N supplies [12], identified multiple sources of physiological variation, such as those leading to the variations of shoot biomass, nitrogen percentage and free amino-acids content [13]. Several studies showed decreases in total N%, in NO_3 reserves and in total free amino-acids content in plants grown on low regimes [14–16], while soluble proteins content remains unchanged in rosettes [10]. On the other hand, rubisco degradation releases numerous free amino acids available for phloem loading and other interconversions in N depleted plants [2]. One possibility is that amino acids would be interconverted to increase the synthesis of amino acids dedicated to transport, such as glutamine and asparagine. It has also been proposed by Richard-Molard et al. [15] that the initial size of the N storage pool is crucial for the capacity of plants to cope with nitrate starvation, unlike the remobilization dynamics and the composition of the internal N pool [14]. The reduction of growth and photosynthesis, the remobilization of N from old, mature organs to actively growing ones, and the accumulation of abundant anthocyanins, have also been observed in N-starved plants [17]. The complexity of plant responses to temporary or chronic N depletion makes it difficult to give simple explanations, and very often results are contradictory, mainly because of different experimental conditions.

Among individuals belonging to the same plant species, there are great genetic variations for traits that contribute to NUE, including total N uptake, post-anthesis N uptake, N translocation from roots to shoot, and N assimilation [18]. Plant responsiveness to N availability depends on both genotype and the interaction of genotype with N supply level [19]. Root architecture plasticity traits have been investigated in Arabidopsis in relation to N availability by De Pessemier et al. [7]. This study, conducted on a core-collection of 24 accessions, together with the one conducted by Sulpice et al. [20] on 97 Arabidopsis accessions for biomass traits in response to carbon (C) and N nutrition, confirms that a large genetic variation exists in response to nutrient availability within the same species. Results point out the existence of a huge potential to elucidate complex traits such as NUE. Using a natural variation approach, several studies have been conducted in different species in order to individuate the most performing genotypes in response to nutrients availability [21]. Since *Arabidopsis* accessions have been found in a wide range of habitats differing notably in soil richness, this species constitutes a very suitable model for studying genetic variability of plant adaptation to nutrient availability [22]. Recent investigations on N perturbed environments have underlined some major differences between Arabidopsis accessions when they face N fluctuations [14,16]. In particular, Ikram et al. [16] identified accessions showing contrasted responses and different growth adaptive strategies using a core-collection of 23 accessions. Thanks to the comparison of N-limited and N-starved plants to control ones, it was possible to differentiate the adaptive responses of plants by highlighting master traits influencing growth under each nutritional condition. Moreover, it was possible to depict four distinct patterns of adaptation that exist among Arabidopsis accessions that allow the plants to tolerate the imbalance in exogenous N supply.

Despite the efforts to discover genes involved in NUE, few crop-breeding programs today include genomic selection [23]. The first limitation of the selection progress comes from the complexity of the genetic architecture of the NUE process. The second limitation is the existence of a strong genotype $\times$ environment interaction (G $\times$ E) that modulates the number of key genes and their interactions in the network controlling the process. These G $\times$ E interactions require considerable experimental effort to identify the persistent traits that contribute to NUE increases across many environments. Computer-based modelling approaches have recently emerged as a method to save time, labor, and resources, and to infer trait values beyond field experiments [24]. Early models focused on leaf to canopy assimilation, with emphasis on light interception and canopy architecture [25].

Subsequent modelling efforts progressed to develop whole crop models where life cycle prediction, life-long C balance and growth of different organs were emphasized. For example, this has been done for phenological development in soybean [26], leaf elongation rate in maize [27], or fruit quality in peach [28]. All of these studies pointed out the necessity for linking model parameters with easily measurable physiological traits [29]. Among the processes that are needed for improving plants models, Boote et al. [25] highlight the importance of mechanistic methods for predicting allocation of C and N assimilates among plant organs and a better linkage to soil nutrients and soil fertility. An eco-physiological model simulating relations between leaf area and root N retrieval has been developed in *Medicago truncatula* [30], another pointed out the relationship between yield-related traits, N uptake and assimilates with drought stress in rice [31], while in wheat, nine nitrogen treatments have been put in correlation with the grain protein composition [32].

In our study, we investigate the Genotype by Nutrition interaction by describing responses of four Arabidopsis accessions to six different N environments. We first characterize the plant response to N stress following variations of the morphological and metabolic traits measured both in shoot and roots. We then present an integrated genetic and physiological model obtained from the investigation of important traits related to N nutrition management in Arabidopsis, thanks to the description of the responses of four contrasted accessions to six different N environments. The proposed model allows the prediction of relevant biomass values from the environmental N availability thanks to strong correlations between different metabolic and morphological traits. The model can moreover lead to a predictive description of the principal entities that respond to N stress environment as well as to estimate the main variations among accessions that might explain differences in their NUE.

2. Materials and Methods

2.1. Selection of Accessions

Each one of the accessions used in this study, *Bur-0*, *Col-0*, *Cvi-0*, and *Ge-0*, belongs to one of the four classes described in Ikram et al. [16]. They are therefore characterized by contrasted responses to an N-perturbed environment. Seeds were obtained from a Versailles stock center (http://publiclines.versailles.inra.fr/).

2.2. Growth Conditions and Experimental Design

Seeds were surface-sterilized by using ethanol-'bayrochlor' (95–5% v/v) and shaken in this solution shaken for 8 min, then they were rinsed in clear sterile water and allowed to dry. Sowing was done using a toothpick, placing two seeds on the top of one cut Eppendorf tube filled with 0.7% agar and inserted into 96-wells trays filled with distilled water. Seeds were then stratified at 4 °C for three days before transfer to the growth chamber. Growth chamber conditions were as follows: short-day photoperiod of 8 h light at 21 °C, and 16 h darkness at 17 °C, 150 µmol m^{-2} s^{-1} photon flux density and relative humidity was 65%. On the seventh day of growth, seedlings were transferred to six black plastic tanks (1 tank per nutrition) each one hosting 54 plants in a 6 × 9 grid. Each genotype was represented by 12 plants per nutrition regime that where harvested by groups of three to form four biological replicates. Spare holes were filled with additional plants in order to dispose of the highest number of homogeneous plants to compose replicates at harvest. Plastic tanks were filled with 15 L of nutrient solution. The whole set of plants was grown in complete nutrient solution (4 mM NO$_3$) until the 21st day after sowing (das) in order to give them a common background and to allow the formation of N initial stock, then split onto 6 different NO$_3$ regimes for a 14 day period (up to harvest). Solutions were renewed once a week. Nutrition regimes contained respectively NO$_3$ at a concentration of 0.01 mM, 0.2 mM, 1 mM, 4 mM, 10 mM, and 50 mM. One set of plants constituted a control group, since it was grown always at 4 mM NO$_3$ during the whole experiment. Hydroponic culture lasted 35 days in total, in which all plants remained in vegetative stage. Shoots and roots of each plant were separated at harvest. Roots were measured for primary root length and patted dry with a paper towel before weighting. Root thickness values were obtained by simply dividing root weight by primary root length. After weighing, shoots and roots were frozen in liquid nitrogen. The same experimental culture was repeated three times and collected samples were stored at $-80\,^\circ$C until the metabolite extraction procedure.

2.3. Composition of Solutions

Complete nutrient solution contained 4mM nitrate as a sole nitrogen source and it was composed as follows: 3 mM KNO_3, 1.7 mM $CaCl_2$, 2 mM $MgSO_4$, 2 mM KH_2PO_4, 1 mM K_2SO_4, and 0.5 mM $Ca(NO_3)_2$. In 0.01 mM, 0.2 mM, and 1 mM NO_3 nutrition, only KNO_3 was present to constitute the desired N concentration. Deficits of potassium and calcium compared to the complete nutrition solution were compensated by using K_2SO_4 at a concentration of 2.5 mM, 2.4 mM, and 2 mM respectively, and $CaCl_2$ at a concentration of 2.2 mM. In 10 mM and 50 mM NO_3 solutions, KNO_3 was added at a concentration of 3 mM and 5 mM respectively, while $Ca(NO_3)_2$ was added at a concentration of 3 mM and 22.5 mM respectively. $CaCl_2$ was decreased to 0.3 mM in both solutions while K_2SO_4 was eliminated in 50 mM NO_3 solution. All nutrient solution contained microelements in the same amount as follows: 22 μM Na_2EDTA, 0.01 μM $CoCl_2$, 0.9 μM $CuSO_4$, 0.2 μM Na_2MoO_4, 0.5 μM KI, and a solution of NaFeEDDHA in final concentration of 1 mL$\cdot$L^{-1}.

2.4. Extraction of Metabolites

Samples were ground with the help of steel bullets in a shaking grinder, and an aliquot of the obtained powder was weighed and used for extraction of metabolites. A two-step ethanol–water extraction was used, as described in Loudet et al. [13]. The first step consisted in a 25-min extraction at 80 $^\circ$C using 500 μL of 80% (v/v) ethanol, whereas the second step completed the extraction by using 500 μL of double-distilled water at 80 $^\circ$C for 20 min. Supernatants obtained from the two extractions were collected and put together in a well of a 2 mL-96 well plate and dried overnight in a speed-vacuum machine. Samples were then dissolved in 600 μL of double-distilled water and frozen at $-20\,^\circ$C before analysis. Pellets obtained after removing supernatants were dried for one night at 40 $^\circ$C and used for starch extraction.

2.5. Starch Content

Pellets were mixed with 50 mM 3-(N-morpholino)propanesulfonic acid (MOPS) (pH 7) and an amylase solution (15 U) was added to each sample. Samples were incubated at a temperature of 100 $^\circ$C for 6 min. Later, amyloglucosidase (35 U) dissolved in 0.2 M sodium acetate (pH 4.8) was added. After agitation and incubation for 5 hours at a temperature of 50 $^\circ$C, samples were spun for 10 min at 14,000 r/min. Supernatants were collected and conserved at $-20\,^\circ$C before dosage. Starch content was quantified from extracts using the Roche analysis kit, using glucose as a standard. The results were expressed in nmol/mg of dry matter.

2.6. Nitrate Content

Extracts were evaporated and diluted in water before analysing them for nitrate contents. The method used was as described by Miranda et al. [33]. The reactant was prepared by dissolving vanadium III chloride (0.5 g), N-(1-naphthyl)ethylenediamine (0.01 g), and sulphanilamide (0.2 g) in HCl (0.5 M); 1 mM $NaNO_3$ was used as standard. After loading the plate with samples (100 μL), an equal volume of reactant was added to each well and the reaction was carried out at room temperature for 5–6 h. The absorbance at 540 nm was measured using a spectrophotometer (Labsystem iEMS Reader MF) and used to estimate the nitrate content in nmol mg^{-1} dry matter (DM).

2.7. Amino Acids Content

The same extracts were subjected to an evaluation of free amino acid content using glutamine as a standard Rosen [34]. Briefly, ninhydrin colour reagent was first made by dissolving 0.3 g of ninhydrin (Sigma-Aldrich Chimie, Lyon, France) in 10 mL of methyl cellosolve (Sigma-Aldrich Chimie, Lyon, France). Cyanide acetate reagent was then prepared by mixing Na-acetate buffer (25 mL, 2.5 M, pH 5.2) with the KCN (10 mM) solution 2/100 (*v/v*) just prior to reaction. In a 96-well (2 mL) plate containing 200 µL of sample (diluted), 100 µL of ninhydrin colour reagent followed by 100 µL of cyanide acetate reagent were added. The plate was shaken and heated for 15 min at 100 °C. After cooling, 1 mL of isopropanol (50% *v/v*) was added to the wells of the plate and mixed well. Absorbance was read at 570 nm on a spectrophotometer. This result was used to calculate the amino acid content in nmol mg^{-1} DM using a glutamine 4 mM dilution set as calibration standard.

2.8. Ammonium Content

Ammonium was determined adding a solution of 2% 5-sulfosalicylic acid according to the Berthelot method to samples extracted in a hydro-alcoholic procedure as previously described. Calibration curve was obtained using a 1 mM $(NH_4)_2SO_4$ dilution series.

2.9. Nitrogen and Carbon Percentage

N% and C% were determined using lyophilized plant powder (1 mg) and the Dumas combustion method with an NA1500CN Fisons instrument (Thermoquest, Runcorn, Cheshire, UK) analyzer.

2.10. Soluble Proteins Content

Soluble proteins were extracted from 30 mg of fresh shoots material, and ground in a shaker with Eppendorf safe-lock 2 mL tubes. In each tube, one spatula of Fontainebleau sand was added to plant material in order to facilitate a homogeneous grinding procedure, together with 250 µL of buffer composed as follows: 250 mM Tris HCl at pH 7.6, 10 mM Na-EDTA, 10 mM MgCl$_2$, 28.6 µM β-mercaptoethanol and 2 µM leupeptine. Tubes were kept at low temperature by cooling them in liquid nitrogen before the grinding process and keeping them on ice afterward. After extraction, samples were centrifuged and 200 µL from each tube were transferred to a 96-well plate which was then stored at −20 °C. Protein concentration was determined using a commercially available kit (Coomassie Protein assay reagent, BioRad, Hercules, CA, USA) and bovine serum albumin (BSA) as a standard.

2.11. Shoot Projected Area

The shoot projected area was obtained by area calculation of rosettes pictures of each genotype, with the software ImageJ (https://imagej.nih.gov/ij/index.html); the selected surface area was then divided by the number of plants grown in the same tank to give an average estimation for a single plant.

2.12. Determination of Nc and Nutrition Indexes

Nc was obtained by choosing the average N% in shoots in plants growing at 4 mM. Nutrition indexes were calculated by the following formula usually used to compute the Nitrogen Nutrition Index (NNI) [35], but generalized to others N pools:

$$\text{Nutrition Index} = (\text{Pool}_{\text{observed}} \times \text{SFM}_{\text{observed}})/(\text{Pool}_{\text{control}} \times \text{SFM}_{\text{control}})$$

where:
SFM$_{\text{control}}$ is the shoot fresh matter (SFM) value at 4 mM NO$_3$,
SFM$_{\text{observed}}$ is the SFM value observed,

Pool$_{control}$ is the N pool value at 4 mM (N% for NNI, SAA and RAA for AANI, and SNO3 and RNO3 for NO3NI),

Pool$_{observed}$ is the N pool observed.

2.13. Statistical Analyses

Type III analysis of variance (ANOVA) analyses was carried out using XLSTAT. The ANOVA model for the phenotypical and metabolic data included three main effects (Genotype, Nutrition and Experiment) and the interactions combining all the effects. The ANOVA model for nutrition indexes included only the three main effects (Genotype, Nutrition and Experiment), the interaction effects were generally not significant. Effects were significant when the *p*-value was less than 0.001 for phenotypical and metabolic data and 0.05 for nutrition indexes.

The pairwise differences of slope of the linear regression of SFM against NNI among accessions were computed following the following adapted *T*-test:

$$T_{test} = \frac{(b_1 - b_2)/\sqrt{(n1 + n2 - 2)}}{\sqrt{SRR.\left[\frac{1}{\sum_{n1}(NNI_{1,n1} - NNI_1)^2} + \frac{1}{\sum_{n2}(NNI_{2,n2} - NNI_2)^2}\right]}}$$

where:

b_1 is the slope of accession 1,

b_2 is the slope of accession 2,

$n1$ and $n2$ are the number of observation for the two accessions,

SSR is the sum of squared of the residuals,

NNI_1 is the average *NNI* of accession 1,

NNI_2 is the average *NNI* of accession 2,

$NNI_{1,n1}$ is the nth observed *NNI* of accession 1,

$NNI_{2,n2}$ is the nth average *NNI* of accession 2.

2.14. Statistical Modelling

Pairwise correlations between phenotypical and metabolic traits were calculated for all with a Pearson's correlation method. Correlation coefficients were tested for significance (*p* value < 0.01) by Excel spreadsheet at http://www.stat-help.com/notes.html with DeCoster utility for Applied Linear Regression [36]. For each correlation pair selected in the model, linear, logarithmic, and power regression were run on the whole data by using least squares methods in Excel spreadsheet. For each regression tested, the R-squared was recorded. Regression with the best R-squared was selected to fit the curve to the whole data. The same regressions were used on subsets concerning a single accession to fit the curve and determine the smooth parameters for the considered accessions.

3. Results

In order to investigate the effect of N environmental availability on Arabidopsis plants, six different supplies were tested with four accessions (*Bur-0*, *Col-0*, *Cvi-0* and *Ge-0*) belonging to different classes on the basis of their response to NO_3 nutrition [16]. All plants were grown at a control nutrition regime (4 mM NO_3) for three weeks, to give them a common background and to allow the formation of an initial N stock. Plants were then moved onto different N treatments. In nutritive solution, NO_3 was present at a concentration of 0.01 mM, 0.2 mM, 1 mM, 4 mM, 10 mM, and 50 mM. It constituted the only N source provided to the plants, which were harvested in their vegetative growth period, 35 days after sowing (das). The response to available N was evaluated by measuring 6 morphological and 13 metabolic traits at harvest, commonly used to give a description of the status of plants in nutritional studies [13,15,16,19]. The morphological traits were SFM, root fresh matter (RFM), the ratio between shoot and roots fresh matter (SRFM), primary root length (RL), root thickness (RT), and the shoot projected area (SPA). The metabolic traits were shoot nitrate content (SNO3), roots nitrate content (RNO3), their ratio (SRNO3), shoot free amino-acids content (SAA), root free amino acid content (RNO3), their ratio (SRAA), total shoot carbon (C%), total shoot nitrate (N%), their ratio (C/N), shoot ammonium content (SNH4), root ammonium content (RNH4), shoot protein content (SPC), and shoot starch content (SStarch). All measurements were done on three independent and identical cultures. The raw data for the three cultures are in listed in Supplemental Table S1.

In order to determine the proportion of variance explained by genetic and environmental effects, ANOVA were performed for all studied traits under the six different nutrition regimes. Nutrition, accession (i.e., genotype), and culture (i.e., experiment) were considered as the main effects with potential interactions. ANOVA results are shown in Figure 1 and the proportion of explained variance for each effect are listed in Supplemental Table S2. Genotype has a significant impact on the variance of all the investigated traits, even when the percentage of explained variance was weak (Supplemental Table S2), expect for N%. Morphological traits variation were confirmed to be mainly driven by genotype constraints, as already stated in Ikram et al. [16], with only one exception for SRFM whose variation was affected mainly by nutrition (40% of explained variation). Nutrition explained most of the variation for all metabolite traits, with a different percentage of explained variation depending upon each trait. However, RNH4 and SRAA were two exceptions, in which genotype effect has a higher impact than nutrition effect on the observed variation. ANOVA results indicated that metabolite traits were more affected by N status than by the genetic origin of accessions.

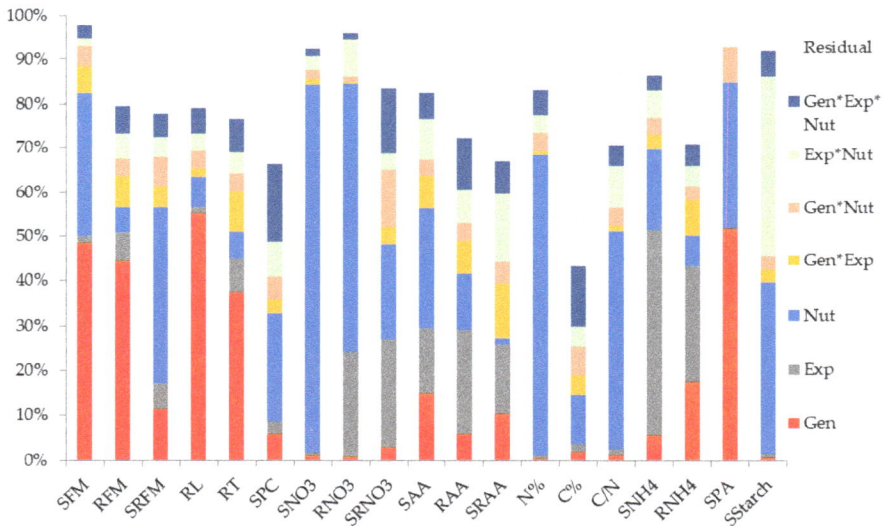

Figure 1. ANOVA of six morphological and 13 metabolic traits in four Arabidopsis accessions cultivated on six different nutrient supplies in three independent experiments. The morphological traits are: shoot fresh matter (SFM), roots fresh matter (RFM), shoot-to-root fresh matter ratio (SRFM), primary root length (RL) and root thickness (RT). The metabolic traits are: shoot protein content (SPC), shoot nitrate content (SNO3), roots nitrate content (RNO3), shoot-to-root nitrate content ratio (SRNO3), shoot amino acids content (SAA), roots amino acids content (RAA), shoot-to-root amino acids content ratio (SRAA), total nitrogen percentage (N%), total carbon percentage (C%), carbon to nitrogen percentages ratio (C/N), shoot ammonium content (SNH4), roots ammonium content (RNH4), shoot projected area (SPA) and shoot starch content (SStarch). Histograms show the effects due to: genotype (Gen), experiment (Exp), nutrition (Nut), interactions and residual (Residual), as percentages of the total variation.

3.1. Global Response of Arabidopsis Plants to a WIDE Range of NO₃ Supplies

The average morphological traits on the four accessions among the six nutrient regimes are presented in Figure 2.

SFM and SPA, two highly correlated traits ($r^2 = 0.88$), responded negatively to the extreme N supplies, both decreasing for the lowest as well as for the highest NO_3 concentrations (Figure 2A,B). SRFM showed the highest values at 4 mM, 10 mM, and 50 mM (Figure 2C). For root-related traits, responses to nitrate availability were the opposite of the responses of shoot-related traits. RFM showed the highest levels at low N (0.01 mM, 0.2 mM and 1 mM) and control (4 mM) conditions, while the lowest RFM were observed at 10 mM and 50 mM NO_3 (Figure 2D). RT showed medium values in the control condition (4 mM), with a significantly increased and decreased level in the lowest and in the highest NO_3 supplies respectively (Figure 2E). RL was at its maximum in the control condition (4 mM). It fell with decreasing NO_3 concentration in the media, with a significant change detectable at 0.01 mM as well as at 10 mM and 50 mM supplies (Figure 2F).

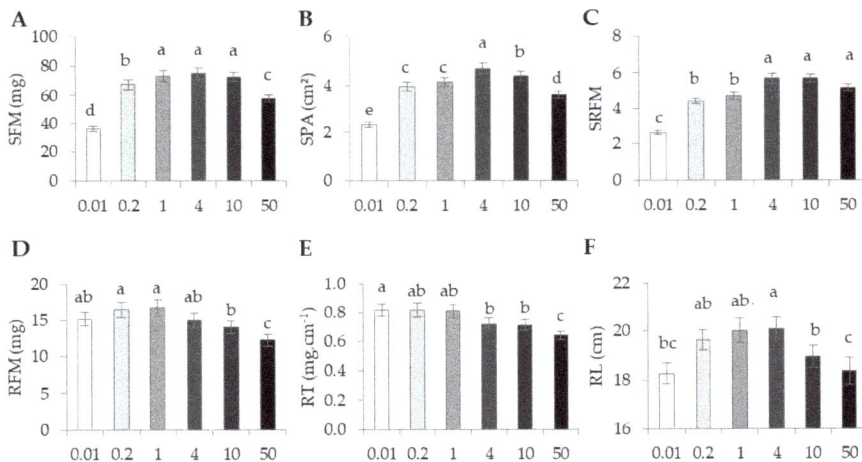

Figure 2. Average of morphological traits on the four Arabidopsis accessions grown in six different NO_3 supplies. On the *x* axis, nutrition regimes are reported in bars of different shades of grey, according to NO_3 concentration (mM). On the y axis, shoot fresh matter ((**A**)-SFM), shoot projected area ((**B**)-SPA), shoot-to-root fresh matter ratio ((**C**)-SRFM), root fresh matter ((**D**)-RFM), root thickness ((**E**)-RT) and primary root length ((**F**)-RL) are reported. Different letters indicate values that are significantly different at $p < 0.05$.

Average Arabidopsis metabolic traits for the six nutrition regimes are reported in Figure 3. By far, metabolic traits had a larger percentage of variation explained by nutrition compared to morphological traits (Figure 1). Interestingly, SNO3, RNO3, SAA, RAA, SNH4, and RNH4 steadily increased following the increase of NO_3 in solution (Figure 3A–F). This was also observed for N% from 0.01 mM, but it remained fairly constant between 0.2 mM and 50 mM (Figure 3G). C% had average values in the control condition with significantly higher and lower values at 0.01 mM and 50 mM respectively (Figure 3H). SPC showed a different trend in which the highest content was found at 1 mM and a gradual, but significant, decrease was observed at higher N supplies and at 0.01 mM (Figure 3L). The C/N ratio did not vary significantly except at 0.01 (Figure 3I). The nitrate ratio SRNO3 remained stable in supplies between 1 mM to 50 mM, but it increased at supplies below 1 mM (Figure 3J). In contrast, the amino acid ratio SRAA remained constant in all conditions (Figure 3K). The average starch contents in shoot were higher than in the control condition in extreme supplies, at 0.01 mM and 50 mM (Figure 3M).

Beyond the potential offered by genotype for further investigations, a G × E interaction represented the specific genetic response to perturbed N environment. This was a significant source of variation for all the investigated traits to different extents, as SRNO3 and SRFM explained variations of 15.6% and 8.5% respectively (Supplemental Table S2).

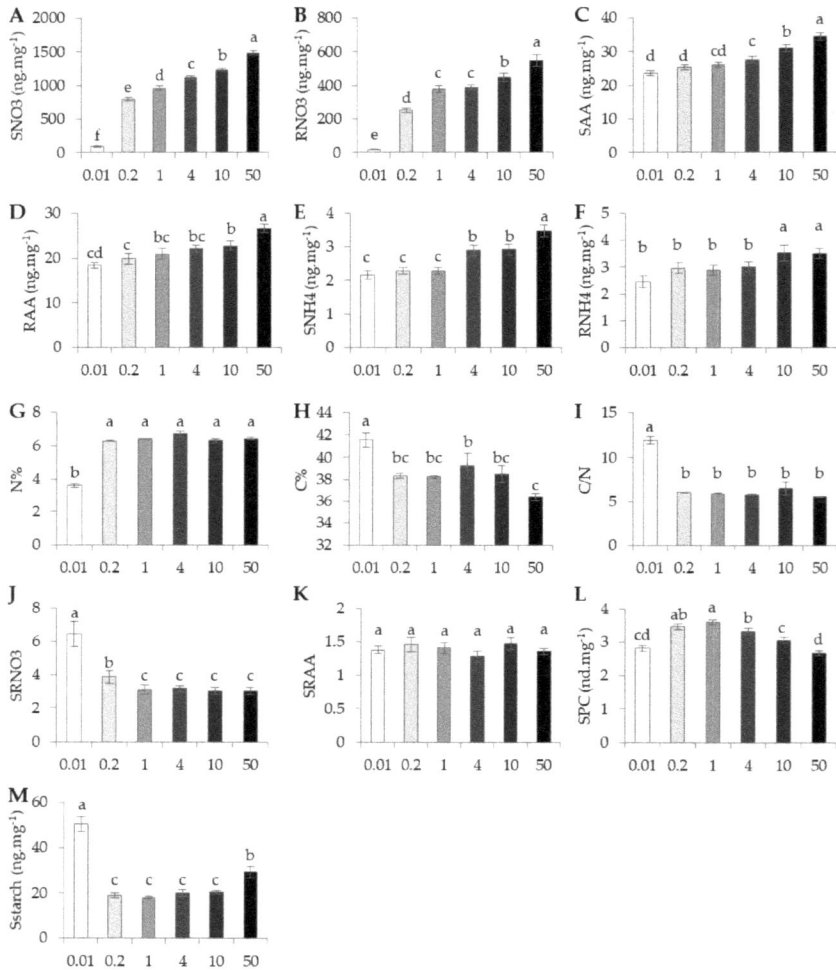

Figure 3. Average of metabolic traits on four Arabidopsis accessions grown in six different NO_3 supplies. On the x axis nutrition regimes are reported in bards of different shades of grey, according to NO_3 concentration (mM). On y axis are reported shoot nitrate ((**A**)-SNO3), roots nitrate ((**B**)-RNO3), shoot amino acids ((**C**)-SAA), roots amino acids ((**D**)-RAA), shoot ammonium ((**E**)-SNH4), roots ammonium ((**F**)-RNH4), nitrogen percentage ((**G**)-N%), carbon percentage ((**H**)-C%), carbon to nitrogen percentages ratio ((**I**)-C/N), shoot-to-root nitrate ratio ((**J**)-SRNO3), shoot-to-root amino acids ratio ((**K**)-SRAA), shoot protein content ((**L**)-SPC) and shoot starch ((**M**)-Sstarch). Different letters indicate values significantly different at $p < 0.05$.

The average Arabidopsis variation for SRNO3 showed the highest value at 0.01 mM NO_3, an intermediate value at 0.2 mM, and lower ones from 1 mM to 50 mM. This general trend was observed in *Bur-0* and *Cvi-0* (Figure 4A). In *Col-0*, SRNO3 remained steady for any nutrition regime and SRNO3 varied significantly only for the two extreme NO_3 supplies (0.01 mM and 50 mM) in *Ge-0*. Among the morphological traits, SRFM showed the highest variation driven by G × E interaction (6% of explained variation). For *Bur-0* and *Cvi-0*, SRFM decreased significantly only at 0.01 mM, whereas in the two last accessions, SRFM was significantly different among low N conditions (0.01 mM, 0.2 mM and 1 mM) compared to high N conditions (4 mM, 10 mM and 50 mM, Figure 4B).

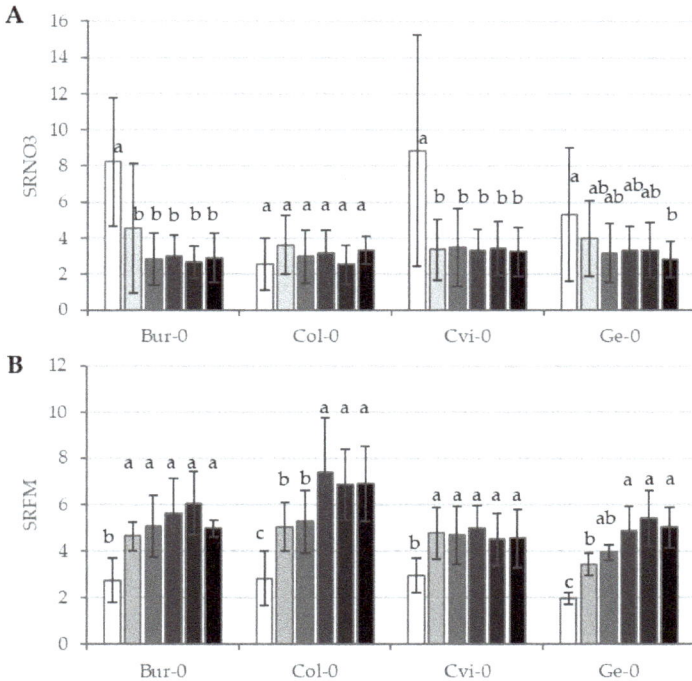

Figure 4. Genotype by nutrition interaction on trait variation. Shoot-to-root NO_3 content ratio ((**A**)-SRNO$_3$) and shoot-to-root fresh matter ratio ((**B**)-SRFM). The six different NO_3 supplies are reported in bars of different shades of grey. Different letters indicate values significantly different at $p < 0.05$.

3.2. Genetic Response to N Stress Environment

When we investigated morphological traits in the different conditions compared to the control at 4 mM, we distinguished clear differences among the 4 investigated accessions. A synthesis of the response of accessions to nutritional change compared to the control condition together with a plot for the average response are shown in Figure 5.

Compared to the situation at 4 mM, SFM and RFM decreased when plants were grown in other supplies in *Bur-0*, *Col-0*, and *Ge-0*, but the two traits increased in *Cvi-0* at 0.2 mM and 1 mM. *Col-0* showed a greater reduction of shoots and roots compared to *Bur-0* and *Ge-0*. These results suggested that *Cvi-0* was the most adapted accession to low N supplies, whereas *Col-0* was the accession the most sensitive accession to N fluctuation. The dramatic sensitivity of *Col-0* was observable on its root architecture. RL was reduced between −10% and −31% in the different N supplies for *Col-0*, whereas RL varied between 4% and −17% in others accessions. Interestingly, *Bur-0* did not show any significant difference for RL compared to the control condition. SRFM, the morphological trait for which the highest impact of G × E interaction was found, tended to decrease for all accessions in N concentrations lower than the control, reaching significantly lower values at 0.01 mM, 0.2 mM, and 1 mM NO_3 for *Col-0* and *Ge-0*, while only at 0.01 mM NO_3 for *Bur-0* and *Cvi-0*. Among metabolic traits, we observed similarly various responses to N environment among accessions. For instance, SAA increased at 50 mM compared to the control condition (4 mM), but this trend was lower in *Col-0* compared to the other accessions (+12% in *Col-0* and +25% in average for the others). We noticed a great reduction of SAA in *Ge-0* at 0.01 mM although the reduction was weaker for the 3 other accessions (−26% for *Ge-0* and on average −0.14% for the others). At the roots level, the amino acids tended to increase with NO_3 availability (Figure 3E). However, RNO3 was higher in all N conditions compared to the control condition in *Col-0* whereas in the three other accessions RNO3 was reduced in low N conditions (0.01 mM, 0.2 mM and 1 mM) and increased at 50 mM. Compared to the control condition, SRNO3 remained stable in other NO_3 conditions except at 0.01 mM where SRNO3 increased for *Bur-0*, *Cvi-0*, and *Ge-0*. *Col-0* did not show any significant change at 0.01 mM NO_3 regime. These variations in the accession response to NO_3 availability suggested the existence of a natural variation for dynamics of nitrogen pools and sensitivity to external NO_3 concentration.

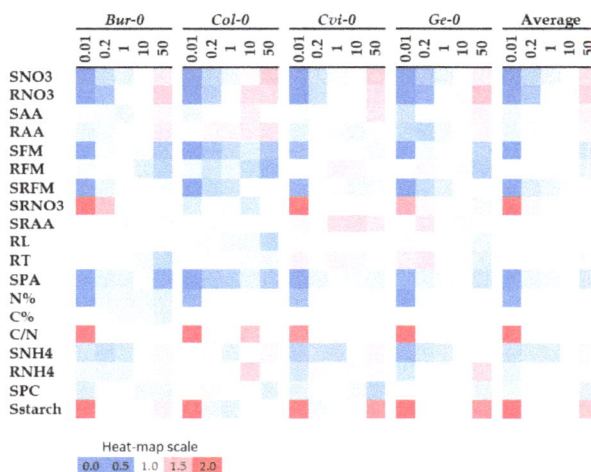

Figure 5. Heat-map of accession changes according to N fluctuations. Values measured in the different N condition are expressed as a ratio to control values (4 mM NO_3) for the four accessions (*Bur-0*, *Col-0*, *Cvi-0* and *Ge-0*) and for the total average (Average). Decreased and increased values in comparison to control are shown in blue and red shades, respectively.

Critical N concentration (Nc), i.e., the minimum N concentration in plants needed for the maximum growth rate [35], was estimated to be 6.4 N% in shoots for all accessions at the control condition (4 mM NO_3). Nc was stable at all N supplies above 0.2 mM NO_3, but fell when plants were grown below this nitrate threshold. We then calculated the Nitrogen Nutrition Index (NNI) as the ratio between the amounts of N in plants growing in a specific supply compared to the N amount in plants growing in the control condition (Supplemental Table S3). This index allowed agronomists to diagnose nitrogen deficiency in crops [35]. NNI can vary from 0 in very deficient conditions, to values greater than 1 for ad lib consumption of N. The four studied accessions showed variation in their NNI under stress conditions (Figure 6A). *Cvi-0* seemed to have a good capacity to store N, since its NNI was greater than 1 when it grew between 1 mM and 10 mM NO_3. We tested the effect of genotype on observed variation of NNI using an ANOVA with three main effects: genotype, nutrition, and experiment. The results of ANOVA, recorded in Supplemental Table S4, showed that the NNI of *Cvi-0* was significantly higher than the values of the three other accessions. For instance, the NNI of

Cvi-0 reached a maximum at 1 mM NO_3. In contrast, the three other accessions always showed a NNI below 1 in conditions different from control (Figure 6A). At 0.01 mM, 38% of N present at the control condition (4 mM) was in *Cvi-0*, while other accessions had only between 22% and 27% of the N needed for the normal growth. In the same manner for NNI, we estimated the variation in the NO_3 pool and the amino acid pool respectively by calculating the nitrate nutrition index (NO3NI) and amino acids nutrition index (AANI—Supplemental Table S3). NO3NI is the ratio between the amounts of nitrate stored in plants growing in a particular supply, divided by the nitrate amount stored in plants growing at the control condition. AANI is the ratio between the amount of free amino acids in plants growing in particular supply divided by the amount stored in plants growing at the control condition, both in shoots and roots for each accession (Figure 6B–E). For the two nutrition indexes, *Cvi-0* showed significant higher indexes than the others accessions (supplementary Table S4). For instance, in shoot at 1 mM, around 75% of the nitrate pool and 80% of the free amino acid pool were filled for *Bur-0*, *Col-0* and *Ge-0* plants (NO3NI = 0.75 and AANI = 0.80) whereas the nitrate pool and the free amino acid pool in *Cvi-0* were optimal (NO3NI = 0.96 and AANI = 1.12, Figure 6B,C). Similarly, in roots at 0.02 mM, around 60% of the nitrate pool 70% of the free amino acid pool were filled for *Bur-0*, *Col-0*, and *Ge-0* plants (NO3NI = 0.54 and AANI = 0.71) whereas the nitrate pool and the free amino acid pool in *Cvi-0* were less reduced (NO3NI = 0.80 and AANI = 0.98, Figure 6D,E). The results suggested *Cvi-0* was less sensitive to NO_3 availability than the three other accessions since its pools of nitrogen remained full in N stress condition unlikely the common pattern in others accessions.

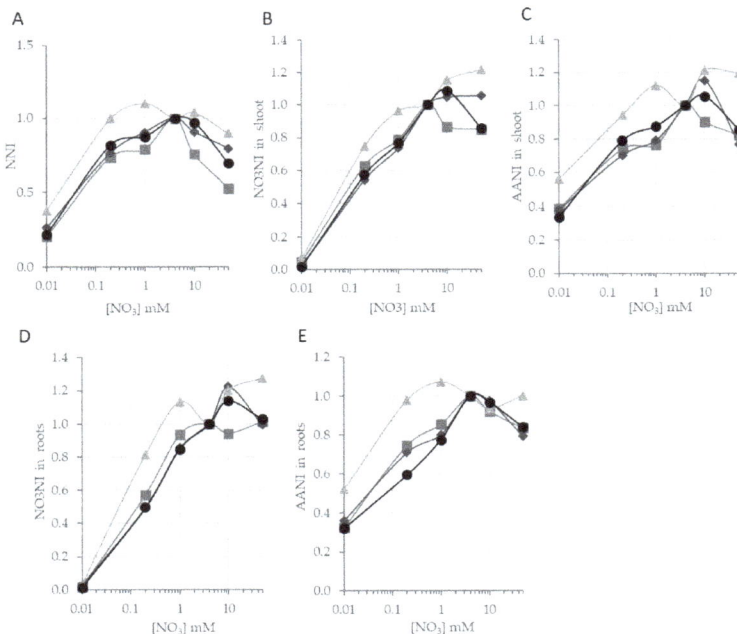

Figure 6. Nutrition Index of different N pools in four Arabidopsis accessions. On the *y* axis, Nitrogen Nutrition Index (NNI, (**A**), Nitrate Nutrition Index (NO3NI, (**B**) in shoot, (**D**) in roots) and Amino Acids Nutrition Index (AANI, (**C**) in shoot, (**E**) in roots) are reported, plotted against environmental NO_3 concentration on a logarithmic scale. Accessions are shown by squares, diamonds, triangles, and circles in different shades of grey for *Bur-0*, *Col-0*, *Cvi-0* and *Ge-0*, respectively.

To give an idea of natural variation for NUE, we plotted the SFM of each accession against their NNI (Figure 7). We observed a strong positive correlation between the two traits, suggesting a linear response of plant growth to the N nutrition status in all accessions. However, the slopes of the linear regression line among accessions were significantly different among the four accessions (Supplemental Figure S1). These slopes allowed us to distinguish three groups of accession: *Bur-0* with the highest slope (77.6), *Cvi-0* and *Ge-0* with intermediate slopes (57.2 and 50.4, respectively) and *Col-0* with the lowest slope (28.8). Interestingly, we noticed that the order of efficiencies between accessions was the same when we considered the NO3NI and AANI in the shoots (Supplemental Figure S2A,B). In contrast, the relationship between RFM and NO3NI and AANI in roots was not linear (Supplemental Figure S2C,D). Taking together these results suggested that for a same nitrogen nutrition index, *Bur-0* was an efficient accession to produce shoot biomass whereas *Col-0* was an inefficient accession.

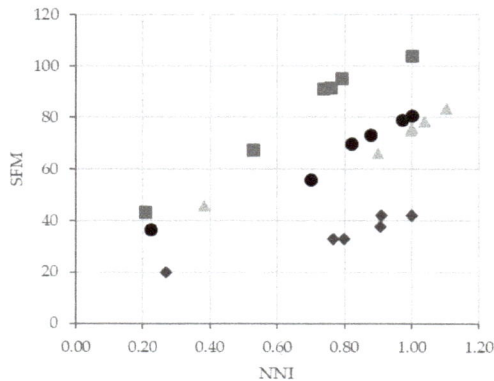

Figure 7. Nutrition Use Efficiency in four Arabidopsis accessions. Shoot fresh matter (SFM) of the four accessions are plotted against their Nitrogen Nutrition Index. Accessions are showed by squares, diamonds, triangles and circles in different shades of grey for *Bur-0*, *Col-0*, *Cvi-0*, and *Ge-0*, respectively.

3.3. Strong Correlations Drive the Construction of a Genetic-Physiologic Model of NUE in Arabidopsis

Because there was a strong genetic variation between plants, N nutrition did not correlate globally with easy-to-measure morphological traits such as SFM and SPA ($r^2 = -0.05$ and $r^2 = 0.01$, respectively). N nutrition only slightly correlated with RFM ($r^2 = -0.19$). It was thus not possible to predict plant biomass by only knowing the nutritional regime applied. Using Pearson's correlation matrix (Supplemental Table S5), we recorded the two highest correlations for each trait to build up a descriptive model starting from the traits that correlated with N nutrition the most, such as SNO3 and RNO3 ($r^2 = 0.60$ and $r^2 = 0.51$ respectively), towards biomass traits (Figure 8). Analyzing the correlation matrix, we identified four groups of traits corresponding to defined pools that were strongly correlated among them (Figure 8). The nitrate pools in shoot and roots were correlated directly to N regime. Traits linked to elements such as N% and C% followed the group of nitrate pools, and in particular SNO3, by which they were linked by strong correlations. In a similar manner, a third group of traits related to amino acids and ammonium pools were well correlated to the group of nitrate pools. Lastly, traits related to biomass allocation, which includes SFM and RFM, were connected to the amino acids and ammonium pool, in view of the good correlation existing between RAA and RT ($r^2 = -0.61$). Interestingly, SPC represents a completely independent trait, showing weak correlations to others traits. From Figure 8, we drew the shorter pathway to connect the N nutrition of plants to SPA, in accordance to the arrow orientations. The model then consisted of six steps, connecting N nutrition to RNO3, RAA, RT, RFM, SFM, and finally SPA (Table 1).

Plotting on *x* and *y* axis the observed values of traits that are linked by the black dotted line in Figure 8, we tested different equations to fit a regression curve to the whole data. For each step, we choose the curve that fit the better with the observed values, i.e., the highest *R*-squared value

(Table 1). Nutrition-RNO3 and RAA-RT plots fitted with a logarithmic curve, RNO3-RAA, RT-RFM, and SFM-SPA with a linear line, RFM-SFM with an exponential curve. Each parameter of these curves was first computed from the best fitted equation from all the available data (Table 1). We then estimated smooth parameters from datasets corresponding to each accession and each experiment (Supplemental Table S6). Finally, we estimated SPA using the six steps of the statistical model (Table 1), the smooth parameters estimated for each accessions and the initial NO_3 concentration. The computed SPA values were close to the observed SPA values ($R^2 = 0.55$), supporting the idea that the model simulated well the process of NUE in the studied plants.

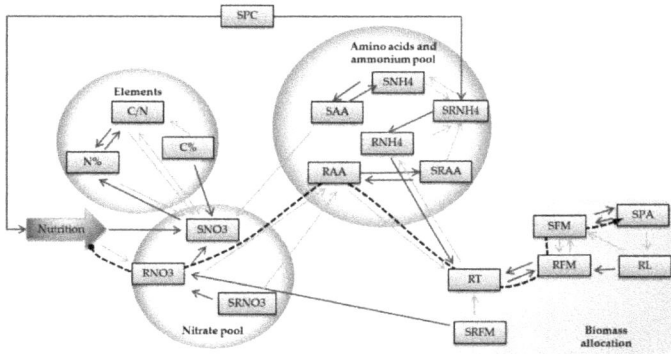

Figure 8. Descriptive model of morphological and metabolic responses to N nutrition. Traits are linked to each other by strong correlative relationships. Dark and light grey arrows link traits with the first and second highest correlated traits, respectively, according to Pearson's correlation matrix (Supplemental Table S5). The black dotted line indicates the pathway used in the model to estimate RFM, SFM and SPA from the nutrition regime. Four groups of correlated traits are classified as elements, nitrate pool, free amino acids-ammonium pool, and biomass allocation.

Table 1. Formulas and parameters used by the model to estimate morphological traits according to N nutrition status. In the first three columns, the traits included in the descriptive model are listed, and the formulas and model parameters that fit with all the data. Model parameters refined for each genotype in the three experiments are listed in the Supplemental Table S6.

Model	Formula	Model Parameters	ANOVA		
			Par.	G	E
EnvNO3→RNO3	$RNO_3 = A1.ln(EnvNO_3) + A2$	A1 = 60 A2 = 325	A1 A2	* (*)	** **
RNO3→RAA	$RAA = B1.(RNO3) + B2$	B1 = 0.018 B2 = 15.87	B1 B2	Ns Ns	(*) (*)
RAA→RT	$RT = C1.ln(RAA) + C2$	C1 = 0.58 C2 = 2.51	C1 C2	* **	ns ns
RT→RFM	$RFM = D1.(RT) + D2$	D1 = 22.14 D2 = −1.76	D1 D2	Ns Ns	ns ns
RFM→SFM	$SFM = E1.(RFM)^{E2}$	E1 = 11.72 E2 = 0.62	E1 E2	Ns Ns	(*) (*)
SFM→SPA	$SPA = F1.(SFM) + F2$	F1 = 0.05 F2 = 0.7	F1 F2	Ns Ns	ns ns

Significance of the variation due to genotype (G) and to experience (E) for the refined parameters (Par.) in ANOVA is reported by "**" for highly significant ($p < 0.01$), "*" for significant ($p < 0.05$), "(*)" for nearly significant ($0.05 < p < 0.06$) and "ns" for not significant.

The ANOVA of these smooth parameters allowed to identifying the ones that respond to experimental and/or genotype factors. Significance of experimental and genotype factors for each parameter were displayed in Table 1. The parameters A1 and A2 of the first equation connecting RNO3 to the NO_3 environment were significantly genotype-dependent (p-value < 0.05 and p-value < 0.06, respectively). In addition, the parameters C1 and C2 in the equation linking RAA to RT were explained by a genotype factor (p-value < 0.05 and p-value < 0.01, respectively). Genetic variation for the A1 parameter revealed a higher and faster nitrate accumulation in roots depending on nutrition regimes in *Col-0* compared to the other accessions. Genetic variation for C1 and C2 parameters revealed a difference in the relationship linking RAA to RT between accessions. Characterized by the highest C1 and C2 values, *Bur-0* was the accession showing the highest increase of its root thickness when RAA decreased. In contrast, following the statistical model, *Col-0* showed the lowest modulation of RT when RAA varied. Significant variation was observed between experiments for A1 and A2 parameters (p-value < 0.01 and p-value < 0.055, respectively), suggesting a continuum environmental differences between experiments.

4. Discussion

Natural variation provides an excellent framework to investigate plant adaptation to simultaneous genetic changes and environmental variations. Plant response to N availability has been widely studied in Arabidopsis as well as in crop species, most often focusing only on ample, limiting, and scarcity conditions [14,16,19,37,38]. With this study, we quantified the Arabidopsis response to an ampler set of nutrition regimes, ranging from 0.01 mM up to 50 mM NO_3, taking into account genotype features. We investigated both induced increases and decreases of N concentration compared to a control nutrition regime, a likely scenario that describes what plants face in an agricultural context. This allowed an accurate description of the dynamics of N pools (total N%, NO_3 pools, free amino acid and ammonium pools) and the formulation of a statistic model to correlate nutrition-responsive traits in Arabidopsis.

The present study gives the global response of Arabidopsis to a wide range of N regimes, computed from the average of contrasted genotypes, and completes some previous studies [13–16]. We observed first that the expected positive correlation existing between NO_3 abundance in media and certain metabolic traits such as the total N percentage and free amino acid content [13,15] was only partially confirmed in our study. Total N percentage indeed remained substantially stable from 0.2 mM to 50 mM nutrition, while we clearly observed an increase of nitrate, amino acids, and ammonium in both shoots and roots towards the highest NO_3 media concentrations (Figure 3). The break happens when total N percentage decreases under the critical N value (the amount of supplied nitrogen beyond which plants stop biomass production). This threshold is reached when the external NO_3 is between 0.2 mM and 1 mM in Arabidopsis (Figure 6). We observed a decrease of protein content in shoots (SPC) under both low and high nutrition regimes (Figure 3). This result is in contrast with previous descriptions [10,13,15] where no changes were detected between low and high N regimes for SPC. We hypothesized a change in the dynamics of N pools when the plants are growing at high NO_3 concentrations. Total N percentage remains stable, but the amount of proteins decreases in favor of accumulation of free amino acids, ammonium, and nitrate. A negative impact at extreme NO_3 concentrations was remarked in our investigation on shoot biomass, a trait which is usually known to be highly impacted by N scarcity [16,18,39]. A decrease in SFM was expected and found at a NO_3 availability lower than the control. In this case, all N-related traits decreased, due the scarcity of the N resource. Interestingly, we also detected a clear decrease of shoot biomass at high NO_3 concentrations, in which plants dispose of excess amounts of NO_3. This may represent a symptom of impairment of carbon inclusion in the GS/GOGAT cycle to form amino acids. Indeed, plant growth and development are highly dependent on the interaction between C and N metabolism [40]. As large amounts of N are invested in the photosynthetic machinery, optimal CO_2 assimilation through photosynthesis and, as a consequence, biomass production requires an adequate N supply. On the other hand, it has also been long established that during the assimilation of inorganic N, significant amounts of fixed C are required to provide the C skeletons that act as acceptors during N assimilation and use, and significant amounts of ATP and NAD(P)H are required to drive these processes. At very high NO_3 media availability, plants continue to uptake N, and they stock it in vacuoles, reaching very high N concentrations [41]. However, we have evidence that increases in shoot biomass become impossible in this case, and at the same time, carbohydrates are driven towards starch production. SStarch was indeed extremely increased at low N regimes, confirming the well-known response to N starvation [16,19,42], but it was also accumulated at high N regimes, where it accompanied the decrease of SFM (Figure 2). Starch, if incompletely remobilized, can therefore sequester carbon that could otherwise be invested in leaf and root growth [42]. To some extent we found the usual attempt of plants to scavenge for nutrients at low NO_3 supplies [11], with a slight increase of RFM essentially due to lateral roots emergence and increase of their architecture complexity. We could in fact even detect a decrease of primary root length in all nutrient regimes compared to control, while root thickness was increased at the lowest supply and decreased at the highest NO_3 supply (Figure 2).

Our investigation provides a statistical model to estimate biomass allocation from the initial NO_3 availability, taking in account the genotype specificities. From the correlation analysis, we can notice that shoot biomass and other morphological traits are more connected to root metabolic traits than to shoot metabolic traits (Figure 8). RT plays a key role in connecting the group of amino acids and ammonium pools to the group of biomass traits. It represents the size and architecture of secondary roots. The negative correlation between RAA and RT was strong in all experiments and for all genotypes. It suggests that plants that present a low level of free amino acids in roots, are able to acquire a higher root biomass or vice-versa, and consequently support the growth of aerial parts, since RFM and SFM are positively correlated to RT. Indeed, *Col-0* is the only genotype that presented a higher RAA content in all conditions compared to control, and did not increase its root size or root architecture shape to respond to N limitation. This is in agreement with results from Krapp et al. [8] that detected stable levels of root amino acids after several days of N shortage accompanied by huge secondary root development in *Col-0*. In contrast, other studies stated that RAA and RT have a positive correlation in limiting N in several accessions [16] and some genotypes that better resist an N-perturbed environment are characterized by a stronger accumulation of RAA and RFM compared to sensitive lines [15]. The negative correlation reported here might have arisen because of the wider range of N supplies in which plants have been grown. RAA has a negative correlation with RNO3 and a positive one with SRNO3 (Table S3), suggesting that the key for adaptation to N stress is the ability of a plant to transport NO_3 towards shoots where it will be reduced and assimilated into amino acids, avoiding RNO3 accumulation [43]. Starch is described by Calenge et al. [39] as a likely contributor to dry matter accumulation according to the positive correlation between these two traits both in normal and limiting N supplies, but the present study revealed a negative correlation between SStarch and SFM (data not shown). With extreme supplies of 0.01 mM and 50 mM NO_3, conditions in which plants undergo a sensitive decrease in SFM, we could indeed observe a significant increase in SStarch, suggesting that starch accumulation may not be a specific response to low N availability but rather a plant response to nutrient conditions unsuitable for optimal growth.

Our study gives some clues on the strategies used by plants when faced with a N stress environment. The four accessions used in our investigation have been previously classified according to their different reactions to N limitations [16]. Briefly, *Bur-0* was described as an accession well adapted to N limitation. *Ge-0* belonged to a group of accessions which supported a complete N starvation. *Cvi-0* was an accession less adapted to N limitation and *Col-0* showed a low growth and a low adaptation to N stress. In the present study, we characterized the response of these four accessions to different N stress environments. Evaluating the status of the different N pools in plants, we showed that the accessions varied for their different nutrition indexes (NNI, NO3NI and AANI). NNI stayed high in *Cvi-0* growing in low or high NO_3 concentrations, whereas the index decreased in the three other accessions (Figure 6). This means that the N pools stay filled in this accession, even under N stress conditions. This result suggested that *Cvi-0* is relatively insensitive to N availability. Its weak response might explain why this genotype is less adapted to N limitation [16]. The shoot biomass (SFM) was positively correlated to NNI (Figure 7). The estimation of the slope of the regression line for each genotype revealed genetic specificities. The slopes of the regression line were high in *Bur-0*, intermediate in *Cvi-0* and *Ge-0*, and low in *Col-0*, indicating that the biomass produced at the same NNI will be higher in *Bur-0* than in *Col-0*, and revealing a higher NUE in *Bur-0* than in *Col-0*. These differences in NUE are congruent with previous observations [15,16,19]. Finally, a genetic variation was observed for the relationship connecting amino acid content in roots, RAA, and root thickness, RT (Table 1). In the highly efficient accession *Bur-0*, the two root traits are well opposed; the reduction of amino acids in roots is associated with a development of root architecture. In contrast, RAA remains high and RT varied slightly in the poorly efficient accession *Col-0*. Taking together, the results suggest that the capacity of the plant to transform amino acids into secondary and lateral roots is an interesting and surprising part of NUE in plants.

Supplementary Materials: The following are available online at http://www.mdpi.com/2077-0472/8/2/28/s1, Table S1: Raw dataset corresponding to all measurements made on four Arabidopsis accessions in six N supplies for three independent experiments, Table S2: Percentage of explained variance of investigated traits, Table S3: Nutrition indexes of different N pools in the four Arabidopsis accessions, Table S4: ANOVA of the five nutrition indexes in four *Arabidopsis* accessions growing in six different N supplies in three independent experiments., Table S5: Pearson's correlation matrix of investigated traits, Table S6: Parameters of the descriptive model for the estimation of morphological traits according to N nutrition, Figure S1: Slope (A) and intercept (B) of linear regression of SFM against NNI for the 4 *Arabidopsis* accessions, Figure S2: Nutrition Use Efficiency in four Arabidopsis accessions.

Acknowledgments: We thank David Lawrence for English editing this work was supported by the Marie Curie Initial Training Network, project No 264296, BIONUT. The IJPB benefits from the support of the LabEx Saclay Plant Sciences-SPS (ANR-10-LABX-0040-SPS).

Author Contributions: F.C. and S.C. conceived and designed the experiments; G.C. and F.C. performed the experiments; G.C., C.M.-D. and F.C. analyzed the data; M.B. and A.M. contributed reagents/materials/analysis tools; G.C. and F.C. wrote the paper.

Conflicts of Interest: The authors declare no conflict of interest.

References

1. Bouguyon, E.; Gojon, A.; Nacry, P. Nitrate sensing and signaling in plants. *Semin. Cell Dev. Biol.* **2012**, *23*, 648–654. [CrossRef] [PubMed]
2. Diaz, C.; Lemaître, T.; Christ, A.; Azzopardi, M.; Kato, Y.; Sato, F.; Morot-Gaudry, J.-F.; Le Dily, F.; Masclaux-Daubresse, C. Nitrogen recycling and remobilization are differentially controlled by leaf senescence and development stage in Arabidopsis under low nitrogen nutrition. *Plant Physiol.* **2008**, *147*, 1437–1449. [CrossRef] [PubMed]
3. Kant, S.; Bi, Y.M.; Rothstein, S.J. Understanding plant response to nitrogen limitation for the improvement of crop nitrogen use efficiency. *J. Exp. Bot.* **2011**, *62*, 1499–1509. [CrossRef] [PubMed]
4. Masclaux-Daubresse, C.; Chen, Q.; Havé, M. Regulation of nutrient recycling via autophagy. *Curr. Opin. Plant Biol.* **2017**, *39*, 8–17. [CrossRef] [PubMed]
5. Kim, J.; Woo, H.R.; Nam, H.G. Toward systems understanding of leaf senescence: An integrated multi-omics perspective on leaf senescence research. *Mol. Plant* **2017**, *9*, 813–825. [CrossRef] [PubMed]
6. Tegeder, M.; Masclaux-Daubresse, C. Source and sink mechanisms of nitrogen transport and use. *New Phytol.* **2018**, *217*, 35–53. [CrossRef] [PubMed]
7. De Pessemier, J.; Chardon, F.; Juraniec, M.; Delaplace, P.; Hermans, C. Natural variation of the root morphological response to nitrate supply in *Arabidopsis thaliana*. *Mech. Dev.* **2013**, *130*, 45–53. [CrossRef] [PubMed]
8. Krapp, A.; Berthomé, R.; Orsel, M.; Mercey-Boutet, S.; Yu, A.; Castaings, L.; Elftieh, S.; Major, H.; Renou, J.-P.; Daniel-Vedele, F. Arabidopsis roots and shoots show distinct temporal adaptation patterns toward nitrogen starvation. *Plant Physiol.* **2011**, *157*, 1255–1282. [CrossRef] [PubMed]

9. Krapp, A.; David, L.C.; Chardin, C.; Girin, T.; Marmagne, A.; Leprince, A.S.; Chaillou, S.; Ferrario-Méry, S.; Meyer, C.; Daniel-Vedele, F. Nitrate transport and signalling in arabidopsis. *J. Exp. Bot.* **2014**, *65*, 789–798. [CrossRef] [PubMed]

10. Lemaître, T.; Gaufichon, L.; Boutet-Mercey, S.; Christ, A.; Masclaux-Daubresse, C. Enzymatic and metabolic diagnostic of nitrogen deficiency in Arabidopsis thaliana Wassileskija accession. *Plant Cell Physiol.* **2008**, *49*, 1056–1065. [CrossRef] [PubMed]

11. Walch-Liu, P.; Ivanov, I.I.; Filleur, S.; Gan, Y.B.; Remans, T.; Forde, B.G. Nitrogen regulation of root branching. *Ann. Bot.* **2006**, *97*, 875–881. [CrossRef] [PubMed]

12. Loudet, O.; Chaillou, S.; Camilleri, C.; Bouchez, D.; Daniel-Vedele, F. Bay-0 x Shahdara recombinant inbred line population: A powerful tool for the genetic dissection of complex traits in Arabidopsis. *Theor. Appl. Genet.* **2002**, *104*, 1173–1184. [CrossRef] [PubMed]

13. Loudet, O.; Chaillou, S.; Merigout, P.; Talbotec, J.; Daniel-Vedele, F. Quantitative trait loci analysis of nitrogen use efficiency in Arabidopsis. *Plant Physiol.* **2003**, *131*, 345–358. [CrossRef] [PubMed]

14. North, K.A.; Ehlting, B.; Koprivova, A.; Rennenberg, H.; Kopriva, S. Natural variation in Arabidopsis adaptation to growth at low nitrogen conditions. *Plant Physiol. Biochem.* **2009**, *47*, 912–918. [CrossRef] [PubMed]

15. Richard-Molard, C.; Krapp, A.; Brun, F.; Ney, B.; Daniel-Vedele, F.; Chaillou, S. Plant response to nitrate starvation is determined by N storage capacity matched by nitrate uptake capacity in two Arabidopsis genotypes. *J. Exp. Bot.* **2008**, *59*, 779–791. [CrossRef] [PubMed]

16. Ikram, S.; Bedu, M.; Daniel-Vedele, F.; Chaillou, S.; Chardon, F. Natural variation of Arabidopsis response to nitrogen availability. *J. Exp. Bot.* **2012**, *63*, 91–105. [CrossRef] [PubMed]

17. Peng, M.; Hudson, D.; Schofield, A.; Tsao, R.; Yang, R.; Gu, H.; Bi, Y.M.; Rothstein, S.J. Adaptation of Arabidopsis to nitrogen limitation involves induction of anthocyanin synthesis which is controlled by the NLA gene. *J. Exp. Bot.* **2008**, *59*, 2933–2944. [CrossRef] [PubMed]

18. Chardon, F.; Noël, V.; Masclaux-Daubresse, C. Exploring NUE in crops and in Arabidopsis ideotypes to improve yield and seed quality. *J. Exp. Bot.* **2012**, *63*, 3401–3412. [CrossRef] [PubMed]

19. Chardon, F.; Barthélémy, J.; Daniel-Vedele, F.; Masclaux-Daubresse, C. Natural variation of nitrate uptake and nitrogen use efficiency in Arabidopsis thaliana cultivated with limiting and ample nitrogen supply. *J. Exp. Bot.* **2010**, *61*, 2293–2302. [CrossRef] [PubMed]

20. Sulpice, R.; Nikoloski, Z.; Tschoep, H.; Antonio, C.; Kleessen, S.; Larhlimi, A.; Selbig, J.; Ishihara, H.; Gibon, Y.; Fernie, A.R.; et al. Impact of the carbon and nitrogen supply on relationships and connectivity between metabolism and biomass in a broad panel of Arabidopsis Accessions. *Plant Physiol.* **2013**, *162*, 347–363. [CrossRef] [PubMed]

21. Chietera, G.; Chardon, F. Natural variation as a tool to investigate nutrient use efficiency in plants. In *Nutrient Use Efficiency in Plants: Concepts and Approaches*; Hawkesford, M.J., Kopriva, S., De Kok, L.J., Eds.; Springer International Publishing: Cham, Switzerland, 2014; pp. 29–50.

22. Weigel, D. Natural variation in Arabidopsis: From molecular genetics to ecological genomics. *Plant Physiol.* **2012**, *158*, 2–22. [CrossRef] [PubMed]

23. Sinclair, T.R.; Purcell, L.C.; Sneller, C.H. Crop transformation and the challenge to increase yield potential. *Trends Plant Sci.* **2004**, *9*, 70–75. [CrossRef] [PubMed]

24. Picheny, V.; Casadebaig, P.; Trépos, R.; Faivre, R.; Da Silva, D.; Vincourt, P.; Costes, E. Using numerical plant models and phenotypic correlation space to design achievable ideotypes. *Plant Cell Environ.* **2017**, *40*, 1926–1939. [CrossRef] [PubMed]

25. Boote, K.J.; Jones, J.W.; White, J.W.; Asseng, S.; Lizaso, J.I. Putting mechanisms into crop production models. *Plant Cell Environ.* **2013**, *36*, 1658–1672. [CrossRef] [PubMed]

26. Stewart, D.W.; Cober, E.R.; Bernard, R.L. Modeling genetic effects on the photothermal response of soybean phenological development. *Agron. J.* **2003**, *95*, 65–70. [CrossRef]

27. Reymond, M.; Muller, B.; Leonardi, A.; Charcosset, A.; Tardieu, F. Combining quantitative trait Loci analysis and an ecophysiological model to analyze the genetic variability of the responses of maize leaf growth to temperature and water deficit. *Plant Physiol.* **2003**, *131*, 664–675. [CrossRef] [PubMed]

28. Quilot, B.; Kervella, J.; Génard, M.; Lescourret, F. Analysing the genetic control of peach fruit quality through an ecophysiological model combined with a QTL approach. *J. Exp. Bot.* **2005**, *56*, 3083–3092. [CrossRef] [PubMed]

29. Bertin, N.; Martre, P.; Génard, M.; Quilot, B.; Salon, C. Under what circumstances can process-based simulation models link genotype to phenotype for complex traits? Case-study of fruit and grain quality traits. *J. Exp. Bot.* **2010**, *61*, 955–967. [CrossRef] [PubMed]

30. Salon, C.; Lepetit, M.; Gamas, P.; Jeudy, C.; Moreau, S.; Moreau, D.; Voisin, A.S.; Duc, G.; Bourion, V.; Munier-Jolain, N. Analysis and modeling of the integrative response of Medicago truncatula to nitrogen constraints. *C. R. Biol.* **2009**, *332*, 1022–1033. [CrossRef] [PubMed]

31. Gu, J.; Yin, X.; Zhang, C.; Wang, H.; Struik, P.C. Linking ecophysiological modelling with quantitative genetics to support marker-assisted crop design for improved yields of rice (*Oryza sativa*) under drought stress. *Ann. Bot.* **2014**, *114*, 499–511. [CrossRef] [PubMed]

32. Martre, P.; Porter, J.R.; Jamieson, P.D.; Triboï, E. Modeling grain nitrogen accumulation and protein composition to understand the sink/source regulations of nitrogen remobilization for wheat. *Plant Physiol.* **2003**, *133*, 1959–1967. [CrossRef] [PubMed]

33. Miranda, K.M.; Espey, M.G.; Wink, D.A. A rapid, simple spectrophotometric method for simultaneous detection of nitrate and nitrite. *Nitric Oxide* **2001**, *5*, 62–71. [CrossRef] [PubMed]

34. Rosen, H. A modified ninhydrin colorimetric analysis for amino acids. *Arch. Biochem. Biophys.* **1957**, *67*, 10–15. [CrossRef]

35. Lemaire, G.; Jeuffroy, M.-H.; Gastal, F. Diagnosis tool for plant and crop N status in vegetative stage: Theory and practices for crop N management. *Eur. J. Agron.* **2008**, *28*, 614–624. [CrossRef]

36. Cohen, J.; Cohen, P.; West, S.G.; Aiken, L.S. *Applied Multiple Regression/Correlation Analysis for the Behavioral Sciences*, 3rd ed.; Laurence Erlbaum Associates, Inc.: Mahwah, NJ, USA, 2003.

37. Cañas, R.A.; Yesbergenova-Cuny, Z.; Simons, M.; Chardon, F.; Armengaud, P.; Quilleré, I.; Cukier, C.; Gibon, Y.; Limami, A.M.; Nicolas, S.; et al. Exploiting the genetic diversity of maize using a combined metabolomic, enzyme activity profiling, and metabolic modelling approach to link leaf physiology to kernel yield. *Plant Cell* **2017**, *29*, 919–943. [CrossRef] [PubMed]

38. Dai, Z.; Plessis, A.; Vincent, J.; Duchateau, N.; Besson, A.; Dardevet, M.; Prodhomme, D.; Gibon, Y.; Hilbert, G.; Pailloux, M.; et al. Transcriptional and metabolic alternations rebalance wheat grain storage protein accumulation under variable nitrogen and sulfur supply. *Plant J.* **2015**, *83*, 326–343. [CrossRef] [PubMed]

39. Calenge, F.; Saliba-Colombani, V.; Mahieu, S.; Loudet, O.; Daniel-Vedele, F.; Krapp, A. Natural variation for carbohydrate content in Arabidopsis. Interaction with complex traits dissected by quantitative genetics. *Plant Physiol.* **2006**, *141*, 1630–1643. [CrossRef] [PubMed]

40. Nunes-Nesi, A.; Fernie, A.R.; Stitt, M. Metabolic and signaling aspects underpinning the regulation of plant carbon nitrogen interactions. *Mol. Plant* **2010**, *3*, 973–996. [CrossRef] [PubMed]

41. De Angeli, A.; Monachello, D.; Ephritikhine, G.; Frachisse, J.M.; Thomine, S.; Gambale, F.; Barbier-Brygoo, H. The nitrate/proton antiporter AtCLCa mediates nitrate accumulation in plant vacuoles. *Nature* **2006**, *442*, 939. [CrossRef] [PubMed]

42. Stitt, M.; Zeeman, S.C. Starch turnover: Pathways, regulation and role in growth. *Curr. Opin. Plant Biol.* **2012**, *15*, 282–292. [CrossRef] [PubMed]

43. Masclaux-Daubresse, C.; Daniel-Vedele, F.; Dechorgnat, J.; Chardon, F.; Gaufichon, L.; Suzuki, A. Nitrogen uptake, assimilation and remobilization in plants: Challenges for sustainable and productive agriculture. *Ann. Bot.* **2010**, *105*, 1141–1157. [CrossRef] [PubMed]

agriculture

MDPI

Review

Sustainable Agriculture—Enhancing Environmental Benefits, Food Nutritional Quality and Building Crop Resilience to Abiotic and Biotic Stresses

Daniel P. Roberts and Autar K. Mattoo *

United States Department of Agriculture, Agricultural Research Service, The Henry A Wallace Beltsville
Agricultural Research Center, Sustainable Agricultural Systems Laboratory, Building 001, Beltsville,
MD 2705-2350, USA; dan.roberts@ars.usda.gov
* Correspondence: autar.mattoo@ars.usda.gov; Tel.: +1-301-504-6622

Received: 9 December 2017; Accepted: 27 December 2017; Published: 1 January 2018

Abstract: Feeding nutrition-dense food to future world populations presents agriculture with enormous challenges as estimates indicate that crop production must as much as double. Crop production cannot be increased to meet this challenge simply by increasing land acreage or using past agricultural intensification methods. Food production doubled in the past through substantial use of synthetic fertilizer, pesticides, and irrigation, all at significant environmental cost. Future production of nutrition-dense food will require next-generation crop production systems with decreased reliance on synthetic fertilizer and pesticide. Here, we present three case studies detailing the development of cover crops and plant-beneficial microbes for sustainable, next-generation small grain, tomato, and oilseed rape production systems. Cover crops imparted weed and pathogen control and decreased soil erosion and loss of soil nitrogen, phosphorus and carbon, while plant-beneficial microbes provided disease control and phosphorus fertility. However, yield in these next-generation crop production systems at best approximated that associated with current production systems. We argue here that to substantially increase agricultural productivity, new crop germplasm needs to be developed with enhanced nutritional content and enhanced tolerance to abiotic and biotic stress. This will require using all available technologies, including intensified genetic engineering tools, in the next-generation cropping systems.

Keywords: next generation cropping systems; plant beneficial microbes; nutritional quality; environmental stresses; plant biotic stress; antimicrobial peptides; genetic engineering; future agriculture

---

1. Introduction/Future Challenges Confronting Agriculture

The global population is expected to increase in number and affluence by the middle of this century, with estimates of the world's population in 2050 varying between 8 and 10 billion [1]. Feeding the world's future population will place unprecedented demands on agriculture. We will need to increase food production while at the same time decreasing the negative impacts of agriculture on land, water, and climate [2]. Estimates indicate agricultural production must as much as double to meet projected demands for food [2]. More food is needed and food quality must be improved, particularly regarding nutrient content [3].

Overarching the challenge of producing more food is global climate change [1,3]. Global climate change is expected to bring increased temperatures and increased concentrations of CO_2 and ozone in the atmosphere. Also predicted are altered patterns of weather and drought. Rising CO_2 concentrations may increase yields of certain crops such as wheat, although the extent of the benefit of CO_2 fertilization to crop yield is the subject of debate [3,4]. Offsetting any potential benefits due to global climate change are associated negative impacts leading to concerns about our ability to increase, or even maintain, crop yields [3]. Increased temperatures will impact yield, as yields of most crops decline dramatically at temperatures much above 30 °C; the optimum temperature for photosynthesis being between 20 °C and 25 °C [5]. For example, yields of the major US crops, corn, soybean, and cotton, increase up to 29 °C, 30 °C, and 32 °C, respectively, with temperatures above these thresholds being very damaging [6]. Drought, salinity stress, and higher ozone levels due to global climate change, as well as the development of new pest and pathogen problems, are predicted to be a drag on crop yields [3,4]. Also of concern is that plants develop more quickly at higher temperatures, leaving less time to accumulate human nutrients such as sugars, fat, and protein [5]. Rising CO_2 levels are also predicted to lower the nutritional quality of certain crops, lowering mineral and protein content [7].

It is unlikely that crop production can be increased to meet the demands of future populations simply by increasing land acreage devoted to agriculture. Competition for land use with urbanization, and the loss of land to salination and desertification will reduce land available for conversion to agricultural production [5]. Additionally, conversion of land to agricultural production has serious environmental consequences. Transforming natural ecosystems to land in agricultural production impacts the global carbon and hydrological cycles, habitat biodiversity, and soil conditions [2,5,8–12]. Among other things, repurposing natural ecosystems for crop production can have substantial impacts on greenhouse gas emissions by releasing carbon stored in vegetation and soil biomass to the atmosphere [12].

It is also unlikely that we can increase productivity of current production systems to meet the demands of future populations by using the agricultural intensification methods of the past. Food production doubled worldwide over the past 35 years; largely due to the use of synthetic fertilizer, pesticides, and irrigation [2]. Nitrogen (N) and phosphorus (P) fertilizer inputs increased dramatically [13]. From 1960 to 2000, the use of N fertilizers increased approximately 800%, with corn, rice, and wheat accounting for about 50% of current fertilizer use. These crops typically have N use efficiencies below 40% leading to loss of N into the environment [8,11,13]. Worldwide, nearly 90% of N fertilizer is NH_4^+ which is converted to NO_3^- by the soil microflora and easily leached into water systems. Under anoxic conditions, excess N in soil from fertilizer is transformed mainly into N_2, but also into the potent greenhouse gas N_2O [10,13]. Overuse of P fertilizer also has negative environmental impacts, causing eutrophication of water systems [11]. Pesticide use increased dramatically (15 to 20 times) over the past 40 years as well [14]. Pesticides used in agriculture are generally hazardous to human health and that of other species, with some pesticides accumulating in food chains [11]. Irrigated land doubled globally over the past 50 years to the point where 70% of freshwater withdrawals are now used to irrigate cropland [2]. Irrigation can result in nutrient loading into water systems and salinization of arable land. Clearly, agricultural intensification over the past decades has had negative impacts, such as increased soil erosion and decreased soil fertility, pollution of ground water and eutrophication of rivers, lakes, and coastal ecosystems, and increased atmospheric constituents that lead to global climate change and water resources with dramatic consequences for food production [2,10,11].

2. Current Cropping Systems

Crops are currently grown globally using various conventional production systems which use synthetic fertilizer and pesticides, and to a lesser extent with organic production systems that use 'natural' sources for maintaining soil fertility and pest control. Conventional production systems can be further categorized into those which incorporate tillage and those relying on a no-till or reduced-tillage strategy. Conventional no-till systems are typically considered more sustainable than conventional tillage systems as they sequester more carbon, have better soil erosion prevention, improve water and fertilizer use efficiency, have better soil nutrient cycling, enhance soil biological activity, and reduce energy, labor, and machinery inputs [15,16]. Organic production systems are designed and perceived to be less detrimental to humans and the environment than the conventional systems as organic production systems place a greater emphasis on managing ecological processes and eliminating inputs that are, or perceived to be, harmful to humans and the environment such as synthetic fertilizers and synthetic pesticides [17–20]. A major principle of organic agriculture is building soil organic matter via the use of cover crops (non-cash crops grown for their environmental benefits) and animal manures and by-products. Organic soils with higher organic matter levels often have higher capacity to mineralize, capture, and store essential nutrients, such as N, and water resources [17,21,22]. Higher soil organic matter also leads to higher soil aggregate stability and is associated with richer food webs and higher biological activities that drive beneficial soil processes [23].

Conventional and organic cropping systems have been compared in numerous studies and have been found to result in different crop yields, impacts on the environment, and levels of sustainability [24]. Yield averages associated with organic production systems were found to be lower. However, these differences in yield were contextual, varying between 5% and 34% lower depending on crop, conditions, and management practices [19]. Yield differences between organic and conventional systems also differed dramatically with different regions of the world [19]. Some have argued that environmental benefits associated with organic production are diminished as lower crop yields lead to greater deforestation and loss of biodiversity when land is converted to agricultural use to maintain crop production at a certain level [19,25]. Results from one meta-analysis showed that organic farming generally had less negative impacts on the environment per unit land area, but not necessarily with respect to per unit product due to lower yields [25]. For example, organic farms tended to have higher soil organic matter and lower nutrient loss (N leaching, N_2O emissions, NH_4 emissions) per unit field area but higher NH_4 emissions, N leaching, and N_2O emissions per unit product [25]. Organic production systems also had higher eutrophication potential per unit product.

3. Development/Refining Crop Production Systems for Sustainable Intensification of Crop Production

3.1. Development of Next-Generation Cropping Systems

Next-generation cropping systems should include a combination of high yield potential and low negative environmental impacts drawing on the most sustainable aspects from organic and conventional crop production systems [12,24,25]. The emphasis on building soil health in organic production systems using cover crops and other organic materials will need to be combined with the development of new crop cultivars using traditional breeding or genetic engineering techniques. New crop cultivars need to be developed that have increased tolerance of abiotic stress; that offer higher yields but use less water, fertilizer, and other inputs; and have higher nutritional quality. More sustainable methods for pathogen, pest, and weed management also need to be developed [1]. The techniques used to increase crop yield and environmental sustainability will depend on the crop in question. The next sections illustrate development of next-generation sustainable crop production systems for the row crops corn, wheat, and soybean; the horticultural crop, tomato; as well as the development of environmentally friendly disease control methods. These next-generation cropping systems place emphasis on the use of cover crops and beneficial microbes for enhancing soil fertility (N, P), weed and disease control, and decreasing soil erosion. These case studies are being presented to illustrate the approaches as well as challenges encountered during development of these next-generation cropping systems. It should be noted that this review is not exhaustive in nature as only a limited number of applications of cover crops and plant-beneficial microbes are discussed. Finally, examples of the power of genetic engineering for development of crop cultivars with enhanced nutritional content, abiotic stress tolerance, and biotic stress resistance are included.

3.2. Development of Next-Generation Sustainable Grain Cropping Systems

The farming systems project (FSP) is a long-term agroecological research project that was established at the Beltsville Agricultural Research Center in Beltsville, Maryland, USA, in 1996 to evaluate the sustainability of the conventional and organic grain cropping systems currently being used in the mid-Atlantic region of the United States. The five cropping systems being evaluated at FSP are a three-year conventional no-till corn-rye cover crop/soybean-wheat/soybean rotation (NT), a three-year conventional chisel-till corn-rye cover crop/soybean-wheat/soybean rotation (CT), a two-year organic hairy vetch/corn-rye/soybean rotation (Org2), a three-year organic hairy vetch/corn-rye/soybean-wheat rotation (Org3), and a six-year organic corn–soybean–wheat–alfalfa–alfalfa–alfalfa rotation (Org6). These five cropping system plots are co-located and incorporate large field-scale plots to address field variability and long timeframes, as some soil processes being compared occur very slowly [26]. Conventional systems were managed with herbicide and synthetic fertilizer programs and current GMO cultivars. Organic systems were managed using USDA National Organic Program Standards [27].

Crop yields were considerably lower with the organic systems, with yields 31% less for corn and 20% less for soybean than with conventional cropping systems in a ten-year analysis [27]. Certain environmental benefits were greater with the organic cropping systems than the CT or NT cropping systems. Soil organic carbon (SOC), measured to a depth of 1 m, was 11% greater in the Org3 than the NT cropping system indicating that tilling organic materials (manure) into soil may be a more effective means of increasing SOC than eliminating tillage [28]. SOC was greater in the NT cropping system at the 0–5 cm soil depth but greater in the Org3 cropping system at the 5–10 cm and the 20–25 cm depths. This increased SOC in the organic cropping systems led to greater N mineralization potential and biodiversity [27]. Increased SOC also decreased global warming potential of the organic cropping systems by sequestering carbon in soil [28]. Global warming potential was negative for Org3, indicating it was a net sink for CO_2 equivalents, and positive for NT and CT production systems. These differences in global warming potential were also driven by lower energy usage with the Org3 production system. Global warming potential per unit of grain yield was negative and significantly lower for the Org3 systems than the NT or CT systems despite lower grain yields [20].

Other environmental benefits associated with organic systems, such as minimizing soil erosion and sediment-bound nutrient transport from fields, were mixed when compared with the CT and NT cropping systems. It was found that predicted soil sediment loss was 33% less with Org3 than CT. N, P, and soil carbon loss was similarly less with Org3 than CT as N, P, and soil carbon are absorbed to the soil sediments lost in run-off [29,30]. However, when NT and Org3 cropping systems were compared, soil erosion and loss of N, P, and soil carbon were 80% less with NT than Org3. Tillage used in Org3 diminishes soil aggregate size and these smaller soil particles are more susceptible to erosion than macroaggregates [30,31]. Using cover crops and animal manure in Org3 built SOC, and had the associated enhancement of soil aggregation, which probably offset the negative impact of tillage on soil aggregation [32], resulting in soil erosion being less with Org3 than CT but greater than with NT. Meteorological influences on corn and soybean grain yields over an 18-year period were compared among the cropping systems at the FSP site as well [27]. Efficiency of grain yield per unit precipitation was greater for conventional than organic cropping systems. Precipitation and heat stress had a significant impact on organic production systems during this study with weed cover playing a significant role due to competition between weeds and the grain crops for water.

3.2.1. Increasing Organic Cropping System Complexity to Enhance Environmental Benefits and Yield

Comparisons of the organic production systems at FSP indicated that increasing cropping system complexity increased grain yield by decreasing weed competition [33]. Corn yield losses due to weed competition were estimated to be 35% in Org2 and decreasing to 14% in Org6, and for sake of comparison, 7% with the NT treatment. In Org2, weed mortality events occur at the same time each year as the two cash crops, corn and soybean, are sown at the similar times in the calendar year. This tends to favor the establishment of summer annual weeds in Org2 plots. With Org3, wheat is added to the rotation, resulting in decreased seed set by the summer annual weeds due to cutting at wheat harvest, or weed seed mortality, due to preparation of the soil for the cover crop used in the Org3 rotation [28]. With Org6, the addition of alfalfa to the rotation adds an additional layer of complexity as alfalfa is cut three to five times a year, incorporating more weed mortality events into the cropping system [28]. All measures of nitrogen availability increased as cropping system complexity increased from Org2 to Org3 to Org6 [22], also possibly contributing to the increased grain yield associated with the Org6 rotation. Preliminary results also suggested that increased complexity decreased predicted soil erosion [28]. Although this approach enhances yield and environmental benefits associated with organic grain cropping systems it does not come close to the estimated doubling in yield needed to feed the future world population.

3.2.2. Organic No-Till Cover Crop-Based Small Grains Cropping System

A second approach for a next-generation sustainable grain cropping system is an organic no-till cover crop-based production system [34]. This strategy integrates the soil conservation benefits of no-till grain production [35–37], with the soil organic matter building practices of organic systems. Features of this system are the use of cover crop mulches for weed suppression and the contribution of N-fixing legume cover crops towards meeting N demand of the subsequent cash crop. Use of cover crop mulch for weed control allows the system to move away from the use of tillage for weed control, and its deleterious impacts on soil aggregation, biology, and erosivity [16,37–39]. These organic no-till systems rely on mechanical termination of winter annual cover crops with roller-crimper devices and no-till planting corn and soybean into the resulting cover crop mulches. Simulation models indicate that this approach has the potential to increase environmental benefits compared to current tillage-based organic grain production systems [40]. However, more research is needed [34].

Cereal rye is typically used as the cover crop preceding soybean in the rotation. Small grains cover crops, such as cereal rye, produce substantial biomass and provide reliable weed suppression as a living cover crop and after termination as a surface mulch [41,42]. These cover crops also prevent erosion and build soil organic matter [41]. Legume cover crops, such as hairy vetch, are used prior to the corn phase of the rotation as corn requires more exogenous N than soybean. These legume cover crops fix atmospheric N and release N during decomposition. However, hairy vetch grows slowly in the fall and decomposes rapidly after termination making it less effective than cereal rye for weed suppression. Recent research has switched focus to mixtures of hairy vetch and cereal rye, or another winter grain, since hairy vetch alone does not provide sufficient weed control. Mixtures of hairy vetch and cereal rye cover crops have been shown to provide greater above ground biomass and weed suppression than hairy vetch monocultures and release greater N content to the soil than cereal rye monocultures [42–44]. However, significant challenges remain regarding management of cover crop and cover crop mixtures to enhance weed control and soil fertility and minimize the cover crop itself competing for resources as a weed with the subsequent cash crop [34,45]. As with the preceding approach, this organic no-till cover crop-based system does not appear to come close to doubling grain yield.

3.3. Next-Generation Cover Crop-Based Sustainable Tomato Production System

Fresh-market tomatoes, like other vegetables in the US, are grown using high-input production systems to maximize yield and product quality. For tomatoes, these production systems use raised beds and depend heavily on synthetic fertilizers, black polyethylene plastic mulch, and tillage. N from synthetic fertilizer is essential to maintain fertility levels, while the black polyethylene plastic is used primarily to control weeds [46]. Fertilizer, black plastic, and tillage comprise a large proportion of the production costs for large-scale production of fresh-market tomatoes. Additionally, N recovery by the tomato plant from synthetic fertilizer is low, causing some growers to apply excess N to maximize yield, driving up production costs and potentially contributing to surface and groundwater pollution [46].

To increase sustainability of this, and potentially other vegetable production systems, the legume cover crop hairy vetch has been substituted for the black polyethylene plastic in the next-generation tomato production system. This system features no-tillage planting of tomato transplants into a killed, hairy vetch cover crop grown on raised soil beds. Importantly, in research conducted over a ten-year period, this no-till hairy vetch mulch system resulted in greater yield and economic return than tomatoes grown using the conventional black polyethylene plastic system [46,47]. Economic advantage with the hairy vetch mulch system was greatest in years with abundant rainfall and less during droughty years.

Intended environmental benefits provided by this no-till hairy vetch system were a reduced need for synthetic nutrient inputs such as N, reduced soil erosion resulting from the no-tillage system, and increased soil water holding capacity [48]. As with the hairy vetch cover crop used in the no-till production system discussed above, hairy vetch was intended to utilize N in soil leftover from the previous crop and fix N from the atmosphere. Both the intercepted soil N and fixed atmospheric N were then expected to be converted into hairy vetch biomass and released to the soil upon decomposition for the subsequent tomato crop [49]. In a three-year field study, exogenous N requirements were reduced for tomato grown with the hairy vetch system compared with those grown using black polyethylene plastic. Minimum N rates necessary to achieve maximum yield were 80 lb/acre with the hairy vetch system and 170 lb/acre with the black plastic system [49]. As expected, soil erosion was less when hairy vetch was used. With the black plastic system, 50% to 75% of the field is covered with the water-impervious plastic enhancing run-off and lessening water retention in field soil. Losses of two to four times more water and five to fifteen times more soil sediment were observed with the black plastic system over a field season. Pesticide loads released from fields were also greater as pesticides intended for the tomato plant collected on the plastic surface and were loaded into run-off during rainfall events [50,51].

Another intended environmental benefit of the no-till hairy vetch production system was reduced weed competition and hence reduced need for herbicides. However, mixed results were obtained for weed control. As with the no-till small grains production system discussed above, better weed control may result from cover crop mixtures where the hairy vetch is mixed with cereal rye or another cover crop that provides more biomass for a weed-suppressive mulch and decomposes more slowly than hairy vetch. The hairy vetch system did, however, reduce disease on tomato due to *Alternaria solani*, the causal agent of tomato early blight [52,53]. Infective propagules of this fungal pathogen in soil are spread by splash dissemination of infested soil into the tomato canopy. Soil sediment detected in the tomato canopy was significantly lower with the hairy vetch production system than the black plastic production system after rainfall events. Development of tomato early blight disease was also slower on tomatoes grown with the hairy vetch production system. It was thought that more complete coverage of the soil surface with the hairy vetch mulch (hairy vetch mulch covers the entire field while black plastic covers the tomato beds but not the interspaced rows) physically obstructed splashed soil near the soil surface preventing soil infested with pathogen inoculum from entering the tomato canopy. The tomato crop grown in the hairy vetch mulch had negligible loss to early blight in the absence of fungicide relative to the fungicide treated controls [52,53]. There was also greater resistance of the tomato crop grown in the hairy vetch mulch to invasion and damage by the Colorado potato beetle [54]. Possibly contributing to disease and pest reduction was induction of disease resistance in tomato by the hairy vetch production system. As discussed below, certain genes functioning in plant defense had an altered expression profile in tomato grown in hairy vetch relative to these genes in tomato grown with the black plastic [55].

3.3.1. Impact of Hairy Vetch Cropping System on Tomato Physiology

This no-till hairy vetch tomato cropping system had a broad and beneficial impact on tomato plant physiology. In field and greenhouse experiments gene transcripts of proteins involved in diverse processes that regulate metabolism and growth were found to be differentially up-regulated in leaves of hairy vetch-grown tomato plants relative to those grown under black plastic [55–57]. Up-regulated genes included the most abundant chloroplast protein, ribulose bisphosphate carboxylase/oxygenase (Rubisco), important for carbon fixation; nitrogen-responsive glutamine synthase, regulating carbon/nitrogen signaling; nitrogen utilizing and nitrite toxicity reducing nitrite reductase; nitrogen-use efficiency protein glucose-6-phosphate dehydrogenase; chaperone proteins (HSP70 and ER protein BiP) that stabilize native proteins; cytokinin- and gibberellin-related regulatory proteins; and plant defense anti-fungal proteins chitinase and osmotin. Up-regulation of these plant defense genes may contribute to disease resistance as, in these and other field experiments, hairy vetch-grown tomato had less disease than tomato grown under black plastic [52,53,55].

Other genes/proteins that promote senescence and aging, the ethylene biosynthesis gene ACC synthase and the senescence-regulated SAG12 gene, were more down-regulated in leaves from hairy vetch-grown tomato. Consistent with this, hairy vetch grown tomato had higher levels of the cytokinin indicator gene, cytokinin receptor protein kinase (CRK) [55,58]. A continued supply of cytokinin from the roots to the upper parts of a plant should delay senescence as cytokinin inhibits accumulation of senescence-enhancing gene transcripts [59]. Interestingly, senescence in hairy vetch-grown tomato was delayed relative to tomato grown with black plastic. Cytokinin signaling has also been found to regulate plant-microbe interactions [60]. Engineered accumulation of cytokinins led to the upregulation of defense-related genes including basic chitinase [61] and osmotin [62]. Therefore, accumulation of transcripts and protein of these two anti-fungal defense proteins, chitinase and osmotin, in hairy vetch-grown tomato [55] indicated that cytokinin signaling may regulate disease resistance as well as senescence. Implications of these results in the organism-to-organism interactions in the ecosystem have been proposed (www.glfc.forestry.ca/frontline/bulletins/bulletin_no.18_e.html).

A field pot experiment was conducted to determine the impact of N released during decomposition of hairy vetch on hairy vetch-grown tomato physiology [63]. In this experiment, one treatment consisted of tomato grown in soil where hairy vetch was grown as the winter cover crop (including hairy vetch residue on the soil surface) while a second consisted of bare soil (no hairy vetch cover crop). Additional treatments consisted of the bare soil and hairy vetch treatments supplemented with varying amounts of inorganic N fertilizer. Tomato fruit yield, plant biomass, and photosynthesis were found to be higher in plants grown with the hairy vetch treatment than the bare soil treatment. Additionally, a parabolic response to inorganic N in the bare soil treatments containing supplementary N fertilizer was evident for tomato growth and photosynthesis, suggesting N toxicity in pots with the highest rates of supplemental N. There was also a decline in expression of several genes such as nitrate reductase and PEP carboxylase that regulate nitrogen and carbon metabolism associated with the high (200 N kg ha^{-1}) rate of supplemental inorganic N. Surprisingly, these parabolic responses were mitigated in the hairy vetch-grown plants, where higher photosynthetic rates were maintained at high supplemental inorganic N rates. Hairy vetch also mitigated the decline in expression of the genes regulating nitrogen and carbon metabolism. Consistent with prior experiments, the plant defense-related gene, osmotin, was up-regulated in tomato plants grown in the hairy vetch treatment relative to those grown in bare soil.

Results from this field pot experiment suggest that physiological cues released from the decomposing cover crop, other than N, have beneficial impacts on tomato plant physiology. These studies also suggest that tomato can distinguish between organic and inorganic sources of N, and that N management by on-site production of legume cover crops such as hairy vetch in sustainable cropping systems offer additional physiological advantages to cash crops than cropping systems utilizing inorganic N fertilizer alone [63]. Pertinent to these findings is a recent study that showed soil organic N influences plant growth as well as nitrogen use efficiency in plants, and that carbon cost of organic N assimilation, with its carbon content, into proteins is lower than with inorganic N [64]. Factors that contribute to higher nitrogen use efficiency in crops grown on organic N include nitrogen-based productivity and higher root:shoot ratio.

Another aspect of the physiology of tomato plants altered due to growth under different cropping systems (hairy vetch, black plastic, rye cover crop) compared to bare soil was the metabolome of the tomato fruit. Since metabolomics data for the fruit is relevant to the nutritional quality these studies affirmed that interaction between nutritional quality and growth environment [65].

3.3.2. Impact of Hairy Vetch Cropping System on the Soil Microbiome

Cover cropping with hairy vetch increased microbial biomass in the bulk soil as well as tomato rhizosphere when compared with the black plastic and bare soil treatments. The hairy vetch treatment also altered the microbial community structure [66,67]. Microbes are associated with healthy soil, nutrient cycling, soil organic matter, disease suppression, and therefore this altered microbial community structure may be at least partially responsible for the impact of hairy vetch on tomato physiology described above. There was indirect evidence suggesting that more readily available carbon in the vetch treatment may have caused the change in microbial community structure.

3.4. Use of Plant-Beneficial Microbes for Sustainable Crop Production

Biological control agents, plant-beneficial microbes that control plant pathogens and pests (invertebrates, pathogens, and weeds), are being developed to replace, or to be used in combination with reduced levels of, synthetic pesticides to enhance sustainability of crop production systems. Agricultural production systems are heavily reliant on pesticides for control of invertebrates, pathogens, and weeds as estimated losses to these pathogens and pests were 48% to 83% in the absence of some form of crop protection [14]. When there is no effective plant resistance, biological controls must be developed if pesticide inputs are to be reduced. Cultural practices such as crop rotation, alteration of planting date, etc. certainly play a role in pathogen and pest management but the level of control is often inadequate or economically nonviable [68]. For example, to decrease fungicide use in oilseed rape production an effective biological control strategy must be developed for the important soil-borne pathogen *Sclerotinia sclerotiorum* as the other forms of disease control tactics for this pathogen can be inadequate. Traditional breeding strategies for plant resistance to *S. sclerotiorum* are challenging due to limited gene pools and the need for multigenic resistance [69]. Crop rotation is limited in effectiveness as *S. sclerotiorum* has a broad host range and lengthy persistence [70].

Toward this end, biological control agents have been developed that can be applied at strategic points during the disease cycle of *S. sclerotiorum* on oilseed rape. *S. sclerotiorum* overwinters as sclerotia in soil, which upon germination, produce apothecia or directly produce mycelia. Ascospores produced from apothecia are the primary inoculum for most diseases of *S. sclerotiorum* and typically germinate on senescing flower petals. The pathogen then infects healthy leaf and stem tissue ultimately killing the plant [71,72]. Application points for biological control agents in this disease cycle are as a seed treatment, where the biological control agent is expected to colonize the developing oilseed rape plant and persist for the growing season, as a foliar spray at oilseed rape flowering to position the biological control agent so that it can prevent infection by germinating ascospores in the plant canopy, and as a spray on the field prior to planting oilseed rape so that the biological control agent can colonize and kill sclerotia of this pathogen prior to production of apothecia. Two seed treatment formulations of *Bacillus subtilis* BY-2 and of *B. subtilis* Tu-100, and spray applications of these two isolates at flowering, resulted in significantly lower disease incidence than the non-treated control and a significantly greater yield than this control in field trials conducted at a few locations. *B. megaterium* A6 was also shown to control *S. sclerotiorum* on oilseed rape at a few field locations when applied as a seed treatment formulation [73–75]. Control by these strains at these field locations compared favorably to that provided by the carbendazim chemical spray control applied at flowering. Fungal mycoparasites of sclerotia of *S. sclerotiorum* have also been isolated and developed for control of this pathogen on oilseed rape [76,77]. Spray application of one mycoparasite, *Aspergillus aculeatus* Asp-4, to the soil prior to sowing rice in a rice-oilseed rape rotation resulted in a significant reduction in incidence of Sclerotinia stem rot on oilseed rape compared with the non-treated control in two field trials. This application of *A. aculeatus* Asp-4 also resulted in a significant reduction in formation of apothecia on sclerotia relative to the non-treated control in these field trials, suggesting that colonization and degradation of sclerotia by Asp-4 and subsequent reduction in sclerotial germination led to disease control [77].

However, for these beneficial microbes to be widely acceptable to farmers as an alternative to pesticides they must be improved in reliability and efficacy so that they compare favorably to synthetic pesticide compounds over a large range of fields [78,79]. Unfortunately, biological control efficacy is often inconsistent due to the inherent complexity of the interaction between the biological control agent, the pathogen, and the environment where this interaction occurs [80]. For example, *A. aculeatus* Asp-4 up-regulated numerous genes and proteins involved in mitigation of environmental stress and dissolution of sclerotial compounds during colonization and degradation of sclerotia of *S. sclerotiorum* [81]. Omic studies of other biological control interactions revealed equally complex gene and/or protein expression profiles by biological control agents during interactions with pathogens or the environment [82–84]. In turn, soils and plant surfaces, where biological control agents are expected to effect disease control, are highly heterogeneous regarding compounds and conditions that impact expression of genes important to biological control; ultimately impacting expression of these genes and reliability and efficacy of disease control (reviewed in [80]). One approach to improving reliability and efficacy of disease control is to combine biological control microbes. These microbes can be combined in individual formulations or through multiple treatments targeting strategic points in the disease cycle. It is thought that a combination of microbes, with different ecological adaptations and mechanisms of disease control, are more likely to express traits important for disease control over a wider range of environmental conditions than an individual microbe.

This concept for improving efficacy and reliability shows promise. For example, when isolates BY-2, Tu-100, and A6 were applied as oliseed rape seed treatment formulations in various combinations there was increased seed yield and decreased disease incidence with increased number of isolates in seed treatment preparations in pot and field experiments conducted in four different soils (Hu et al., unpublished). Further, the treatment containing the three isolates BY-2, Tu-100, and A6 resulted in an incidence of disease on oilseed rape that was significantly lower than that associated with seed treatments containing individual isolates most of the time. Pieces are in place to develop and test an integrated disease control strategy to increase performance (level of disease suppression, reliability of disease suppression) of biological control of *S. sclerotiorum* on oilseed rape. Application of seed treatment formulations containing multiple *Bacillus* isolates, can be combined with application of formulations of the mycoparasites of sclerotia of *S. sclerotiorum*. These two application methods would reinforce each other regarding disease control. The application of mycoparasites prior to planting oilseed rape would reduce initial inoculum of *S. sclerotiorum*, resulting in less disease to be controlled with the *Bacillus* seed treatment or spray formulations. Likewise, the follow-up use of the *Bacillus* seed treatments or spray formulations would minimize disease caused by pathogen inoculum that escaped the treatment with the mycoparasite.

A second approach to increasing sustainability of disease control is to combine microbial biological control agents with reduced rates of synthetic pesticides. For example, formulations of the mycoparasite *Trichoderma* sp. Tri-1 were tested in combination with reduced application rates of the chemical pesticide carbendazim for control of *S. sclerotiorum* on oilseed rape [77]. The treatment containing the recommended rate of carbendazim provided the greatest reduction in disease when compared with treatments containing individual applications of lower rates of this pesticide or the formulated Tri-1 treatment in all field experiments. Encouragingly, treatments containing formulated Tri-1 combined with carbendazim applied at 75% the recommended rate reduced incidence of disease to levels to those obtained with the treatment containing carbendazim applied at the recommended rate in field trials.

Plant-beneficial microbes are also being developed to replace, or be used in combination with reduced levels of, synthetic phosphatic fertilizers to increase cropping system sustainability. The ability to solubilize P from compounds in soil is fairly wide-spread amongst soil microbes and certain biological control agents have been shown to control disease and solubilize P [85]. All three *Bacillus* isolates, *B. subtilis* Tu-100, *B. subtilis* BY-2, and *B. megaterium* A6 solubilized phosphate from inorganic and organic sources (Hu et al., unpublished; [86]. In addition to controlling disease, combinations of all three isolates in treatments of oilseed rape seeds resulted in promotion of growth relative to the nontreated control in five different soils (Hu et al., unpublished) indicating that these plant-beneficial microbes contribute multiple environmental benefits. As with biological control, environmental conditions influence the ability of these microbes to solubilize P [85]. More research is needed regarding colonization and mode of action of these plant-beneficial microbes to facilitate their use in enhancing sustainability of crop production systems.

4. Development of New Crop Cultivars for Sustainable Intensification in Agriculture

4.1. Improving Nutritional Quality of Crops

Until recently, classical breeding strategies were focused on developing crop cultivars for mechanical harvesting, yield, size, and disease control while improving nutritional quality was mostly unexplored. Lack of knowledge regarding metabolic pathways and their genetic components likely contributed to slow progress on developing strategies to improve nutritional quality of crops. Lately, attention has focused on developing nutrient-rich crops in response to consumer demand for high nutritional quality foods. Although subject to debate [87]; and references therein], a number of nutrition studies have linked diet with certain human health maladies–cancer, osteoporosis, diabetes, hypertension, cardiovascular health, and age/lifestyle-related diseases [88–91]. Diet fortification with off-the-shelf vitamins and antioxidants is a solution to enhancing human nutrition, but it is the bioavailability of a nutrient in food that determines the extent to which the potential of a nutrient is realized [92,93].

Improving nutritional quality of food will require new strategies and approaches. Currently available crop cultivars developed with classical breeding approaches have been diminished in certain health-promoting and flavor-enhancing molecules while the important agronomic attributes, yield and time to harvest, have been optimized [94–98]. Significant reductions in minerals, protein content, and vitamins over time were found in 43 garden crops in a US Department of Agriculture (USDA) survey conducted between 1950 and 1999 [94]. Also, a recent study combining biochemical and genomic analyses with consumer tasting panel data of approximately 400 varieties of tomatoes showed that modern commercial varieties were significantly reduced in many flavor chemicals relative to older varieties [99]. These and other studies clearly illustrate the need to use new approaches to recover not only the flavor-imparting genes/compounds but also other important nutritional traits that may have been lost due to intense classical breeding and selection.

Biotechnological approaches, using refined genomics, proteomics, and metabolomics tools are available and capable of engineering crops for enhanced nutritional content. In particular, progressive genetic engineering allows specific and controlled introduction of more efficient and novel genes while RNA-programmable genome CRISPR/Cas9 technology has great potential for editing inefficient or unwanted genes for development of crop germplasm with higher concentrations of nutrients [100]. For example, engineered 'golden' rice that synthesizes and accumulates pro-vitamin A, β-carotene, and protein is playing an important role in conquering malnutrition in the developing world, particularly with children [101,102]. Similarly, transgenic multivitamin corn was developed by introducing four different cDNAs encoding enzymes in the biosynthetic pathways of the vitamins β-carotene, ascorbate, and folate. The resulting transgenic corn endosperm was shown to accumulate β-carotene (~59.32 μg/gDW), the anti-cancerous lycopene (22.78 μg/gDW), vitamin C (ascorbate) (106.94 μg/gDW), and zeaxanthin (35.76 μg/gDW) [103]. Additionally, tomato fruit with enriched nutritional content (choline, lycopene, amino acids, sugars and organic acids) and enhanced shelf-life was developed using ripening-specific expression of the polyamines spermidine and spermine [104,105]. Unanticipated enrichment in lycopene levels in the engineered tomato by 2- to 3.5-fold as compared to the conventional tomatoes was substantial, exceeding that achieved by classical breeding methods [104]. Transcriptomic analysis of high polyamine-accumulating tomato fruits has shown that increased content of lycopene and flavonoids is consistent with upregulation in the transcription profiles related to carotenoid and flavonoid biosynthesis pathways (Fatima T. et al., unpublished). Other examples of genetic engineering-mediated enhancement of specific food crops, including fruits and vegetables, with higher or novel doses of nutritional and disease-preventing molecules have been collated and reviewed [91,98,106–108].

It is understood that genetic events, as well as environmental and crop production system factors, impact both the type and quantity of cellular metabolites in crops [109,110]. In this regard, cross talk between carbon metabolism and nitrogen sensing was detected in the polyamine-accumulating, transgenic tomatoes described above when these plants were grown with hairy vetch as a cover crop but not when grown with black plastic or in bare soil [105]. In these polyamine-accumulating, transgenic hairy vetch-grown tomatoes expression of genes for PEP carboxylase (PEPC) and cytosolic isocitrate dehydrogenase (ICDHc) was upregulated; these gene transcripts generally being activated in response to nitrogen assimilation [111,112]. Increased high polyamine levels in tomato fruits grown in hairy vetch were associated with a higher respiration rate, decreased concentrations of sucrose and glucose, and increases in the aspartate family of amino acids, as well as malate, fumarate and citrate relative to the other treatments [105] suggesting that additional nutritional benefit may be obtained by combining transgenic tomato with a cropping system tailored to optimize nutritional content of tomato fruit. There is a good probability that approaches used in this tomato model system can be applied towards enhancing the levels and composition of health-related nutrients into other crops such as grains [113].

4.2. Improving Abiotic Stresses Tolerance in Crop Cultivars

Crops are exposed to one or more abiotic stresses daily that impact crop yield and quality. These include drought, temperature extremes, high light intensities and UV radiation, and salinity. For example, high temperatures are known to reduce the grain filling period in wheat, oat, and field corn [114–117]. Also, suboptimal environmental conditions resulting from abiotic stress are a major cause for crops not achieving their full genetic potential for yield and crop quality, with yield losses being as high as 50% [118–120]. Clearly, the development of stress-tolerant crops is important if we are to increase agricultural productivity.

Developing stress-tolerant crops will require new strategies and approaches as classical breeding techniques have been largely unsuccessful, possibly due to the involvement of multigenic traits. Genetic transformation of crops by introducing genes that provide tolerance to one or more abiotic stresses is likely needed. A medley of genes, including specific transcription factors that interact directly or indirectly with genes associated with the abiotic stress signaling, have been identified by techniques such as comparative transcriptomics and validated [108,121,122]. Some transcription factors, belonging to a number of transcription factor families, enabled protection against multiple stresses, including cold, drought, and excess salt [120,122–124]. Additional, important candidate molecules that directly or indirectly impart tolerance to abiotic stresses include proteins with protective functions (dehydrins, heat shock proteins [HSPs], late embryogenesis abundant proteins), osmolytes (proline/trehalose/sugars), glycine betaine, signaling molecules (polyamines), and hormones (abscisic acid, ethylene, and methyl jasmonate). With regard to transcription factors, it is important to characterize which transcription factors among a family of transcription factors actually confer stress tolerance, rather than simply being stress responsive, in order to successfully utilize them for genetic transformation of a crop for improving resistance against one or more abiotic stresses.

Plants have been successfully enhanced in stress tolerance using genetic engineering. Overexpression of wheat transcription factor TaHsfA2d, similar to rice OsHsfA2d, infused tolerance of the transgenic Arabidopsis plant to a number of abiotic stresses—elevated temperatures, salinity and drought [125]. Also, gene shuffling improved the function of Rubisco activase, a critical protein in carbon fixation, and the resulting transgenic Arabidopsis plants performed better at moderately high temperatures [126]; the chloroplast chaperone, a HSP cpn60b chaperone with Rubisco activase acclimatized photosynthesis to higher temperatures [127]. HSPs accumulate in diverse plant cells in response to even a short exposure to high temperature, and are normally synthesized to maintain homeostasis of plant processes during these conditions. HSPs maintain functional conformation of other proteins and moonlight as chaperones in the assembly and transport of nascent proteins, normally as well as in response to abiotic stresses [118,128–131]. Additionally, engineering crop plants has resulted in resistance against various abiotic stresses and industry has successfully generated drought resistant germplasm for farmers.

Genetic evidence has validated that plants synthesize small molecules for protection against extreme environmental conditions. Small molecules that act as osmoprotectants include glycine betaine, sarcosine, trimethylamine-*N*-oxide, glycine, proline, glutamate, mannitol, and trehalose. These compounds correct the cytosolic imbalance caused by stress exposure [132,133], and references therein]. For instance, salt-tolerant alfalfa plants roots accumulate two-fold more proline than the salt-sensitive plants [134,135]. Additionally, different strategies used to develop transgenic rice plants that accumulate proline and soluble sugars resulted in plants that were protected against drought and osmotic stress [136–138].

Transgenic approaches have confirmed that polyamines, small molecules discussed above with regard to their role in the fruit metabolome, play an important role in plant responses to various abiotic stresses [120,139–142]. Polyamine pathway-deficient Arabidopsis mutants were more sensitive to salt stress [143–147] while heterologous overexpression of polyamine genes generated transgenic plants—Arabidopsis, rice, tobacco, and tomato, which were tolerant to abiotic stresses [148–154]. Also, transgenic tobacco, rice, and tomato plants engineered to express heterologous *S*-adenosylmethionine decarboxylase gene, a rate-limiting enzyme/gene in polyamine biosynthesis [100], showed tolerance against salt, osmotic, and heat stresses [139,155–158], while overexpression of spermidine synthase gene—in Arabidopsis, pear and potato plants—made these transgenic plants tolerant against drought, salt and oxidative stresses due to the higher polyamine levels [152,153,159].

Thus, the success in developing novel germplasm/genotypes against abiotic stresses provides a positive spin and encouragement for conducting bon-a-fide field trials of the engineered crop plants. Such validation will catalyze producing super new high-yielding crops with durable resistance to harsh environmental stresses to meet the food demands of world population.

4.3. Plant Biotic Stress Resistance and Antimicrobial Peptides (AMPs)

Traditional breeding strategies for plant resistance to pathogens have successfully mined 'R' (resistance) genes from related plant hosts and incorporated them into high yielding cultivars [68,160]. Research carried out to understand plant-pathogen interactions helped the discovery of plant 'R' genes that contained pathogen attack via 'innate' immunity. Elite breeding lines became sources of disease-resistant germplasm that helped contain crop diseases with additional help from chemical pesticide application. Such strategies are inherently time consuming in nature, pathogen species-specific, and tend to be of short duration as new race(s) of pathogens develop/evolve continuously, eventually overcoming R-gene plant resistance.

Antimicrobial peptides (AMPs) are a promising alternative to the use of 'R'-genes for plant resistance to plant pathogens [161,162]. AMPs are structurally diverse, small proteins consisting of 20–100 amino acids with the potential of imparting durable immunity to plants against pathogens. Being small molecules, changing their chemical structure to enhance potency is relatively easy. The combination of hydrophobic regions and net positive charge of AMPs allows electrostatic interactions with the negatively charged polar heads and hydrophobic core of microbial pathogen membranes [163]. Structural parameters other than conformation and charge include hydrophobicity, hydrophobic moment, amphipathicity, and polar angle that make them toxic and target specific. After binding their target, they permeabilize the pathogen membrane and impair pathogen cellular functions. AMPs establish interactions at the surface of these microbes in spite of the fact that prokaryotic organisms have significant structural differences, while their broad spectrum toxicity also involves targeting intracellular components necessary for the survival and proliferation of the pathogen. Structural disparity of prokaryotic and eukaryotic membranes contributes towards the

AMP selectivity [164–166]. They are also amenable to alteration for higher potency via production in higher amounts in heterologous systems using genetic biotechnology.

In vitro studies and in vivo expression of AMPs through transgenic approaches suggest that their antimicrobial activity is taxonomically very broad, likely because they bind to commonly specific domains that occur across great phylogenetic distances. AMPs are not cytotoxic to the host cells, thus enhancing their scope and application in disease management. Natural selection favors a defensive apparatus with a minimal maintenance cost plus a high deterrence value, which make AMPs fitting tools in plant defense strategies. It is noted here that AMPs are also a tool used by microorganisms including bacteria and fungi, likely for their defense [167].

AMPs Have Moonlighting Functions

Advancing research toward defining roles of AMPs in plant responses to abiotic and biotic stress has led to new information that suggests they also catalyze other important functions related to plant development. Some of these include roles in self-incompatibility—OCP-A1 and SP11 AMPs [168,169], pollination—LUREs, DEFL, ZmES-1 and DEF2 AMPs [170–172], and root nodule-specific secretary pathway [173,174]. Lipid transfer proteins, LTPs, are a class of cysteine-containing AMPs that function in the synthesis of cuticular wax [175], pollen adhesion [176], guiding pollen tube towards fertilization [177]—enabling enhanced tolerance to bacterial pathogens [178]. Molecular genetic manipulation of some AMPs confirmed their other cellular function(s). Such manipulation of DEF2 demonstrated its role in pollen viability, seeding and morphology of tomato [179] and inhibiting endogenous production of snakin-1 impaired potato plant development [180].

The promise and potential of employing AMPs not only in defense against pathogens but also in modulating and enhancing plant developmental processes can be immense in contributing to global food security. A test case that involved specific modification of the structure of the AMP, *msrA3*, and its introduction into potato via genetic engineering [181] brought to light important features about such a strategy which bear not only on disease-resistance phenomenon but also crop production. *msrA3*-transformed potato lines were found to be resistant to the necrotrophic pathogen *Fusarium solani*. Surprisingly, these transgenic potato plants mitigated normal plant defense responses such as the hypersensitive response (HR) and production of reactive oxygen species (ROS). Other characteristics of these transgenic potato lines included delayed leaf senescence and altered timing of bud development. Additionally, these transgenic plants did not elevate ROS (H_2O_2) levels in response to temperature stress or upon wounding, compared with the non-transformed wild-type potato. These properties were accompanied by dampened levels of gene markers for HR, ROS and senescence in the *msrA3*-transgenic plants [181]. Also apparent was that these transgenic potato plants were mitigated in normal jasmonic acid and H_2O_2 signaling observed in the control plants. It was concluded that the lack of oxidative burst, reduced H_2O_2 levels, and early suppression of gene transcription in response to different stressors is an indication that *msrA* functions upstream of these processes, consistent with the suggestion that the stress response pathways converge downstream of the stress recognition patterns [182]. Coincident to these observations was the finding that the potato vegetative phase was prolonged and bud development delayed. Of important agronomic consequence was the fact that the *msrA3* transgenic plants had significantly higher yield (52–57%) compared to the control plants.

The expression of *msrA3* in potato suppressed ROS (and HR) and prevented the induction of a number of gene transcripts, characteristics associated with extended vegetative growth, delayed floral development, and higher tuber yield. By extrapolation to studies in the literature, the delayed allocation of resources for reproductive growth seemed to translate into an increased tuber yield in the transgenic potato. Therefore, a dual action of *msrA* involving containing pathogen growth and maintaining a lower basal oxidative stress may contribute to enhanced productivity in plants. Since resource reallocation involves a global shift in the levels of hormones IAA and GA and/or nutrient balance, one can assume that *msrA* function may influence these processes. A stress environment induces a higher threshold of ROS, which in plants modulates development, signaling the stressed plant to grow rapidly, flower early and even shorten the grain filling period in field crops to complete the life cycle. Such a redirection of nutrient flow from vegetative organs to reproductive growth seems to be the norm during a plant's transition from vegetative to reproductive growth. It is also known that generation of ROS-mediated HR (as a response to a stress or a pathogen attack) causes a shift in cellular metabolism for resource re-allocation, involving global changes in gene expression. Thus, a heightened defense response of a plant contributes to the fitness cost, as seen during JA-dependent defense against herbivores and pathogenesis.

5. Conclusions

Feeding the world's future population will place unprecedented demands on agriculture, necessitating a dramatic increase in food production while at the same time decreasing the negative impacts of agriculture on land, water, and climate [2]. Next-generation cropping systems employing sustainable aspects of existing conventional and organic cropping systems can decrease the impact of agriculture on the environment. Cover crops, in addition to environmental benefits such as reducing erosion, can be used for weed and pathogen control decreasing the need for synthetic pesticides and tillage. Cover crops can also be used for managing soil fertility decreasing the need for synthetic N fertilizer inputs. Likewise, beneficial microbes can be used for control of pathogens and for improving soil fertility. As illustrated here, next-generation cropping systems can also have beneficial impacts on crop plant physiology. Vetch-grown tomatoes were delayed in senescence, producing fruit for extended periods, and induced in expression of genes involved in plant disease resistance. Unfortunately, the use of these cover crops and plant-beneficial microbes only leads to yields that approximate, or are attempting to approximate, yields already obtained using synthetic pesticides and fertilizer. To increase yields we will need to develop better performing crop cultivars using all available technologies, including genetic engineering.

Genetic improvement of plants through biotechnology to tolerate, or be resistant to, abiotic and biotic stresses will be a key component of future global food security as these stresses can result in substantial yield losses globally each year [183]. Current high-yielding crop varieties only yield well under ideal environmental conditions with high pesticide and fertilizer inputs. A second Green Revolution is needed where crops are developed that yield well under environmental extremes with low input of pesticide and fertilizer [184] for use in sustainable crop production systems. Genetic improvement of crops via genetic engineering and other biotechnology approaches will be needed to complement, or replace, traditional breeding efforts as they make available a broader range of genes such as AMPS, etc. and can be used in a more precise manner than traditional breeding [184,185]. In addition to being more time consuming and limited by diminished gene pools, traditional crop breeding methods are subject to linkage drag. As illustrated here, novel approaches are available using genetic engineering of crop plants to enhance abiotic stress tolerance and increase resistance to biotic stress. Examples of commercial success of genetic engineering for crop resistance to stress include Bt cotton, corn, and other crops [186].

More nutritious foods are also needed. Although there appears to be an impact of cropping system on nutritional content of food the perception that organic foods are more nutritious than conventionally produced food is the subject of debate. Some reviews and meta-analyses found some evidence of organic food being more nutritious in certain cases, however, there were questions as to whether the differences in nutritional content were meaningful. Other studies concluded that there were no meaningful differences in nutritional content between organically produced and conventionally produced food [24,187]. Therefore, substantial improvement of the nutritional content of food will likely require new germplasm generated using advanced biotechnological approaches. As pointed out with the tomato-hairy vetch system added nutritional benefit can be obtained combining genetically modified crops with the appropriate sustainable crop production strategy.

It should be noted that approaches for sustainable intensification described here relate primarily to regions of the world with sophisticated agricultural production systems. Countries that have not yet achieved food security face different challenges than the those with sophisticated agricultural production systems. In these countries crops are typically produced by farmers on small parcels of land not suitable for advanced agricultural technologies. However, increasing yield may be as simple as obtaining improved seed and fertilizer, or outreach providing informed advice on appropriate rates of fertilizer [188]. Clearly, to meet the challenge of sustainably feeding the future world population many different regional-specific approaches need to be developed and implemented.

Author Contributions: Both authors contributed equally to this review.

Conflicts of Interest: The authors declare no conflict of interest.

References

1. Godfray, H.C.J.; Beddington, J.R.; Crute, I.R.; Haddad, L.; Lawrence, D.; Muir, J.F.; Robinson, S.; Thomas, S.M.; Toulmin, C. Food security: The challenge of feeding 9 billion people. *Science* **2010**, *237*, 812–818. [CrossRef] [PubMed]
2. Foley, J.A.; Ramankutty, N.; Brauman, K.A.; Cassidy, E.S.; Gerber, J.S.; Johnston, M.; Mueller, N.D.; O'Connell, C.; Ray, D.K.; West, P.C.; et al. Solutions for a cultivated planet. *Nature* **2011**, *478*, 337–342. [CrossRef] [PubMed]
3. Tester, M.; Langridge, P. Breeding technologies to increase crop production in a changing world. *Science* **2010**, *327*, 818–822. [CrossRef] [PubMed]
4. Long, S.P.; Ainsworth, E.A.; Leakey, A.D.B.; Nösberger, J.; Ort, D.R. Food for thought: Lower-than-expected crop yield stimulation with rising CO_2 concentrations. *Science* **2006**, *312*, 1918–1921. [CrossRef] [PubMed]
5. Fedoroff, N.V.; Battisti, D.S.; Beachy, R.N.; Cooper, P.J.M.; Fischhoff, D.A.; Hodges, C.N.; Knauf, V.C.; Lobell, D.; Mazur, B.J.; Molden, D.; et al. Radically rethinking agriculture for the 21st Century. *Science* **2010**, *327*, 833–834. [CrossRef] [PubMed]
6. Schlenker, W.; Roberts, M.J. Nonlinear termperature effects indicate severe damages to U.S. crop yields under climate change. *Proc. Natl. Acad. Sci. USA* **2009**, *106*, 15594–15598. [CrossRef] [PubMed]
7. Myers, S.S.; Zanobetti, A.; Kloog, I.; Huybers, P.; Leakey, A.D.B.; Bloom, A.J.; Carlisle, E.; Dietterich, L.H.; Fitzgerald, G.; Hasegawa, T.; et al. Increasing CO_2 threatens human nutrition. *Nature* **2014**, *510*, 139–142. [CrossRef] [PubMed]
8. Cassman, K.G.; Dobermann, A.; Walters, D.T. Agroecosystems, nitrogen-use efficiency, and nitrogen management. *Ambio* **2002**, *31*, 132–139. [CrossRef] [PubMed]
9. Foley, J.A.; DeFries, R.; Asner, G.P.; Barford, C.; Bonan, C.; Carpenter, S.R.; Chapin, F.S.; Coe, M.T.; Daily, G.C.; Gibbs, H.K.; et al. Global consequences of land use. *Science* **2005**, *309*, 570–574. [CrossRef] [PubMed]
10. Matson, P.A.; Parton, W.J.; Power, A.G.; Swift, M.J. Agricultural intensification and ecosystem properties. *Science* **1997**, *277*, 504–509. [CrossRef] [PubMed]
11. Tilman, D.; Fargione, J.; Wolff, B.; D'Antonio, C.; Dobson, A.; Howarth, R.; Schindler, D.; Schlesinger, W.H.; Simberloff, D.; Swackhamer, D. Forecasting agriculturally driven global environmental change. *Science* **2001**, *292*, 281–284. [CrossRef] [PubMed]
12. West, P.C.; Gibbs, H.K.; Monfreda, C.; Wagner, J.; Barford, C.C.; Carpenter, S.R.; Foley, J.A. Trading carbon for food: Global comparison of carbon stocks vs. crop yields on agricultural land. *Proc. Natl. Acad. Sci. USA* **2010**, *107*, 19645–19648. [CrossRef] [PubMed]
13. Canfield, D.E.; Glazer, A.N.; Falkowski, P.G. The evolution and future of Earth's nitrogen cycle. *Science* **2010**, *330*, 192–196. [CrossRef] [PubMed]
14. Oerke, E.-C. Crop losses to pests. *J. Agric. Sci.* **2005**, *144*, 31–43. [CrossRef]
15. Peigné, J.; Ball, B.C.; Roger-Estrade, J.; David, C. Is conservation tillage suitable for organic farming? *Soil Use Manag.* **2007**, *23*, 129–144. [CrossRef]
16. Triplett, G.B., Jr.; Dick, W.A. No-tillage crop production: A revolution in agriculture! *Agron. J.* **2008**, *100*, S153–S165. [CrossRef]
17. Mattoo, A.K.; Teasdale, J.R. Ecological and genetic systems underlying sustainable horticulture. *Hortic. Rev.* **2010**, *37*, 331–362.
18. Gomeiro, T.; Pimentel, D.; Paoletti, M.G. Environmental impact of different agricultural management practices: Conventional vs. organic agriculture. *Crit. Rev. Plant Sci.* **2011**, *30*, 95–124. [CrossRef]

19. Seufert, V.; Ramankutty, N.; Foley, J.A. Comparing the yields of organic and conventional agriculture. *Nature* **2012**, *485*, 229–234. [CrossRef] [PubMed]
20. Cavigelli, M.A.; Djurickovic, M.; Rasmann, C.; Spargo, J.T.; Mirsky, S.B.; Maul, J. Global warming potential or organic and conventional grain cropping systems in the mid-Atlantic region of the US. In Proceedings of the Farming Systems Design 2009 International Symposium, Monterey, CA, USA, 23–26 August 2009.
21. Marriott, E.E.; Wander, M.M. Qualitative and quantitative differences in particulate organic matter fractions in organic and conventional farming systems. *Soil Biol. Biochem.* **2006**, *38*, 1527–1536. [CrossRef]
22. Spargo, J.T.; Cavigelli, M.A.; Mirsky, S.B.; Maul, J.E.; Meisinger, J.J. Mineralizable soil nitrogen and labile soil organic matter in diverse long-term cropping systems. *Nutr. Cycl. Agroecosyst.* **2011**, *90*, 253–266. [CrossRef]
23. Mäder, P.; Flieback, A.; Dubois, D.; Gunst, L.; Fried, P.; Nigglli, U. Soil fertility and biodiversity in organic farming. *Science* **2002**, *296*, 1694–1697. [CrossRef] [PubMed]
24. Reganold, J.P.; Wachter, J.M. Organic agriculture in the twenty-first century. *Nat. Plants* **2016**, *2*, 15221. [CrossRef] [PubMed]
25. Tuomisto, H.L.; Hodge, I.D.; Riordan, P.; Macdonald, D.W. Does organic farming reduce environmental impacts?—A meta-analysis of European research. *J. Environ. Manag.* **2012**, *112*, 309–320. [CrossRef] [PubMed]
26. Drinkwater, L.E. Cropping systems research: Reconsidering agricultural experimental approaches. *HortTechnology* **2002**, *12*, 355–361.
27. Teasdale, J.R.; Cavigelli, M.A. Meteorological fluctuations define long-term crop yield patterns in conventional and organic production systems. *Sci. Rep.* **2017**, *7*, 688. [CrossRef] [PubMed]
28. Cavigelli, M.A.; Mirsky, S.B.; Teasdale, J.R.; Spargo, J.T.; Doran, J. Organic grain cropping systems to enhance ecosystems services. *Renew. Agric. Food Syst.* **2013**, *28*, 145–159. [CrossRef]
29. Boesch, D.F.; Brinsfield, R.B.; Magnien, R.E. Chesapeake Bay eutrophication: Scientific understanding, ecosystem restoration, and challenges for agriculture. *J. Environ. Qual.* **2001**, *30*, 303–320. [CrossRef] [PubMed]
30. Green, S.V.; Cavigelli, M.A.; Dao, T.H.; Flanagan, D.C. Soil physical properties and aggregate-associated C, N, and P distributions in organic and conventional cropping systems. *Soil Sci.* **2005**, *170*, 822–831. [CrossRef]
31. Foster, G.R.; Young, R.A.; Neibling, W.H. Sediment composition for nonpoint source pollution analysis. *Trans. ASAE* **1985**, *28*, 133–139. [CrossRef]
32. Miksha, M.M.; Rice, C.W. Tillage and manure effects on soil and aggregate-associated carbon and nitrogen. *Soil Sci. Soc. Am. J.* **2004**, *68*, 809–816. [CrossRef]
33. Teasdale, J.R.; Cavigelli, M.A. Subplots facilitate assessment of corn yield losses from weed competition in a long-term systems experiment. *Agron. Sustain. Dev.* **2010**, *30*, 445–453. [CrossRef]
34. Wallace, J.M.; Williams, A.; Liebert, J.A.; Ackroyd, J.J.; Vann, R.A.; Curran, W.S.; Keene, C.L.; Van Gessel, M.J.; Ryan, M.R.; Mirsky, S.B. Cover crop-based, organic rotational no-till corn and soybean production systems in the mid-Atlantic United States. *Agriculture* **2017**, *7*, 34. [CrossRef]
35. Mahboubi, A.A.; Lal, R.; Faussey, N.R. Twenty-eight years of tillage effects on two soils in Ohio. *Soil Sci. Soc. Am. J.* **1993**, *57*, 506–512. [CrossRef]
36. Weil, R.R.; Magdoff, F. Significance of soil organic matter to soil quality and health. In *Soil Organic Matter in Sustainable Agriculture*; Magdoff, F., Weil, R.R., Eds.; CRC Press: Boca Raton, FL, USA, 2004; pp. 1–43.
37. Grandy, A.S.; Robertson, G.P.; Thelen, K.D. Do productivity and environmental trade-offs justify periodically cultivating no-till cropping systems? *Agron. J.* **2006**, *98*, 1377–1383. [CrossRef]
38. Reicosky, D.C.; Kemper, W.D.; Langdale, G.W.; Douglas, C.L.; Rasmussen, P.E. Soil organic matter changes resulting from tillage and biomass production. *J. Soil Water Conserv.* **1995**, *50*, 253–261.
39. Camargo, G.G.T.; Ryan, M.R.; Richard, T.L. Energy use and greenhouse gas emissions from crop production using the farm energy analysis tool. *Bioscience* **2013**, *63*, 263–273. [CrossRef]
40. Mirsky, S.B.; Ryan, M.R.; Curran, W.S.; Teasdale, J.R.; Maul, J.E.; Spargo, J.T.; Moyer, J.; Grantham, A.M.; Weber, D.; Way, T.R.; et al. Conservation tillage issues: Cover crop-based organic rotational no-till grain production in the mid-Atlantic region, USA. *Renew. Agric. Food Syst.* **2012**, *27*, 31–40. [CrossRef]
41. Snapp, S.S.; Swinton, S.M.; Labarra, D.; Mutch, D.; Black, J.R.; Leep, R.; Nyiraneza, J.; O'Neil, K. Evaluating cover crops for benefits, costs and performance within cropping system niches. *Agron. J.* **2005**, *97*, 322–332.
42. Poffenbarger, H.J.; Mirsky, S.B.; Weil, R.R.; Maul, J.E.; Kramer, M.; Spargo, J.T.; Cavigelli, M.A. Biomass and nitrogen content of hairy vetch-cereal rye cover crop mixtures as influenced by species proportions. *Agron. J.* **2015**, *107*, 2069–2082. [CrossRef]

43. Hayden, Z.D.; Brainard, D.C.; Henshaw, B.; Ngouajio, M. Winter annual weed suppression in rye-vetch cover crop mixtures. *Weed Technol.* **2012**, *26*, 818–824. [CrossRef]
44. Hayden, Z.D.; Ngouajio, M.; Brainard, D. Rye-vetch mixture proportion tradeoffs: Cover crop productivity, nitrogen accumulation, and weed suppression. *Agron. J.* **2014**, *106*, 904–914. [CrossRef]
45. Mirsky, S.B.; Ryan, M.R.; Teasdale, J.R.; Curran, W.S.; Reberg-Horten, C.S.; Spargo, J.T.; Wells, M.S.; Wells, S.M.; Keene, C.L.; Moyer, J.W. Overcoming weed management challenges in cover crop-based organic rotational no-till soybean production in the eastern United States. *Weed Technol.* **2013**, *27*, 31–40. [CrossRef]
46. Abdul-Baki, A.A.; Teasdale, J.R.; Korcak, R.F.; Chitwood, D.J.; Huettel, R.N. Fresh-market tomato production in a low-input alternative system using cover-crop mulch. *J. Am. Soc. Hortic. Sci.* **1996**, *31*, 65–69.
47. Kelly, T.C.; Lu, Y.-C.; Abdul-Baki, A.A.; Teasdale, J.R. Economics of a hairy vetch mulch system for producing fresh-market tomatoes in the mid-Atlantic region. *J. Am. Soc. Hortic. Sci.* **1995**, *120*, 854–860.
48. Abdul-Baki, A.A.; Teasdale, J.R. A no-tillage tomato production system using hairy vetch and subterranean clover mulches. *HortScience* **1993**, *28*, 106–108.
49. Abdul-Baki, A.A.; Teasdale, J.R.; Korcak, R.F. Nitrogen requirements of fresh-market tomatoes on hairy vetch and block polyethylene mulch. *HortScience* **1997**, *32*, 217–221.
50. Rice, P.J.; McConnell, L.L.; Heighton, L.P.; Sadeghi, A.M.; Isensee, A.R.; Teasdale, J.R.; Abdul-Baki, A.A.; Harman-Fetcho, J.A.; Hapeman, C.J. Runoff loss of pesticides and soil: A comparison between vegetative mulch and plastic mulch in vegetable production systems. *J. Environ. Qual.* **2001**, *30*, 1808–1821. [CrossRef] [PubMed]
51. Rice, P.J.; McConnell, L.L.; Heighton, L.P.; Sadeghi, A.M.; Isensee, A.R.; Teasdale, J.R.; Abdul-Baki, A.; Harman-Fetcho, J.A.; Hapeman, C.J. Comparison of copper levels in runoff from fresh-market vegetable production using polyethylene mulch or a vegetative mulch. *Environ. Toxicol. Chem.* **2002**, *21*, 24–30. [CrossRef] [PubMed]
52. Mills, D.J.; Coffman, B.; Teasdale, J.R.; Everts, K.L.; Abdul-Baki, A.A.; Lydon, J.; Anderson, J.D. Foliar disease in fresh-market tomato grown in differing bed strategies and fungicide spray programs. *Plant Dis.* **2002**, *86*, 955–959. [CrossRef]
53. Mills, D.J.; Coffman, B.; Teasdale, J.R.; Everts, K.L.; Anderson, J.D. Factors associated with foliar disease of staked fresh market tomatoes grown under differing bed strategies. *Plant Dis.* **2002**, *86*, 356–361. [CrossRef]
54. Teasdale, J.R.; Abdul-Baki, A.A.; Mills, D.J.; Thorpe, K.W. Enhanced pest management with cover crop mulches. *Acta Hortic.* **2004**, *638*, 135–140. [CrossRef]
55. Kumar, V.; Mills, D.J.; Anderson, J.D.; Mattoo, A.K. An alternative agriculture system is defined by a distinct expression profile of select gene transcripts and proteins. *Proc. Natl. Acad. Sci. USA* **2004**, *101*, 10535–10540. [CrossRef] [PubMed]
56. Kumar, V.; Abdul-Baki, A.; Anderson, J.D.; Mattoo, A.K. Cover crop residues enhance growth, improve yield and delay leaf senescence in greenhouse-grown tomatoes. *HortScience* **2005**, *40*, 1307–1311.
57. Mattoo, A.K.; Abdul-Baki, A. Crop genetic responses to management: Evidence of root-to-shoot communication. In *Biological Approaches to Sustainable Soil Systems*; Uphoff, N., Ball, A.S., Palm, C., Fernandes, E., Pretty, J., Herren, H., Sanchez, P., Husson, O., Sanginga, N., Laing, M., et al., Eds.; Taylor & Francis: Boca Raton, FL, USA, 2006; pp. 221–230.
58. Papon, N.; Clastre, M.; Andreu, F.; Gantet, P.; Rideau, M.; Creche, J. Expression analysis in plant and cell suspensions of CrCKR1, a cDNA encoding histidine kinase receptor homologue in *Catharanthus roseus* (L.) G. Don. *J. Exp. Bot.* **2002**, *53*, 1989–1990. [CrossRef]
59. Noh, Y.S.; Amasino, R.M. Identification of a promoter region responsible for the senescence-specific expression of SAG12. *Plant Mol. Biol.* **1999**, *41*, 181–194. [CrossRef] [PubMed]
60. Ryu, C.M.; Farag, M.A.; Hu, C.-H.; Reddy, M.S.; Wei, H.-X.; Pare, P.W.; Kloepper, J.W. Bacterial volatiles promote growth in Arabidopsis. *Proc. Natl. Acad. Sci. USA* **2003**, *100*, 4927–4932. [CrossRef] [PubMed]
61. Memelink, J.; Hoge, J.H.C.; Schilperoort, R.A. Cytokinin stress changes the developmental regulation of several defence-related genes in tobacco. *EMBO J.* **1987**, *6*, 3579–3583. [PubMed]
62. Thomas, J.C.; Smigocki, A.C.; Bohnert, H.J. Light-induced expression of *ipt* from *Agrobacterium tumefaciens* results in cytokinin accumulation and osmotic stress symptoms in transgenic tobacco. *Plant Mol. Biol.* **1995**, *27*, 225–235. [CrossRef] [PubMed]

63. Fatima, T.F.; Teasdale, J.; Bunce, J.; Mattoo, A.K. Tomato response to legume cover crop and nitrogen: Differing enhancement patterns of fruit yield, photosynthesis and gene expression. *Funct. Plant Biol.* **2012**, *39*, 246–254. [CrossRef]

64. Franklin, O.; Cambui, C.A.; Gruffman, L.; Palmroth, S.; Oren, R.; Nashholm, T.Y. The carbon bonus of organic nitrogen enhances nitrogen use efficiency of plants. *Plant Cell Environ.* **2017**, *40*, 25–35. [CrossRef] [PubMed]

65. Fatima, T.; Sobolev, A.; Teasdale, J.; Kramer, M.; Bunce, J.; Handa, A.; Mattoo, A.K. Fruit metabolite networks in engineered and non-engineered tomato genotypes reveal fluidity in a hormone and agroecosystem specific manner. *Metabolomics* **2016**, *12*, 103. [CrossRef] [PubMed]

66. Buyer, J.S.; Teasdale, J.R.; Roberts, D.P.; Zasada, I.A.; Maul, J.E. Factors affecting soil microbial community structure in tomato cropping systems. *Soil Biol. Biochem.* **2010**, *42*, 831–841. [CrossRef]

67. Maul, J.E.; Buyer, J.S.; Lehman, R.M.; Culman, S.; Blackwood, C.B.; Roberts, D.P.; Zasada, I.A.; Teasdale, J.R. Microbial community structure and abundance in the rhizosphere and bulk soil of a tomato cropping system that includes cover crops. *Appl. Soil Ecol.* **2014**, *77*, 42–50. [CrossRef]

68. Vincelli, P. Genetic engineering and sustainable crop disease management: Opportunities for case-by-case decision making. *Sustainability* **2016**, *8*, 495. [CrossRef]

69. Bardin, S.D.; Huang, H.C. Research on biology and control of *Sclerotinia* diseases in Canada. *Can. J. Plant Pathol.* **2001**, *23*, 88–98. [CrossRef]

70. Nelson, B. Biology of *Sclerotinia*. In *Proceedings of the 10th International Sclerotinia Workshop, Fargo, ND, USA, 21 January 1998*; North Dakota State University Department of Plant Pathology: Fargo, ND, USA, 1998; pp. 1–5.

71. Abawi, G.S.; Grogan, R.G. Epidemiology of diseases caused by *Sclerotinia* species. *Phytopathology* **1979**, *69*, 899–904. [CrossRef]

72. Boland, G.J. Fungal viruses, hypovirulence, and biological control of *Sclerotinia* species. *Can. J. Plant Pathol.* **2004**, *26*, 6–18. [CrossRef]

73. Hu, X.; Roberts, D.P.; Maul, J.E.; Emche, S.E.; Liao, X.; Guo, X.; Liu, X.; McKenna, L.F.; Buyer, J.S.; Liu, S. Formulations of the endophytic bacterium *Bacillus subtilis* Tu-100 suppress *Sclerotinia sclerotiorum* on oilseed rape and improve plant vigor in field trials conducted at separate locations. *Can. J. Microbiol.* **2011**, *57*, 539–546. [CrossRef] [PubMed]

74. Hu, X.; Roberts, D.P.; Xie, L.; Maul, J.E.; Yu, C.; Li, Y.; Zhang, S.; Liao, X. *Bacillus megaterium* A6 suppresses *Sclerotinia sclerotiorum* on oilseed rape in the field and promotes oilseed rape growth. *Crop Prot.* **2013**, *52*, 151–158. [CrossRef]

75. Hu, X.; Roberts, D.P.; Xie, L.; Maul, J.E.; Yu, C.; Li, Y.; Liao, X. Formulations of *Bacillus subtilis* BY-2 suppress *Sclerotinia sclerotiorum* on oilseed rape in the field. *Biol. Control* **2014**, *70*, 54–64. [CrossRef]

76. Hu, X.; Roberts, D.P.; Xie, L.; Maul, J.E.; Yu, C.; Li, Y.; Zhang, Y.; Qin, L.; Xing, L. Components of a rice-oilseed rape production system augmented with *Trichoderma* sp. Tri-1 control *Sclerotinia sclerotiorum* on oilseed rape. *Phytopathology* **2015**, *105*, 1325–1333. [CrossRef] [PubMed]

77. Hu, X.; Roberts, D.P.; Xie, L.; Yu, C.; Li, Y.; Qin, L.; Hu, L.; Zhang, Y.; Liao, X. Use of formulated *Trichoderma* sp. Tri-1 in combination with reduced rates of chemical pesticide for control of *Sclerotinia sclerotiorium* on oilseed rape. *Crop Prot.* **2016**, *79*, 124–127. [CrossRef]

78. Fravel, D.R. Commercialization and implementation of biocontrol. *Annu. Rev. Phytopathol.* **2005**, *43*, 337–359. [CrossRef] [PubMed]

79. Glare, T.; Caradus, J.; Gelernter, W.; Jackson, T.; Keyhani, N.; Köhl, J.; Marrone, P.; Morin, L.; Stewart, A. Have biopesticides come of age? *Trends Biotechnol.* **2012**, *30*, 250–258. [CrossRef] [PubMed]

80. Roberts, D.P.; Kobayashi, D.Y. Impact of spatial heterogeneity within spermosphere and rhizosphere environments on performance of bacterial biological control agents. In *Bacteria in Agrobiology: Crop Ecosystems*; Maheshwari, D.K., Ed.; Springer: Berlin, Germany, 2011.

81. Hu, X.; Qin, L.; Roberts, D.P.; Lakshman, D.K.; Gong, Y.; Maul, J.E.; Xie, L.; Yu, C.; Li, Y.; Hu, L.; et al. Characterization of mechanisms underlying degradation of sclerotia of *Sclerotinia sclerotiorum* by *Aspergillus aculeatus* Asp-4 using a combined qRT-PCR and proteomic approach. *BMC Genom.* **2017**, *18*, 674. [CrossRef] [PubMed]

82. Muthumeenakshi, S.; Sreenivasaprasad, S.; Rodgers, C.W.; Challen, M.P.; Whipps, J.M. Analysis of cDNA transcripts from *Coniothyrium minitans* reveals a diverse array of genes involved in key processes during sclerotial mycoparasitism. *Fungal Genet. Biol.* **2007**, *44*, 1262–1284. [CrossRef] [PubMed]

83. Seidl, V.; Song, L.; Lindquist, E.; Gruber, S.; Koptchinskiy, A.; Zeilinger, S.; Schmoll, M.; Martinez, P.; Sun, J.; Grigoriev, I.; et al. Transcriptomic response of the mycoparasitic fungus *Trichoderma atroviride* to the presence of a fungal prey. *BMC Genom.* **2009**, *10*, 567. [CrossRef] [PubMed]

84. Kidarsa, T.A.; Shaffer, B.T.; Goebel, N.C.; Roberts, D.P.; Buyer, J.S.; Johnson, A.; Kobayashi, D.Y.; Zabriskie, T.M.; Paulsen, I.; Loper, J.E. Genes expressed by the biological control bacterium *Pseudomonas protegens* Pf-5 on seed surfaces under the control of the global regulators GacA and RpoS. *Environ. Microbiol.* **2013**, *15*, 716–735. [CrossRef] [PubMed]

85. Alori, E.T.; Glick, B.R.; Babolola, O.O. Microbial phosphorus solubilization and its potential for use in sustainable agriculture. *Front. Microbiol.* **2017**, *8*, 971. [CrossRef] [PubMed]

86. Hu, X.; Roberts, D.P.; Xie, L.; Maul, J.E.; Yu, C.; Li, Y.; Zhang, S.; Liao, X. Development of a biologically based fertilizer, incorporating *Bacillus megaterium* A6, for improved phosphorus nutrition of oilseed rape. *Can. J. Microbiol.* **2013**, *59*, 231–236. [CrossRef] [PubMed]

87. Newell-McGloughlin, M. Nutritionally improved agricultural crops. *Plant Physiol.* **2008**, *147*, 939–953. [CrossRef] [PubMed]

88. Block, G.; Patterson, B.; Subar, A. Fruit, vegetable and cancer prevention: A review of the epidemiological evidence. *Nutr. Cancer* **1992**, *18*, 1–29. [CrossRef] [PubMed]

89. Potter, J.D.; Steinmetz, K. Vegetables, fruit and phytoestrogens as preventive agents. *IARC Sci. Publ.* **1996**, *139*, 61–90.

90. Botella-Pavía, P.; Rodriguez-Conceptíon, M. Carotenoid biotechnology in plants for nutritionally improved foods. *Plant Physiol.* **2006**, *126*, 369–381. [CrossRef]

91. Mattoo, A.K.; Shukla, V.; Fatima, T.; Handa, A.K.; Yachha, S.K. Genetic engineering to enhance crop-based phytonutrients (nutraceuticals) to alleviate diet-related diseases. In *Bio-Farms for Nutraceuticals: Functional Food and Safety Control by Biosensors*; Giardi, M.T., Rea, G., Berra, B., Eds.; Springer: Georgetown, TX, USA, 2010; pp. 122–143.

92. Rao, A.V.; Agarwal, S. Bioavailability and in vivo antioxidant properties of lycopene from tomato products and their possible role in the prevention of cancer. *Nutr. Cancer* **1995**, *31*, 199–203. [CrossRef] [PubMed]

93. Milner, J.A. Nutrition and cancer: Essential elements for a roadmap. *Cancer Lett.* **2008**, *269*, 189–198. [CrossRef] [PubMed]

94. Davis, D.R.; Epp, M.D.; Riordan, H.D. Changes in USDA food composition data for 43 garden crops, 1950 to 1999. *J. Am. Coll. Nutr.* **2004**, *23*, 669–682. [CrossRef] [PubMed]

95. White, P.J.; Broadley, M.R. Historical variation in the mineral composition of edible horticultural products. *J. Hortic. Sci. Biotechnol.* **2005**, *80*, 660–666. [CrossRef]

96. Garvin, D.F.; Welch, R.M.; Finley, J.W. Historical shifts in the seed mineral micronutrient concentration of US hard red winter wheat germplasm. *J. Sci. Food Agric.* **2006**, *86*, 2213–2220. [CrossRef]

97. Murphy, K.M.; Reeves, P.G.; Jones, S.S. Relationship between yield and mineral nutrient concentrations in historical and modern spring wheat cultivars. *Euphytica* **2008**, *163*, 381–390. [CrossRef]

98. Klee, H.J.; Tieman, D.M. Genetic challenges of flavor improvement in tomato. *Trends Genet.* **2013**, *29*, 257–262. [CrossRef] [PubMed]

99. Tieman, D.; Zhu, G.; Resende, M.F., Jr.; Lin, T.; Nguyen, C.; Bies, D.; Rambla, J.L.; Beltran, K.S.O.; Taylor, M.; Zhang, B.; et al. A chemical genetic roadmap to improved tomato flavor. *Science* **2017**, *355*, 391–394. [CrossRef] [PubMed]

100. Doudna, J.A.; Charpentier, E. The new frontier of genome engineering with CRISPR-Cas9. *Science* **2014**, *346*, 1077. [CrossRef] [PubMed]

101. Ye, X.; Al-Babili, S.; Kloti, A.; Zhang, J.; Lucca, P.; Beyer, P.; Potrykus, I. Engineering the provitamin A (β-carotene) biosynthetic pathway into (carotenoid free) rice endosperm. *Science* **2000**, *287*, 303–305. [CrossRef] [PubMed]

102. Paine, J.A.; Catherine, A.S.; Chaggar, S.; Howells, R.M.; Kennedy, M.J.; Vernon, G.; Wright, S.Y.; Hinchliffe, E.; Adams, J.L.; Silverstone, A.L.; et al. Improving the nutritional value of golden rice through increased pro-vitamin A content. *Nat. Biotechnol.* **2005**, *23*, 482–487. [CrossRef] [PubMed]

103. Naqvi, S.; Zhu, C.; Farre, G.; Ramessar, K.; Bassie, L.; Breitenbach, J.; Conesa, D.P.; Ros, G.; Sandmann, G.; Capell, T.; et al. Transgenic multivitamin corn through biofortification of endosperm with three vitamins representing three distinct metabolic pathways. *Proc. Natl. Acad. Sci. USA* **2009**, *106*, 7762–7767. [CrossRef] [PubMed]

104. Mehta, R.A.; Cassol, T.; Li, N.; Ali, N.; Handa, A.K.; Mattoo, A.K. Engineered polyamine accumulation in tomato enhances phytonutrient content, juice quality, and vine life. *Nat. Biotechnol.* **2002**, *20*, 613–618. [CrossRef] [PubMed]

105. Mattoo, A.K.; Sobolev, A.P.; Neelam, A.; Goyal, R.K.; Handa, A.K.; Segre, A.L. NMR spectroscopy-based metabolite profiling of transgenic tomato fruit engineered to accumulate spermidine and spermine reveals enhanced anabolic and nitrogen-carbon interactions. *Plant Physiol.* **2006**, *142*, 1759–1770. [CrossRef] [PubMed]

106. Christou, P. Plant genetic engineering and agricultural biotechnology 1983–2013. *Trends Biotechnol.* **2013**, *31*, 125–127. [CrossRef] [PubMed]

107. Mattoo, A.K.; Yachha, S.K.; Fatima, T. Genetic manipulation of vegetable crops to alleviate diet-related diseases. In *Improving the Health-Promoting Properties of Fruit and Vegetable Products*; Tomas-Barberan, F.A., Gil, M.I., Eds.; CRC Press, Woodhead Publishing, Ltd.: Cambridge, MA, USA, 2008; pp. 326–345.

108. Kaur, B.; Handa, A.K.; Mattoo, A.K. Genetic engineering of tomato to improve nutritional quality, resistance to abiotic and biotic stresses, and for non-food applications. In *Achieving Sustainable Cultivation of Tomatoes*; Mattoo, A.K., Handa, A.K., Eds.; Burleigh Dodds Science Publishing: Cambridge, UK, 2017; pp. 239–281.

109. Neelam, A.; Cassol, T.; Mehta, R.A.; Abdul-Baki, A.A.; Sobolev, A.; Goyal, R.K.; Abbott, J.; Segre, A.L.; Handa, A.K.; Mattoo, A.K. A field-grown transgenic tomato line expressing polyamines reveals legume cover crop mulch-specific perturbations in fruit phenotype at the levels of metabolite profiles, gene expression and agronomic characteristics. *J. Exp. Bot.* **2008**, *59*, 2337–2346. [CrossRef] [PubMed]

110. Slimestad, R.; Verheu, M. Review of flavonoids and other phenolics from fruits of different tomato (*Lycopersicon esculentum* Mill.) cultivars. *J. Sci. Food Agric.* **2009**, *89*, 1255–1270. [CrossRef]

111. Scheible, W.R.; Gonzales-Fontes, A.; Lauerer, M.; Muller-Rober, B.; Caboche, M.; Stitt, M. Nitrate acts as a signal to induce organic acid metabolism and repress starch metabolism in tobacco. *Plant Cell* **1997**, *9*, 783–798. [CrossRef] [PubMed]

112. Scheible, W.-R.; Morcuende, R.; Czechowski, T.; Fritz, C.; Osuna, D.; Palacios-Rojas, N.; Schindelasch, D.; Thimm, O.; Udvardi, M.K.; Stitt, M. Genome-wide reprogramming of primary and secondary metabolism, protein synthesis, cellular growth processes, and the regulatory infrastructure of Arabidopsis in response to nitrogen. *Plant Physiol.* **2004**, *136*, 2483–2499. [CrossRef] [PubMed]

113. Schijlen, E.; Ric de Vos, C.H.; Jonker, H.; van den Broeck, H.; Molthoff, J.; van Tunen, A. Pathway engineering for healthy phytochemicals leading to the production of novel flavonoids in tomato fruit. *Plant Biotechnol. J.* **2006**, *4*, 433–444. [CrossRef] [PubMed]

114. Mitchell, R.A.C.; Mitchell, V.J.; Driscoll, S.P. Effects of increased CO_2 concentration and temperature on growth and yield of winter wheat at two levels of nitrogen application. *Plant Cell Environ.* **1993**, *16*, 521–529. [CrossRef]

115. Wheeler, T.R.; Hong, T.D.; Ellis, R.H.; Batts, G.R.; Morison, J.I.L.; Hadley, P. The duration and rate of grain growth, and harvest index, of wheat (*Triticum aestivum*) in response to temperature and CO_2. *J. Exp. Bot.* **1996**, *47*, 623–630. [CrossRef]

116. Rosenzwig, C.; Hillel, C. *Climate Change and the Global Harvest: Potential Impacts of the Greenhouse Effect on Agriculture*; Oxford University Press: New York, NY, USA, 1998.

117. Ghaffari, A.; Cook, H.F.; Lee, H.C. Climate change and winter wheat management: A modeling scenario for South-Eastern England. *Clim. Chang.* **2002**, *55*, 509–533. [CrossRef]

118. Wang, W.; Vinocur, B.; Altman, A. Plant responses to drought, salinity and extreme temperatures: Towards genetic engineering for stress tolerance. *Planta* **2003**, *218*, 1–14. [CrossRef] [PubMed]

119. Shao, H.B.; Chu, L.Y.; Jaleel, C.A.; Zhao, C.X. Water-deficit stress-induced anatomical changes in higher plants. *Comptes Rendus Biol.* **2008**, *331*, 215–225. [CrossRef] [PubMed]

120. Shukla, V.; Mattoo, A.K. Developing robust crop plants for sustaining growth and yield under adverse climatic changes. In *Climate Change and Plant Abiotic Stress Tolerance*; Tuteja, N., Gill, S.S., Eds.; Wiley-VCH Verlag GnbH & Co, KGaA: Weinheim, Germany, 2014; pp. 27–56.

121. Zhu, J.-K. Salt and drought stress signal transduction in plants. *Annu. Rev. Plant Biol.* **2002**, *53*, 247–273. [CrossRef] [PubMed]

122. Pandey, G.P. *Elucidation of Abiotic Stress Signaling in Plants. Functional Genomics Perspectives*; Springer: New York, NY, USA, 2015; Volume 2, p. 488, ISBN 978-1-4939-2539-1.

123. Mattoo, A.K. Translational research in agricultural biology—Enhancing crop resistivity against environmental stress alongside nutritional quality. *Front. Chem.* **2014**, *2*, 30. [CrossRef] [PubMed]

124. Mattoo, A.K.; Upadhyay, R.K.; Rudrabhatla, S. Abiotic stress in crops: Candidate genes, osmolytes, polyamines and biotechnological intervention. In *Elucidation of Abiotic Stress Signaling in Plants: A Functional Genomic Perspective*; Pandey, G.K., Ed.; Springer: New York, NY, USA, 2015.

125. Chauhan, H.; Khurana, N.; Agarwal, P.; Khurana, J.P.; Khurana, P. A seed preferential heat shock transcription factor from wheat provides abiotic stress tolerance and yield enhancement in transgenic *Arabidopsis* under hear stress environment. *PLoS ONE* **2013**, *8*, e79577. [CrossRef] [PubMed]

126. Kurek, I.; Chang, T.K.; Bertain, S.M.; Madrigal, A.; Liu, L.; Lassner, M.W.; Zhu, G. Enhanced thermostability of *Arabidopsis* Rubisco activase improves photosynthesis and growth rates under moderate heat stress. *Plant Cell* **2007**, *19*, 3230–3241. [CrossRef] [PubMed]

127. Salvucci, M.E. Association of Rubisco activase with chaperonin-60 beta: A possible mechanism for protecting photosynthesis during heat stress. *J. Exp. Bot.* **2008**, *59*, 1923–1933. [CrossRef] [PubMed]

128. Vierling, E. The roles of heat shock proteins in plants. *Ann. Rev. Plant Physiol. Plant Mol. Biol.* **1991**, *42*, 579–620. [CrossRef]

129. Wang, W.; Vinocur, B.; Shoseyov, O.; Altman, A. Role of plant heat-shock proteins and molecular chaperones in the abiotic stress response. *Trends Plant Sci.* **2004**, *9*, 244–252. [CrossRef] [PubMed]

130. Kazuko, Y.S.; Shinozaki, K. Transcriptional regulatory networks in cellular responses and tolerance to dehydration and cold stresses. *Annu. Rev. Plant Biol.* **2006**, *57*, 781–803.

131. Mu, C.; Zhang, S.; Yu, G.; Chen, N.; Li, X. Overexpression of small heat shock protein LinHSP16.45 in Arabidopsis enhances tolerance to abiotic stresses. *PLoS ONE* **2013**, *8*, e82264. [CrossRef] [PubMed]

132. Hsiao, T.C.; Acevedo, E.; Fereres, E.; Henderson, D.W. Stress metabolism, water stress, growth, and osmotic adjustment. *Philos. Trans. R. Soc. Lond. B* **1976**, *273*, 479–500. [CrossRef]

133. Hayat, S.; Hayat, Q.; Alyemeni, M.N.; Wani, A.S.; Pichtel, J.; Ahmad, A. Role of proline under changing environments: A review. *Plant Signal Behav.* **2012**, *7*, 1456–1466. [CrossRef] [PubMed]

134. Fougere, F.; Le-Rudulier, D.; Streeter, J.G. Effects of salt stress on amino acid, organic acid, and carbohydrate composition of roots, bacteroids, and cytosol of alfalfa (*Medicago sativa* L.). *Plant Physiol.* **1991**, *96*, 1228–1236. [CrossRef] [PubMed]

135. Petrusa, L.M.; Winicov, I. Proline status in salt tolerant and salt sensitive alfalfa cell lines and plants in response to NaCl. *Plant Physiol. Biochem.* **1997**, *35*, 303–310.

136. Ito, Y.; Maruyama, K.; Taji, T.; Kobatashi, M.; Seki, M.; Shinozaki, K.; Yamaguchi-Shinozaki, K.; Katsura K. Functional analysis of rice DREB1/CBF-type transcription factors involved in cold-responsive gene expression in transgenic rice. *Plant Cell Physiol.* **2006**, *47*, 141–153. [CrossRef] [PubMed]

137. Zhang, H.; Liu, W.; Wan, L.; Li, F.; Dai, L.; Li, D.; Zhang, Z.; Huang, R. Functional analyses of ethylene response factor *JERF3* with the aim of improving tolerance to drought and osmotic stress in transgenic rice. *Transgenic Res.* **2010**, *19*, 809–818. [CrossRef] [PubMed]

138. Yang, A.; Dai, X.; Zhang, W.H. A R2R3-type MYB gene, *OsMYB2*, is involved in salt, cold and dehydration tolerance in rice. *J. Exp. Bot.* **2012**, *63*, 2541–2556. [CrossRef] [PubMed]

139. Alcazar, R.; Francisco Marco, F.; Cuevas, J.C.; Patron, M.; Ferrando, A.; Carrasco, P.; Tiburcio, A.F.; Altabella, T. Involvement of polyamines in plant response to abiotic stress. *Biotechnol. Lett.* **2006**, *28*, 1867–1876. [CrossRef] [PubMed]

140. Kusano, T.; Berberich, T.; Tateda, C.; Takahashi, Y. Polyamines: Essential factors for growth and survival. *Planta* **2008**, *228*, 367–381. [CrossRef] [PubMed]

141. Gill, S.S.; Tuteja, N. Polyamines and abiotic stress tolerance in plants. *Plant Signal Behav.* **2010**, *5*, 26–33. [CrossRef] [PubMed]

142. Gupta, V.K.; Scheunemann, L.; Eisenberg, T.; Mertel, S.; Bhukel, A.; Koemans, T.S.; Kramer, J.M.; Liu, K.S.; Schroeder, S.; Stunnenberg, H.D.; et al. Restoring polyamines protects from age-induced memory impairment in an autophagy-dependent manner. *Nat. Neurosci.* **2013**, *16*, 1453–1460. [CrossRef] [PubMed]

143. Watson, M.B.; Emory, K.K.; Piatak, R.M.; Malmberg, R.L. Arginine decarboxylase (polyamine synthesis) mutants of *Arabidopsis thaliana* exhibit altered root growth. *Plant J.* **1998**, *13*, 231–239. [CrossRef] [PubMed]

144. Kasinathan, V.; Wingler, A. Effect of reduced arginine decarboxylate activity on salt tolerance and on polyamine formation during salt stress in Arabidopsis thaliana. *Plant Physiol.* **2004**, *121*, 101–107. [CrossRef] [PubMed]

145. Urano, K.; Yoshiba, Y.; Nanjo, T.; Ito, Y.; Seki, M.; Yamaguchi-Shinozaki, K.; Shinozaki, K. *Arabidopsis* stress-inducible gene for arginine decarboxylase *AtADC2* is required for accumulation of putrescine in salt tolerance. *Biochem. Biophys. Res. Commun.* **2004**, *313*, 369–375. [CrossRef] [PubMed]
146. Yamaguchi, K.; Takahashi, Y.; Berberich, T.; Imai, A.; Miyazaki, A.; Takahashi, T.; Michael, A.; Kusano, T. The polyamine spermine protects against high salt stress in *Arabidopsis thaliana*. *FEBS Lett.* **2006**, *580*, 6783–6788. [CrossRef] [PubMed]
147. Yamaguchi, K.; Takahashi, Y.; Berberich, T.; Imai, A.; Takahashi, T.; Michael, A.; Kusano, T. A protective role for the polyamine spermine against drought stress in *Arabidopsis*. *Biochem. Biophys. Res. Commun.* **2007**, *352*, 486–490. [CrossRef] [PubMed]
148. Capell, T.; Bassie, L.; Christou, P. Modulation of the polyamine biosynthetic pathway in transgenic rice confers tolerance to drought stress. *Proc. Natl. Acad. Sci. USA* **2004**, *101*, 9909–9914. [CrossRef] [PubMed]
149. Roy, M.; Wu, R. *Arginine decarboxylase* transgene expression and analysis of environmental stress tolerance in transgenic rice. *Plant Sci.* **2002**, *160*, 869–875. [CrossRef]
150. Kumria, R.; Rajam, M.V. Ornithine decarboxylase transgene in tobacco affects polyamines, in vitro morphogenesis and response to salt stress. *J. Plant Physiol.* **2002**, *159*, 983–990. [CrossRef]
151. Thu-Hang, P.; Bassie, L.; Safwat, G.; Trung-Nghia, P.; Christou, P.; Capell, T. Expression of a heterologous S-adenosylmethionine decarboxylase cDNA in plants demonstrates that changes in S-adenosyl-L-methionine decarboxylase activity determine levels of the higher polyamines spermidine and spermine. *Plant Physiol.* **2002**, *129*, 1744–1754. [CrossRef] [PubMed]
152. Kasukabe, Y.; He, L.; Nada, K.; Misawa, S.; Ihara, I.; Tachibana, S. Over expression of spermidine synthase enhances tolerance to multiple environmental stresses and up-regulates the expression of various stress-regulated genes in transgenic *Arabidopsis thaliana*. *Plant Cell Physiol.* **2004**, *45*, 712–722. [CrossRef] [PubMed]
153. Wen, X.P.; Pang, X.M.; Matsuda, N.; Kita, M.; Inoue, M.; Hao, Y.J.; Honda, C.; Moriguchi, T. Overexpression of the apple spermidine synthase gene in pear confers multiple abiotic stress tolerance by altering polyamine titers. *Transgenic Res.* **2008**, *17*, 251–263. [CrossRef] [PubMed]
154. Alcázar, R.; Planas, J.; Saxena, T.; Zarza, X.; Bortolotti, C.; Cuevas, J.; Bitrian, M.; Tiburcio, A.F.; Altabella, T. Putrescine accumulation confers drought tolerance in transgenic Arabidopsis plants overexpressing the homologous arginine decarboxylase 2 gene. *Plant Physiol. Biochem.* **2010**, *48*, 547–552. [CrossRef] [PubMed]
155. Roy, M.; Wu, R. Overexpression of S-adenosylmethionine decarboxylase gene in rice increases polyamine level and enhances sodium-chloride-stress tolerance. *Plant Sci.* **2002**, *163*, 987–992. [CrossRef]
156. Waie, B.; Rajam, M.V. Effect of increased polyamine biosynthesis on stress responses in transgenic tobacco by introduction of human S-adenosylmethionine decarboxylase gene. *Plant Sci.* **2003**, *164*, 727–734. [CrossRef]
157. Wi, S.J.; Kim, W.T.; Park, K.Y. Over expression of carnation S-adenosylmethionine decarboxylase gene generates a broad-spectrum tolerance to abiotic stresses in transgenic tobacco plants. *Plant Cell Rep.* **2006**, *25*, 1111–1121. [CrossRef] [PubMed]
158. Cheng, L.; Zou, Y.J.; Ding, S.L.; Zhang, J.J.; Yu, X.L.; Cao, J.; Lu, G. Polyamine accumulation in transgenic tomato enhances the tolerance to high temperature stress. *J. Integr. Plant Biol.* **2009**, *51*, 489–499. [CrossRef] [PubMed]
159. He, L.; Ban, Y.; Inoue, H.; Matsuda, N.; Liu, J.; Moriguchi, T. Enhancement of spermidine content and antioxidant capacity in transgenic pear shoots overexpressing apple spermidine synthase in response to salinity and hyperosmosis. *Phytochemistry* **2008**, *69*, 2133–2141. [CrossRef] [PubMed]
160. Chisholm, S.T.; Coaker, G.; Day, B.; Staskawicz, B.J. Host-microbe interactions shaping the evolution of the plant immune response. *Cell* **2006**, *124*, 803–814. [CrossRef] [PubMed]
161. Goyal, R.K.; Mattoo, A.K. Multitasking antimicrobial peptides in plant development and host defense against biotic/abiotic stress. *Plant Sci.* **2014**, *228*, 135–149. [CrossRef] [PubMed]
162. Goyal, R.K.; Mattoo, A.K. Plant antimicrobial peptides. In *Host Defense Peptides and Their Potential as Therapeutic Agents*; Epand, R.M., Ed.; Springer: Cham, Switzerland, 2016; pp. 111–136.
163. Thevissen, K.; Ferket, K.K.A.; Francois, I.E.J.A.; Cammue, B.P.A. Interactions of antifungal plant defensins with fungal membrane components. *Peptides* **2003**, *24*, 1705–1712. [CrossRef] [PubMed]
164. Zasloff, M. Antimicrobial peptides of multicellular organisms. *Nature* **2002**, *415*, 389–395. [CrossRef] [PubMed]

165. Yeaman, M.R.; Yount, N.Y. Mechanisms of antimicrobial peptide action and resistance. *Pharmacol. Rev.* **2003**, *55*, 27–55. [CrossRef] [PubMed]

166. Yount, N.Y.; Yeaman, M.R. Peptide antimicrobials: Cell wall as a bacterial target. *Ann. N. Y. Acad. Sci.* **2013**, *1277*, 127–138. [CrossRef] [PubMed]

167. Paiva, A.D.; Breukink, E. Antimicrobial peptides produced by microorganisms. In *Antimicrobial Peptides and Innate Immuniyty—Progress in Inflammation Research*; Hiemstra, P.S., Zaat, S.A.J., Eds.; Springer: Basel, Switzerland, 2013; pp. 53–95. [CrossRef]

168. Doughty, J.; Dixon, S.; Hiscock, S.J.; Willis, A.C.; Parkin, I.A.P.; Dickinsin, H.G. PCP-A1, a defensing-like Brassica pollen coat protein that binds the S locus glycoprotein, is the product of gametophytic gene expression. *Plant Cell* **1998**, *10*, 1333–1347. [CrossRef] [PubMed]

169. Takayama, S.; Shimosato, H.; Shiba, H.; Funato, M.; Che, F.S.; Watanabe, M.; Iwano, M.; Isogal, A. Direct ligand- Receptor complex interaction controls Brassica self-incompatibity. *Nature* **2001**, *4513*, 534–538. [CrossRef] [PubMed]

170. Okuda, S.; Tsutsui, H.; Shiina, K.; Sprunnck, S.; Takeuchi, H.; Yui, R.; Kasahara, R.D.; Hamamura, Y.; Mizukami, A.; Susaki, D.; et al. Defensin-like polypeptide LUREs are pollen tube attractants secreted from synergid cells. *Nature* **2009**, *458*, 357–361. [CrossRef] [PubMed]

171. Amien, S.; Kliwer, I.; Marton, M.L.; Debener, T.; Geiger, D.; Becker, D.; Dresselhaus, T. Defensin-like ZmES4 mediates pollen tube burst in maize via opening of the potassium channel KZM1. *PLoS Biol.* **2010**, *8*, e1000388. [CrossRef] [PubMed]

172. Takeuchi, H.; Higashiyama, T. A species-specific cluster of defensing-like genes encodes diffusible pollen tube attractants in Arabidopsis. *PLoS Biol.* **2012**, *10*, e1001449. [CrossRef] [PubMed]

173. Marshall, E.; Costa, L.M.; Gutierrez-Marcos, J. Cysteine-rich peptides (CRPs) mediate diverse aspects of cell-cell communication in plant reproduction and development. *J. Exp. Bot.* **2011**, *62*, 1677–1686. [CrossRef] [PubMed]

174. Penterman, J.; Abo, R.P.; De Nisco, N.J.; Arnold, M.F.; Longhi, R.; Zanda, M.; Walker, G.C. Host plant peptides elicit a transcriptional response to control the *Sinorhizobium meliloti* cell cycle during symbiosis. *Proc. Natl. Acad. Sci. USA* **2014**, *111*, 3561–3566. [CrossRef] [PubMed]

175. DeBono, A.; Yeats, T.H.; Rose, J.K.C.; Bird, D.; Jetter, R.; Kunst, L.; Samuels, L. Arabidopsis LTPG is a glycosylphosphatidylinositol-anchored lipid transfer protein required for export of lipids to the plant surface. *Plant Cell* **2009**, *21*, 1230–1238. [CrossRef] [PubMed]

176. Park, S.Y.; Jauh, G.Y.; Mollet, J.C.; Eckard, K.J.; Nothnagel, E.A.; Walling, L.L.; Lord, E.M. A lipid transfer-like protein is necessary for lily pollen tube adhesion to an in vitro stylar matrix. *Plant Cell* **2000**, *12*, 151–163. [CrossRef] [PubMed]

177. Chae, K.; Kieslich, C.A.; Morikis, D.; Kim, S.C.; Lord, E.M. A gain-of-function mutation of *Arabidopsis* lipid transfer protein 5 disturbs pollen tube tip growth and fertilization. *Plant Cell* **2009**, *21*, 3902–3914. [CrossRef] [PubMed]

178. Molina, A.; Garcia-Olmedo, F. Enhanced tolerance to bacterial pathogens caused by the transgenic expression of barley lipid transfer protein LTP2. *Plant J.* **1997**, *12*, 669–675. [CrossRef] [PubMed]

179. Stotz, H.U.; Spence, B.; Wang, Y. A defensing from tomato with dual function in defense and development. *Plant Mol. Biol.* **2009**, *71*, 131–143. [CrossRef] [PubMed]

180. Nahirnak, V.; Almasia, N.I.; Fernandez, P.V.; Hopp, H.E.; Estevez, J.M.; Carrarii, F.; Vazquez-Rovere, C. Potato Snakin-1 gene silencing affects cell division, primary metabolism, and cell wall composition. *Plant Physiol.* **2012**, *158*, 252–263. [CrossRef] [PubMed]

181. Goyal, R.K.; Hancock, R.E.W.; Mattoo, A.K.; Misra, S. Expression of an engineered heterologous antimicrobial peptide in potato alters plant development and mitigates normal abiotic and biotic responses. *PLoS ONE* **2013**, *8*, e77505. [CrossRef] [PubMed]

182. Fujita, M.; Fujita, Y.; Noutoshi, Y.; Takahashi, F.; Narusaka, Y.; Yamaguchi-Shinozaki, K. Crosstalk between abiotic and biotic stress responses: A current view from the point of convergence in the stress signaling networks. *Curr. Opin. Plant Biol.* **2006**, *9*, 436–442. [CrossRef] [PubMed]

183. Ronald, P. Plant genetics, sustainable agriculture and global food security. *Genetics* **2011**, *188*, 11–20. [CrossRef] [PubMed]

184. Bruce, T.J.A. GM as a route for delivery of sustainable crop protection. *J. Exp. Bot.* **2011**, *63*, 537–541. [CrossRef] [PubMed]

185. Wally, O.; Punja, Z.K. Genetic engineering for increased fungal and bacterial disease resistance in crop plants. *GM Crops* **2010**, *1*, 199–206. [CrossRef] [PubMed]

186. Bennett, A.B.; Chi-Ham, C.; Barrows, G.; Sexton, S.; Zilberman, D. Agricultural biotechnology: Economics, environment, ethics, and the future. *Annu. Rev. Environ. Resour.* **2013**, *38*, 249–279. [CrossRef]

187. Dangour, A.D.; Dodhia, S.K.; Hayter, A.; Allen, E.; Lock, K.; Uauy, R. Nutritional quality of organic foods: A systematic review. *Am. J. Clin. Nutr.* **2009**, *90*, 680–685. [CrossRef] [PubMed]

188. Chen, X.-P.; Cui, Z.-L.; Vitousek, P.M.; Cassman, K.G.; Matson, P.A.; Bai, J.-S.; Meng, Q.-F.; Hou, P.; Yue, S.-C.; Romheld, V.; et al. Integrated soil-crop system management for food security. *Proc. Natl. Acad. Sci. USA* **2011**, *108*, 6399–6404. [CrossRef] [PubMed]

MDPI

St. Alban-Anlage 66

4052 Basel

Switzerland

Tel. +41 61 683 77 34

Fax +41 61 302 89 18

www.mdpi.com

Agriculture Editorial Office

E-mail: agriculture@mdpi.com

www.mdpi.com/journal/agriculture